O9-AID-585

THE
RED
AND THE
BLUE

THE
RED
AND THE
BLUE

★ ★ ★ ★ ★ ★ ★ ★ ★ ★ ★ ★ ★

THE 1990s AND THE BIRTH
OF POLITICAL TRIBALISM

★ ★ ★ ★ ★ ★ ★ ★ ★ ★ ★ ★ ★

STEVE KORNACKI

AN IMPRINT OF HARPERCOLLINSPUBLISHERS

HarperCollins books may be purchased for educational, business, or sales promotional use. For information, please email the Special Markets Department at SPsales@harpercollins.com.

FIRST EDITION

Designed by Suet Chong

Library of Congress Cataloging-in-Publication Data

Names: Kornacki, Steve, 1979- author.
Title: The red and the blue : the 1990s and the birth of political tribalism.
Description: First edition. | New York, NY : Ecco, [2018] | Includes
    bibliographical references and index.
Identifiers: LCCN 2018016635| ISBN 9780062438980 | ISBN 9780062439000 |
    ISBN 9780062865380 | ISBN 9780062865403 | ISBN 9780062865403
    (electronic)
Subjects: LCSH: United States—Politics and government—1989-1993. | United
    States—Politics and government—1993-2001. | Political culture—United
    States—History—20th century.
Classification: LCC E839.5 .K67 2018 | DDC 306.20973/0904—dc23 LC
    record available at https://lccn.loc.gov/2018016635

ISBN 978-0-06-243898-0

18 19 20 21 22 RS/LSC 10 9 8 7 6 5 4 3 2 1

For Mom and Dad

THE
RED
AND THE
BLUE

PROLOGUE

All the momentum was with Al Gore when November 7, 2000, arrived. For much of the fall, he'd lagged a few points behind, stymied by clashing public attitudes toward his boss. The economy was humming and voters were willing to credit Bill Clinton and his administration, but lingering revulsion over the president's affair with a former White House intern left them receptive to Republican candidate George W. Bush and his promise to restore "honor and dignity" to the office.

Through this all, Bush had somehow managed to keep quiet a damaging secret, an old arrest for driving under the influence, back when he was thirty years old. But just days before the vote, it had all come out, reinforcing the doubts about this son of a president and his maturity. Gore closed the gap in the final polls, even grabbed the lead in one, and now as the Election Night returns rolled in, there was suspense that there hadn't been in at least a generation.

The critical call came early, or at least it seemed to. At ten minutes till eight, CBS's Dan Rather cut off the network's elder statesman, Mike Wallace, with an urgent bulletin. "Mike, excuse me one second.

I'm so sorry to interrupt you. Mike, you know I wouldn't do this if it weren't big. Florida goes for Al Gore."

The other networks followed suit, and so did the Associated Press. Florida's twenty-five electoral votes were nearly essential to each candidate's path to 270—a "very, very, very important state," Tim Russert said on NBC. Losing it, CNN's Jeff Greenfield said, amounted to "a roadblock the size of a boulder" for Bush. On screen, though, the tallies continued to show Bush ahead of Gore in Florida, and votes were still being cast in the Panhandle region, part of the Central time zone. Elsewhere in the state, there were reports of long lines and delayed closings, and a flood of absentee ballots.

In Austin, the Bush team rushed network television cameras into the hotel suite where the candidate was watching returns with his family, including his father, the former president. "The networks called this thing awfully early," the candidate said. "But the people who are actually counting the votes are coming up with a little different perspective. And so, we're—I'm pretty darn upbeat about things."

Through the eight o'clock hour and well past nine, Bush maintained the lead, and the networks dug in their heels. On CBS, Rather assured viewers that "if we say somebody's carried the state, you can take that to the bank." But doubt began seeping into coverage. Seated next to Tom Brokaw on NBC, Russert used a pair of dry erase boards to play out Electoral College scenarios. He jotted down nine states, some of them long shots for Bush, and said the Republican would now have to carry all of them to reach 270. Suppose, Brokaw responded, that the Florida call were to be reversed. Russert promptly erased four of the states. "It gets a lot easier," he said.

CNN pulled back first. "Stand by, stand by," anchor Bernard Shaw interjected. "CNN right now is moving our earlier declaration of Florida back to the too-close-to-call column. Twenty-five very big electoral votes, and the home state of the governor's brother, Jeb Bush, are hanging in the balance. This is no longer in the victory column for Vice President Gore."

"Oh waiter!" Greenfield exclaimed, "One order of crow!"

Within a half-hour, everyone had done the same. Said Brokaw: "What the networks giveth, the networks taketh away." Bush still led in the running tally, but his spread dwindled in the next hours. A recount seemed likely until, in the wee hours of November 8, the Bush advantage spiked dramatically. At 2:16 A.M., the Fox News Channel moved to declare Bush the winner of Florida—and the presidency.

Quickly, the other networks followed. On NBC, historian Doris Kearns Goodwin was making a point about the deep divisions being revealed in the election when Brokaw jumped in.

"Stop! Stop!" Doris, Doris, Doris, Doris, Doris . . ."

"Uh oh! Something's happened!"

"George Bush is the president-elect of the United States. He has won the state of Florida, according to our projections."

Cameras captured jubilation in Austin and dejection in Nashville, where Gore placed a concession call to Bush. On the air, there was talk of who might join the new administration and what might be next for Gore, now poised for unemployment. There was speculation about 2004. Would he run again? The First Lady, Hillary Rodham Clinton, the winner on this night of a Senate election in New York, was an obvious prospect too.

They were trying to fill time until the candidates delivered their speeches and ended this bizarrely suspenseful night. Gore's motorcade arrived at the Nashville War Memorial, where his supporters were gathered. Minutes passed without him emerging. Meanwhile, the tally from Florida's secretary of state showed Bush's lead evaporating, a twist at odds with the networks' own count. The waiting continued. Confusion was building. In the national popular vote, Gore was now outpacing Bush. What if more people actually voted for the losing candidate? The Constitution was clear on the supremacy of the Electoral College, but there hadn't been a split like this since 1876. How would Americans react now?

It was nearly four in the morning when Candy Crowley, stationed with the Bush campaign in Austin, came on CNN's air. "Something to report to you here in this very unusual night," she said. The vice president has re-called the governor and retracted his concession, say-ing that Florida is too close right now."

Now all the networks were pulling back their calls—again. "All right," Brokaw said, "we're officially saying that Florida is too close to call. So we take Florida away from George W. Bush. That means that he is short of the 270 electoral votes that he needs to win." Gore, of course, was short too. The scoreboard stood at 267 for Gore and 246 for Bush, with only Florida unresolved. Now the talk moved to a re-count, and military ballots, and a report about a confusing "butterfly" ballot in Palm Beach County, where some Gore voters were claiming they'd mistakenly voted for Pat Buchanan, a third party candidate.

"This is astonishing," Peter Jennings said on ABC.

There would be no winner that night, or the next day, or the next week. Election Night 2000 is remembered for its wild swings of emotion and for the galling errors in exit polling and vote-counting that produced them. It became the trigger point for an even more chaotic drama, the Florida recount, which would proceed in fits and starts for the next thirty-six days until its abrupt termination by a divided Supreme Court. Only then would Bush, courtesy of an of-ficial 537-vote margin in the state, become the next president of the United States.

The 2000 election, though, was far more than the story of one pivotal state. The divisions were deep and vivid. Gore was carrying nine of the ten states north of the Mason-Dixon line, many by crush-ing margins, and cleaning up along the Pacific Coast, with an easy victory in California. Bush was sweeping the South, including Gore's home state of Tennessee (and Bill Clinton's Arkansas), and posting historic gains in traditionally Democratic Appalachia. Outside of Florida, most of the contested turf was in the Rust Belt. It all added up to the closest thing to a tie ever seen in the modern era.

These divisions were geographic, demographic, and cultural, ex-pressed to Americans watching on television in two colors: red for the

Republican states and blue for the Democratic states. The scheme was accidental. In past elections, the colors had been used interchangeably for the two parties, and it hadn't mattered at all, since there hadn't been a close race in a generation. But on this election night all of the networks had landed on the same designations, and in an instant colors that had meant nothing before came to mean everything.

Red America and Blue America as we now know them were born on November 7, 2000, the product of an entire nation torn perfectly in half. But these divisions were created in the decade that preceded them. Many recall the 1990s as a modern answer to the Era of Good Feelings. The economy exploded, the stock market soared, and crime came crashing downward. There was one quick, triumphant war, followed by years of peace and a few surgical interventions that made military force feel like an almost benign instrument. Hollywood was at its peak, with movies like Forrest Gump, Braveheart, and Pulp Fiction hitting that magical sweet spot between artistic style and mass appeal. From *Seinfeld* and *ER* to *NYPD Blue* and *The Fresh Prince of Bel Air*, a renaissance swept through network television. The Internet and cell phones connected Americans to information and to each other without overwhelming their lives—"just the right amount of technology," as Kurt Andersen has argued. America, by and large, seemed to work.

That happy narrative, however, masks the reality of our politics in the 1990s. Here the dominant features were confrontation, gridlock, polarization, and a new kind of tribalism. The source of this disorder—with a partisan fervor and level of sordidness previously unseen in modern politics—was a collision between two titanic personalities.

On the Democratic side, it was Bill Clinton, who offered salvation to a party that had lost three straight presidential elections. For Republicans, it was Newt Gingrich who insisted he could lead them out of the desert of forty years of minority status on Capitol Hill. Their collision unleashed two historic backlashes that shaped the decade. One seemed to fulfill Gingrich's vision—a "Republican Revolution"

that installed him as his party's first Speaker of the House in generations, a transfer of power so jarring that serious political observers
asked openly if the Democratic Party might be facing extinction. The
other, just a few years later, was a backlash against Gingrichism itself,
an awakening of the white collar professional class that found itself
alienated by this new Republican Party.

Some of the most indelible characters in recent history shaped
these battles that defined the decade. There was Ross Perot, dubbed
America's first populist billionaire, who upended the 1992 presidential
campaign and whose Reform Party later in the decade gave Donald
Trump entre to electoral politics. Jesse Jackson, the charismatic civil
rights leader, whose breakthrough White House bids in the 1980s
certified him as a leading force in the Democratic Party, setting up
a climactic confrontation with Clinton that reoriented the '92 race.

Pat Buchanan, one of the original television pundits, whose re-
bellious Republican campaigns tormented George H. W. Bush in
1992 and Bob Dole in 1996 while establishing a market for an edgy
brand of nationalist politics. And Hillary Clinton, offered to voters by
her husband as part of a package deal, seizing a policy-making White
House role without precedent for a first lady, weathering intense criticism and controversy only to emerge through scandal as a sympathetic
figure and electoral star with a political future of her own.

The policy fights of the '90s echo where we are today: clashes in
Washington over budget deficits, spending, and taxes; and healthcare
and broader culture war flare-ups over gay rights, guns, and political
correctness.

The country emerged from the 1990s cleaved into two distinct
political groupings. Red and blue become stand-ins for the two parties and Americans chose sides and dug in. The lines have shifted
here and there, but it's the same basic framework we know today.
This book is the story of how it came to be. It turns out the decade
of good feelings, the Pax Americana nestled between the end of the
Cold War and the start of the War on Terror, is far more tumultuous
and meaningful than ever.

To tell this story properly, though, to take in the larger context of dizzying electoral swings, smashed coalitions, and brand new cultural divisions, requires us to begin a few years before the start of this fateful decade, when the old order seemed locked in place, even as a rising generation of innovative and endlessly ambitious politicians was plotting to upend it.

ONE

Bill Clinton came to San Francisco for the Democratic National Convention hungry for attention. It was July 1984, and the second-term Arkansas governor, not yet forty years old, knew where to find opportunity.

The convention itself would be a morose affair, with party regulars dutifully ratifying former vice president Walter Mondale as their nominee. Four years earlier, Mondale had been number two on the Jimmy Carter–led ticket that surrendered forty-four states to Ronald Reagan. For a while, Democrats had believed Reagan's triumph to be a fluke, especially when a nasty recession pushed unemployment to over 10 percent early in his term. But by the summer of '84, the economy was resurgent, patriotism was in full bloom (Los Angeles would soon host the Summer Olympics), and America's grandfatherly president was enjoying some of the best poll numbers of his tenure. Democrats, a survey of delegates revealed, were significantly less optimistic about Mondale's prospects than they had been about Carter's back in 1980. Even the truest of true believers knew it: Reagan was going to swamp them—again.

This meant the race for the next time around was already on. Once Mondale suffered his inevitable defeat in November, the search would commence for a new leader for the party, someone to carry the torch in 1988, when Reagan would be forbidden from running again and the country might, after eight years of Republican rule, be ready for a break. The conversation was already under way, and Clinton was intent on forcing his name into it.

His ambition had never been parochial or regional. It was national office Clinton coveted, and he'd spent his postadolescent life maneuvering to attain it—Boys Nation delegate, Georgetown undergrad, intern for a powerful senator, Rhodes Scholar, law school at Yale. Those early years had been about making personal connections, mixing with the national power class and those who would one day join it, charming them with his warmth and charisma, dazzling them with his intellect, his sharp political instincts, his potential.

Then, after finishing law school in 1973, the critical decision: to leave it all behind and head back to his native Arkansas, an impoverished state practically invisible to the very people he'd just spent years cultivating. But in Arkansas there was political opportunity. The ancient segregationist order was at last giving way. If he could just get a toehold and win some political office back home, any political office, Clinton knew he could move up and make his way to the top of the Arkansas political ranks. And then he would have it—the springboard to go national, to reconnect with all those people he'd so impressed as a young man; except now they would have matured into positions of power and influence, making them immensely useful to their old pal from Arkansas.

He took his first shot in 1974, and the timing seemed perfect. The Watergate scandal finally took down Richard Nixon that summer, and he left behind a Republican Party in ruins, discredited by the blind eye that too many of its leaders had turned to the president's misdeeds for too long. Clinton, only a year out of law school, set his sights on a congressman named John Paul Hammerschmidt, the only Republican elected to the House from Arkansas since Reconstruction. Surely, in this most anti-Republican of states, in this most anti-Republican of years, Hammerschmidt's party label would doom him—especially since he'd been one of the very last men in Washington to see Nixon for what he was. But the district's voters were also conservative—these were the days before party label and ideology synced up—and so was the incumbent. The national tide was strong for Democrats that November, but Clinton fell three points short.

It was still the closest any Democrat had come to knocking off Hammerschmidt, and the closest any Democrat would ever come. With his near miss, Clinton earned credibility with the Arkansas political world. Here was someone who was going places—the kind of guy you'd want to get in good with today because you'd surely need him tomorrow. When Clinton ran for attorney general two years later, Republicans didn't even bother to put up an opponent. And when Governor David Pryor opted to run for the Senate in 1978, the top job in Little Rock was suddenly open, and Clinton's for the taking. The race wasn't close. At thirty-two, he was his state's youngest governor ever, and the youngest any state had seen since the 1930s.

"His eloquence, often offhand, captivates crowds," read an Associated Press profile just after that election, "and the word 'charisma' shows up often in news articles about him. Friends see the U.S. Senate and maybe more in his future."

That naked ambition nearly undid him. A little over a year after taking office, Clinton accepted an invitation to Washington. Georgetown University wanted to honor this young alum in a hurry, hailing his rapid rise as "dramatic evidence that ambition to be of service coupled with a Georgetown formation quickly bridges the generation gap." The homecoming caught the Washington crowd's attention—that was the point, after all—and led the *Washington Post*'s gardening columnist, of all people, to attend the ceremony and write it up. "He may be president eight years hence," Henry Mitchell wrote of Arkansas's governor.

Clinton gave interviews to the national press, weighing in on domestic politics, the future of the Democratic Party, whatever they wanted to talk about, anything to get his name out there, to remind the political world that he was more than just another good ol' boy pol from a backwater state. Back in Arkansas, the locals weren't exactly blind to all of this, and the notion that their boy wonder governor might be moving a little too fast, getting a bit too big for his britches, started to take hold.

Then there was his wife. People wondered how a woman like Hillary Rodham ended up in Arkansas in the first place. A whip-smart daughter of the Chicago suburbs, she shared Bill's immense ambition and had plenty to show for herself. At just twenty-two, she had made national headlines with her student speech at Wellesley College's 1969 graduation ceremony, using it to present a blistering rebuke of the commencement speaker, Republican senator Edward Brooke. From Wellesley, it was on to Yale Law, where she and Bill began their relationship, and then to D.C., where she landed a gig as a staff attorney for the House Judiciary Committee as it explored Nixon's impeachment.

She faced a dilemma. Her prospects were bright in D.C., at a time when high-level opportunities that had until now been unthinkable for women were suddenly opening up. But her boyfriend wasn't going to be sticking around town and was asking her to come to Arkansas with him and to make a life—and a political career—together. She took the bar exam in D.C. and flunked it, but passed in Arkansas. In the summer of 1974, with one of her Washington mentors along for the ride and pleading with her to reconsider the whole way, Hillary made the long drive south to Arkansas. Within a year, she and Bill were a married couple.

She worked briefly as an advocate for children, then segued into a spot in the state's premier law firm. Here was a driven career woman from up north, the antithesis of the kind of deferential southern political wife Arkansans were accustomed to. And she was adamant: She would not be taking her husband's last name. You could call her Hillary Rodham, thank you very much.

Her stand coincided with a growing backlash in the South against feminism and the Equal Rights Amendment. Rodham's defiance marked her as one of those troublemaking outsiders, a sophisticate looking down on the natives. Her husband defended her use of her maiden name: "She decided to do that when she was nine, long before women's lib came along. People wouldn't mind if they knew how old-fashioned she was in every conceivable way."

It was all too much. The Arkansas governorship still came with a two-year term back then, and when Bill Clinton faced the voters again in 1980, he did so with the presidential race at the top of the ticket. Nationally, Ronald Reagan trounced President Jimmy Carter, piling up victories in every corner of the country. Most stunning was Reagan's strength in the South, a region where the Republican Party had barely existed since the Civil War. The GOP had been gaining strength in Dixie, but Carter, the old Georgia peanut farmer, had seemingly restored his party's grip on the region in 1976. Now, in 1980, the wheels came off and Carter carried only his home state in "the solid South." Arkansas went to Reagan by nine points, and that was the final dagger for Clinton, who lost his job to Republican Frank White. The state's youngest-ever governor was now its youngest-ever former governor.

The ambition still burned, but was there a path back from such a humbling defeat? It was possible Clinton had alienated too many Arkansans, and that stripped of the power—and relevance—that comes with incumbency, his reputation would now decline further. Plus, White himself might prove popular, locking down the governorship for years to come. At just thirty-four years old, had the rising star already been extinguished?

He hunkered down to learn the lessons of his defeat. He'd attempt a comeback in 1982. With a win, he'd be right back in the game, a young governor on the rise. Another loss, though, and he'd be on his way to the discard pile. The political climate was different this time. The Reagan recession was at its peak and Democrats were salivating over a midterm atmosphere nearly as promising as the Watergate year of '74. White, meanwhile, had made some missteps and was vulnerable. Clinton entered with a sense of contrition, running television ads apologizing for his hubris. "I made a young man's mistake," he told a reporter. "I tried to do too much too soon. I spent too much time doing what I wanted to do rather than what people wanted me to do." He asked for a second chance, promised to listen better.

Checking in on the Arkansas race, a pair of *Washington Post* reporters discovered an ex-governor who was "belatedly heeding the advice of those who said his image was out of tune with his conservative state. Clinton's ears show under his newly cropped hair, he has replaced his bearded young aides with clean-shaven, middle-aged men, and Hillary Rodham will not only give up her law practice to campaign full-time, she also will stop using her maiden name and henceforth be known as Mrs. Bill Clinton." He won in November by double digits. Mr. and Mrs. Clinton were heading back to the governor's mansion. He still had a future after all.

This was the Bill Clinton who arrived in San Francisco for the '84 convention. The dark years had sharpened him, instilling in him a new mindfulness of the price to be paid for offending the public's sensibilities. He measured his actions more carefully now, but the ultimate goal never changed, and by now he was beginning to see it. He would win reelection in the fall (this time it really would be a layup) and again in 1986, and then there'd be a wide-open Democratic race for president in 1988. His moment was coming. What he needed was for others to start seeing it too, and the convention was an opportunity to charm the national bigwigs.

Political coverage in 1984 was still driven by major newspapers and broadcast television networks. So it was a coup for Clinton when he won a mention in a *Washington Post* story the Sunday before the convention. It was written by David Broder, one of the foremost shapers of elite opinion, and it explored what Broder portrayed as an identity crisis for a Democratic Party on the verge of its second consecutive national beatdown. Clinton's appearance was a brief cameo in a three-thousand-word story ("I don't know that we've made a lot of progress in honing our message to the country"), but also a signal to Broder's powerful audience that this young governor merited keeping an eye on.

The real coup for Clinton was this: He would get to speak. And not in some dead-of-the-afternoon time slot when no one would be watching. The Mondale campaign was putting him in prime time on opening night—9 P.M. on the East Coast. Preceding Clinton would be a short film narrated by Hal Linden, star of *Barney Miller*, just off the air but still popular to millions of Americans, recalling Harry Truman and his famously improbable victory thirty-six years earlier. It was the best (and only) card convention organizers could play to argue that Mondale wasn't doomed, and Clinton was there to reinforce the message. The audience would be significant. Even in 1984, the networks were scaling back their coverage of conventions, which had evolved into little more than party-produced infomercials, and viewership was shrinking. But this was all relative: as Clinton made his way to the podium, nearly 40 percent of all households watching television were tuned to ABC, CBS, or NBC for the convention.

Clinton hailed the late president's defiant, populist spirit. "If Truman were here," he said, "he'd remind us that in 1948 he was further behind and the Democrats were more divided than is the case today. And he'd tell us to scrap the nostalgia and quit whining about our internal problems and get on with the business of taking our message to the country."

"The real way to honor Harry Truman," he declared, "is to wage a campaign that would make him as proud of us as we are of him."

It was a well-received call to action, crisp and forcefully delivered. The delegates cheered and Clinton beamed. All the big-name anchors—Rather, Jennings, and Brokaw, to say nothing of luminaries like Walter Cronkite, John Chancellor, and David Brinkley—were in the convention hall, along with just about every major political scribe in America. Clinton wanted their attention, needed it, and now he'd gotten it, and he'd made a nice impression, too. It was just the kind of win he needed.

And a few minutes later, it was forgotten forever.

I f Clinton craved validation from political and media elites, the next featured speaker on opening night in San Francisco was his opposite. The delegation breakfasts, the late-night parties, the chance hotel lobby encounters with power brokers, all the priceless networking opportunities that come with a national convention: Mario Cuomo conveyed indifference, even hostility, to all of it. It was a character trait that would, in the years to come, only magnify the impact of what happened this night.

Cuomo was in town at the invitation of Mondale, who'd scored a crucial victory in New York's March primary with an assist from the governor. Now Mondale had deputized Cuomo to deliver the keynote address. But there was no preconvention schmoozing for the New York governor, and there'd be no lingering afterward. He liked his own bed, hated to travel, and was fond of telling anyone who'd listen about the gravity of his responsibilities in Albany. This was a day trip for Cuomo, nothing more.

Expectations were minimal. The fifty-two-year-old Cuomo had only been elected governor two years earlier. It had been a big upset, Cuomo rallying liberals and organized labor to stun his longtime rival, New York City mayor Ed Koch, in the Democratic primary. Even then, Cuomo only won the general election in a squeaker. Back home, he was a formidable figure, but his national profile was limited. At first glance, Cuomo seemed like just another of a familiar type. He was urban, ethnic, and liberal in the tradition of FDR— a friend to labor unions and the proliferating assortment of interest groups now operating within the Democratic Party. It was organization men like Cuomo who'd delivered the nomination to Mondale over Gary Hart and his appeal to a new, more suburban and machine-averse breed of Democrat.

More discerning followers of politics knew something else, though: the guy could talk. Introduced by Mark White, the governor of Texas, Cuomo walked to the podium to polite cheers. Then, the opening, jarring and deliberate: "Please allow me to skip the stories and the poetry and the temptation to deal in nice but vague rhetoric. Let me instead use this valuable opportunity to deal immediately with the questions that should determine this election and that we all know are vital to the American people." The chattering in the hall came to a halt. Now he had their attention.

Cuomo put a new twist on one of Reagan's favorite lines, one he popularized during his 1980 campaign and used throughout his first term. "In many ways," he said, "we are a shining city on a hill. But the hard truth is that not everyone is sharing in this city's splendor and glory." He spoke of another side of this city, a side Reagan might not see "from the portico of the White House and the veranda of his ranch, where everyone seems to be doing well"—a place suffering with unemployment, poverty, crime, and an ever-fraying safety net. "There is despair, Mr. President, in the faces that you don't see, in the places that you don't visit in your shining city."

The Moscone Center was silent. This was not the speech anyone had been expecting. Where were the built-in applause lines, the corny jokes, all the usual clichés? There was an urgency and purpose to Cuomo's cadence that commanded the room to sit up and pay attention. He had them transfixed. What he was describing, Cuomo now told the delegates, was nothing less than a fault line separating their party from Reagan's.

"It's an old story," he said. "It's as old as our history. The difference between Democrats and Republicans has always been measured in courage and confidence. The Republicans—the Republicans believe that the wagon train will not make it to the frontier unless some of the old, some of the young, some of the weak are left behind by the side of the trail. 'The strong'—'The strong,' they tell us, 'will inherit the land.'

"We Democrats believe in something else. We Democrats believe that we can make it all the way with the whole family intact—and we have, more than once. Ever since Franklin Roosevelt lifted himself from his wheelchair to lift this nation from its knees—wagon train after wagon train, to new frontiers of education, housing, peace; the whole family aboard, constantly reaching out to extend and enlarge that family; lifting them up into the wagon on the way; blacks and Hispanics, and people of every ethnic group, and Native Americans—all those struggling to build their families and claim some small share of America.

"For nearly fifty years we carried them all to new levels of comfort, and security, and dignity, even affluence. And remember this: some of us in this room today are here only because this nation had that kind of confidence. And it would be wrong to forget that."

For forty minutes, he continued like this, rarely permitting the audience to interrupt him, as if fanfare would cheapen the message. His party had come to San Francisco aware of its dim November prospects. It was the height of the Reagan era, the tone for which had been set four years earlier when the new president used his inaugural address to declare that "government is not the solution to our problem—it is our problem." Old-style liberalism had never felt so out of date, so politically toxic. But in this moment, Cuomo was making his party forget all of this. He was summoning their pride in the America that their values had built—the America that Reagan and his crew were now threatening to dismantle. It was a valentine to the New Deal legacy of the Democratic Party, and an intensely personal one. Cuomo spoke of his father, an Italian immigrant who'd run a grocery store in the South Jamaica section of Queens.

"I watched a small man with thick calluses on both his hands work fifteen and sixteen hours a day," he said. "I saw him once literally bleed from the bottoms of his feet, a man who came here uneducated, alone, unable to speak the language, who taught me all I needed to know about faith and hard work by the simple eloquence of his example."

The cameras caught delegates in tears. He never did mention the

name of Walter Mondale, not that anyone minded. What he was saying felt bigger than one candidate, one election. He ended with a call
to action: "For the love of God: please make this nation remember
how futures are built!" Then he thanked the crowd, walked offstage,
got on a plane, and flew back home.

H e was all anyone in San Francisco could talk about. For years,
Reagan, a trained actor and master of public performance, had
flummoxed them by forging such an easy, natural connection with
Americans. To these exasperated, beaten-down Democrats, the governor of New York was nothing short of a revelation: at long last they
had discovered their own Great Communicator.

"New York Gov. Mario M. Cuomo," wrote the *Washington Post*'s
Haynes Johnson, "gave Democratic Party power brokers dramatic
reason to wish they could bring back the old smoke-filled room that
in conventions past produced deals to anoint someone other than
[Mondale] as the party's presidential nominee."

Gushed one delegate, "I am overwhelmed by his speech. He's got
great potential nationally. He talks from the heart. He's terrific." Said
another, "He's reaching an emotional chord that hasn't been touched
since John Kennedy." And a third: "That's who should be running
for president." California senator Alan Cranston said it was the best
speech he'd ever heard. Representative Tony Coelho, one of the top
Democrats in the House, called Cuomo "fabulous."

"We've got to get the floating voters who went Republican in
1980," Coelho said. "We need a message that the Democratic Party
is a home for them, and that's what Cuomo is saying."

Those watching at home, wrote *Washington Post* television critic
Tom Shales, "saw yet another star born on TV: Mario Cuomo, the
keynote speaker and New York governor who might this morning lay
claim to the title Son of the Great Communicator.

"Like Ronald Reagan, the absolute standard by which all others are
judged, Cuomo appears to have mastered the art of speaking directly

to the television audience while only appearing to be speaking to an assembled throng."

Mario Cuomo had come to San Francisco just another Democratic governor from the Northeast. He left the next big thing in American politics. It was the kind of sudden, only-in-America turn of fortune that Bill Clinton had been straining to achieve. Now, without even seeming to want it, Cuomo had pulled it off for himself.

Their defeat that fall was worse than even the most pessimistic of Democrats had braced for. Voters went to the polls in all fifty states on November 6, 1984, and in forty-nine of them they sided with Reagan. All that was left for Democrats, besides the eternally loyal District of Columbia, was Mondale's home state of Minnesota, and even there the president came a mere thirty-eight hundred votes from triumphing. Never before had someone come so close to pulling off a clean sweep of all fifty states. All told, Reagan took 59 percent of the national popular vote, his margin of victory a staggering seventeen million votes.

Nothing the Democrats tried in 1984 had worked. Putting a woman, New York congresswoman Geraldine Ferraro, on a national ticket for the first time? Reagan still got 58 percent of the female vote. Mondale's decision to double down on the party's traditional alliance with organized labor? Reagan still grabbed nearly half the votes cast from union households. More than a quarter of self-identified Democrats deserted their party and voted to reelect Reagan. About the only bedrock Democratic constituency left was minority voters. African Americans backed Mondale at a 91 percent clip, while Latinos gave him 66 percent of their votes. But the American electorate of 1984 was 89 percent white—and whites went for the president two to one. As beatings go, this one was thorough.

Worse for Democrats, this was becoming the norm. Once upon a time, from FDR all the way through LBJ, with only the interruption of Eisenhower, they'd been the dominant party in presidential politics. No longer. Since 1968, they'd now lost four of five White House races—three in landslides. Reagan's initial victory in 1980, dismissed as a near impossibility when the campaign began, had shaken them to the core. But now it seemed that everything they'd done in response to it had only made their plight more desperate. The talk now was of a Republican lock on the White House. How had Americans become so allergic to the idea of a Democratic presidency? What could they do about it? Who could rescue them?

Cuomo was a tantalizing possibility. Within days of Mondale's political demise, the venerable columnist Mary McGrory was out with a column heralding the star of the San Francisco convention the party's "last liberal hope." She'd been with him at a watch party in New York on Election Night. When he came to the podium, she noted, there were chants of " '88! '88!"—which, she added, made the governor grimace.

"Cuomo's appeal is obvious," McGrory wrote. "He is the popular governor of the second largest state, and the finest orator in the country. He is an intellectual who loves the meat and potatoes of politics. He is both thoughtful and zestful. And although he says his face is that of 'a tired frog,' he is totally telegenic."

The two-decade decline of liberalism was congealing into despair on the left, a sense that winning back the White House—if it was even possible anymore—would require some kind of traumatic ideological makeover. But now in this darkness, as if on cue, came Cuomo. What if all they'd been missing was him? What if there was nothing wrong with liberalism that couldn't be fixed by one magnetic and morally righteous orator? Revival without compromise. This was the promise of Mario Cuomo.

Not everyone was buying in, especially in Bill Clinton's neck of the woods. In these parts, Democrats tended to see Cuomo as the opposite of a savior, an emblem of every self-defeating choice their party had been making for decades, the recipe for yet more electoral devastation. Here was a very different way of grappling with the wreckage of Mondale, the rise of Reagan conservatism, and the Democratic Party's declining national fortunes. There was a regional aspect that was impossible to ignore, a dramatic political realignment playing out in what Democrats had for nearly a century called "the Solid South."

Reagan won everywhere on the map in '84, but nowhere bigger than in Dixie, where he gobbled up 64 percent of the vote—this in a place where, not long before, any association with the Republican Party was the kiss of death. It was a particular type of southerner with whom Reagan was striking such a resonant chord: white southerners, who cast more than 70 percent of their ballots for him. When it came to blacks in the South, fully 90 percent went for Mondale. Were Democrats headed for oblivion below the Mason-Dixon line? It was a chilling prospect for them, and all too imaginable. Carter, the peanut farmer from Plains, was now the only Democrat in two decades to win multiple southern states, and even his native appeal had its limits. In 1976, he nearly swept his home region. After four years as president, though, Georgia was all he had left; every other southern state went for Reagan in 1980.

Meanwhile, the South was experiencing a population boom with no end in sight. Its electoral muscle was expanding with each new census. Democrats had problems all over the map, but there was no one who believed a return to the White House was possible without a return to competitiveness in the South.

They could take heart in one fact. Even after Mondale's drubbing, they still held nearly every southern governorship and state legislature, plus the lion's share of Senate and House seats; and when it came to county courthouses, wide swaths of the South were still one-party territory, just like in the old days. These were not, by and large, liberal Democrats. "Boll Weevil" was the catchy term of the 1980s for the conservative southern Democrats who were providing pivotal congressional support to Reagan and his program.

The mixed messages spoke to a wider clash, between the national party and its leftward drift and the more moderate, even conservative, sensibility that still drove southern Democratic politics. Even as it rebelled against one, Dixie was clinging to the other. It was a tension fueled by many factors, but its roots could be traced all the way back to the country's original sin.

In the aftermath of the Civil War, the most liberal force in American politics had been Republicans from the North. They'd applied the most pressure on Abraham Lincoln to embrace emancipation and the Thirteenth Amendment and after the war set out to remake the South through Reconstruction—barring Confederate leaders from holding power, enshrining legal protections for blacks, and guaranteeing suffrage for all males. When white southerners mounted a violent resistance through the Ku Klux Klan, a Republican president, Ulysses S. Grant, former general of the Union army, mobilized federal forces to preserve the rule of law.

Thus was the South transformed, for a time, into something akin to a multiracial democracy. In 1868, southerners elected their first black congressmen; more than a dozen others would follow over the next decade. All of them were Republicans. Mississippi's legislature chose a black man, Hiram Revels, to represent the state in the U.S. Senate. He was also a Republican.

This racial progress was short-lived, however. The white resistance strengthened, and when the 1876 election ended in a deadlock, the white South saw a chance to free itself for good from the insult of Reconstruction. They agreed to hand the White House to the Republican candidate, Rutherford B. Hayes; in return, they got a promise: no more federal troops in Dixie. Reconstruction was over, and the white South could again do as it pleased. The last Republican governments vanished, white rule—white Democratic rule—was restored, and laws were drawn up to make sure the vast majority of blacks in the South would never cast a ballot again. The Jim Crow era had begun.

It would endure for generations, well into the twentieth century. From top to bottom, the South became a white Democratic bastion. Among themselves, southerners would wage fierce political battles. To the nation, they presented a united front: a wall of Democratic congressmen and senators bent on preserving "the southern way of life."

It was still an awkward marriage, these conservative southerners and their hostility to federal power making common cause with the urban and industrialized north; small-town segregationists and big-city ward heelers sharing the same tent. With time, the alliance frayed.

It was racial equality that tore it apart. Led by a young Minneapolis mayor named Hubert Humphrey, northern liberals pushed a civil rights plank into the Democratic platform in 1948. "The time has arrived in America for the Democratic Party to get out of the shadow of states' rights and to walk forthrightly into the bright sunshine of human rights," he said. Enraged southern delegations stormed out of the Philadelphia convention and called their own in Birmingham. They launched a rump party—the States' Rights Party officially, but most just called them Dixiecrats—and nominated South Carolina governor Strom Thurmond for president. "All the laws of Washington and all the bayonets of the army," Thurmond declared, "cannot force the Negro into our homes, into our schools, our churches, and our places of recreation and amusement." Thurmond took four southern states in November, but somehow Truman survived the rupture and won anyway.

But the intensity of the civil rights issue only grew, and sixteen years later another Democratic president finally delivered on that old platform plank. Beating back a southern filibuster, Lyndon Johnson signed the Civil Rights Act on July 2, 1964, putting an end to segregated businesses and public spaces.

The next year, though, Johnson signed the Voting Rights Act and African Americans began registering in droves. In the days of Reconstruction, they'd been Republicans. Now they were Democrats. Southern whites, on the other hand, found themselves questioning their attachment to the party of their birth. For the first time since Reconstruction, multiracial democracy was returning to the South.

Never again would either party nominate a civil rights opponent for national office, but now there was an opening for Republicans to plant their flag in a place that had long been off-limits to them. It was a deal with Thurmond that ensured the GOP nomination in 1968 for Nixon, whose "southern strategy" won over many of those aggrieved white voters. Reagan was aiming straight at them, too, when he kicked off his general election campaign in 1980 in Philadelphia, Mississippi, the notorious scene of a triple homicide committed against civil rights activists by the local white resistance just sixteen years earlier. "I believe in states' rights," Reagan told the crowd. Officially, the line had nothing to do with race, but to critics it was a wink as ugly as it was blatant.

Republicans bristled at the charge of race baiting. Civil rights, their party platform proclaimed, was a settled matter, and America was better for it. But race was so intertwined with the most contentious issues of the 1970s and '80s, from school busing and affirmative action to welfare and crime, that it was possible to stoke prejudices in ways that were politically advantageous.

Moreover, Democrats moved in the opposite direction, embracing newly enfranchised blacks in the South and forging tighter bonds with nonwhite voters everywhere. The New Left movement of activists and academics gained influence within the party, pushing it toward a vision of social equality and powering the nomination of George McGovern for president in 1972. Quotas to ensure diversity were imposed starting with that year's convention, and the Democratic Party became associated not just with civil rights for blacks, but with feminism, the fledgling gay rights movement, migrant rights, and a host of other cultural causes.

Black politicians were making their way to prominence, too. By 1984, two of the nation's three biggest cities were led by black Democrats—Harold Washington in Chicago and Tom Bradley in Los Angeles. This was also the year that the Reverend Jesse Jackson, a veteran of the civil rights movement, sought to channel the growing clout of black voters within the party into a full-fledged national campaign for president. He won three primaries and caucuses and more than three million votes, the best showing ever by a black candidate for national office.

Pitted against the Republican Party of Reagan, this Democratic Party—the national one—was facing an existential crisis in the South. It could win the support of black voters, but not nearly enough whites. The math was the math. Not only did Reagan take every southern state in 1984; the margin wasn't within twenty points in any of them.

But then there were the Democrats like Bill Clinton, one step removed from their Washington brethren and thriving with the same voters who were rejecting their party's national candidates. As Mondale lost Arkansas by twenty-two points on Election Day, Clinton was reelected by twenty-six. He personified a new breed of southern Democrat, a younger generation whose political maturation had coincided with integration. Like their parents, they'd been born into the Democratic Party, but the South in which they launched their political careers had been transformed. Segregation was dead, blacks were voting in real numbers, and whites were drifting toward the GOP, a volatile mix of variables the old guard had never had to worry about.

These new southern Democrats strived for a conciliatory tone on race and welcomed African Americans into their party, while still minding the conservative cultural values of their white constituents. They liked to emphasize improvements to public education and economic development and tended to keep their distance from the national party, ducking the fraught cultural issues of the day.

Some lived in fear of their national party and its move to the left. How many more McGoverns and Mondales could they survive? Others saw opportunity. If national Democrats needed to make inroads in the South to win back the White House, well, then maybe they'd come looking for a candidate who could do just that. There was no question which of these camps included Bill Clinton.

TWO

T he incursion of the national GOP was reversing the age-old plight of the ambitious southern Republican. In Dixie's one-party heyday, it was a novelty just to find a Republican listed on a ballot, and it was political and social suicide to be that Republican. Now, there was opportunity not just to break into the game, but to go far, and fast.

The emerging southern Republican Party featured its share of aging Democratic expats, like Thurmond and North Carolina's Jesse Helms, but it lacked a calcified pecking order. Democrats had to worry about waiting their turn; if you were a Republican, the future was already here. Maybe that's what drew in young Newt Gingrich.

The ambition was obvious. He was twenty-seven years old and had been a member of the West Georgia College faculty for less than a year when he applied to be the school's president. It was the spring of 1971 and he was just receiving his Ph.D. in European history. He knew how it looked, and he didn't mind. The quiet, diligent life of an academic wasn't what he'd been seeking when he came to this little agricultural college ten miles from the Alabama border. He was a man of big and urgent ideas, he wanted it known, deserving of a proper perch from which to pursue them. He didn't get the job.

It was politics that Gingrich was really playing. He'd caught the bug early, moved by a teenage visit to a World War I battlefield in France. Even though it was four decades after the fact, the carnage of Verdun was still apparent that day. This stuff's all real, he would recall thinking. It awakened a passion for history, and he drew a connection to politics as the preeminent tool for shaping it. He wanted to change the world.

He was a military brat, a soldier's son born in Pennsylvania, raised for a time on bases in Europe and finally brought to Georgia in time for his senior year of high school. He was eager for adulthood. Only a year after graduation, he married his geometry teacher. He was nineteen, she was twenty-six, and soon they had a young daughter, then a second.

The South agreed with Newt, who graduated from Emory in Atlanta, then took his new family to New Orleans for graduate work at Tulane. Republican politics called to him. When New York governor Nelson Rockefeller ran for president in 1968, Gingrich headed up his campaign in the Southeast. He was still a full-time student, just twenty-four years old. These were the Wild West days of GOP politics in the South. A young man with a little moxie and a lot of self-regard could come to town and make himself pretty important pretty quickly.

Georgia never left his mind, though. It was an enticing place for a hungry Republican.

Like the rest of the South, it was in the throes of realignment. From the end of the Civil War through 1968, it had never once voted for a Republican for president, and only twice in that time did the GOP candidate even break 40 percent. Democratic dominance was common at all levels of the ballot. Georgia's governorship, both of its U.S. Senate seats, its entire House delegation, the state legislature, and county courthouses: each a bastion of single-party rule for nearly a century.

When LBJ broke a southern filibuster that had lasted for generations and signed the '64 Civil Rights Act, Georgians retaliated by voting for Barry Goldwater by eight points in the fall. The GOP nominee had some coattails, too. Calling himself a "Goldwater Republican," Bo Callaway became the first Republican since Reconstruction to win a congressional seat from Georgia. Two years later, in 1966, two more joined him in the delegation. The white backlash was injecting vital energy into the Georgia Republican Party, something that had been missing since the nineteenth century.

In 1972, Richard Nixon earned 75 percent of the vote in Georgia against George McGovern, the antiwar senator whose activist-fueled campaign helped to align the Democratic Party with liberal cultural values. At least in its national form, it was the Republican Party that was more in tune with Georgia's conservative sensibilities. For an up-and-comer like Gingrich to break in, the challenge was to convince these Georgians who were abandoning the Democratic Party in presidential races to do the same for every other office on the ballot.

It sounded simple, but the reality on the ground could be thornier. Gingrich took a leave of absence from West Georgia in 1974 to challenge John J. Flynt, the local congressman. A ten-term incumbent, Flynt was a Democrat of the Old South variety, an ardent segregationist who'd shunned McGovern and embraced Nixon. Good luck attaching him to the growing cultural liberalism of the national Democratic Party,

Gingrich needed a different wedge, and it was probably just as well. He'd been on the side of integration from the beginning, influenced by the integrated military bases of his youth. And his work for Rockefeller, a liberal Republican from the Northeast, had put him in direct conflict with the party's Goldwater-style conservatives. If anything, Gingrich's politics were somewhere to the left of Flynt's. The first Earth Day had been celebrated a few years earlier and he spoke approvingly of the new conservation movement. Invoking Alvin Toffler's *Future Shock*, a bestselling phenomenon of the early '70s, he talked about technology and its potential to unleash vast societal disruption.

There was no question racial politics were feeding the growth of the southern Republican Party, and even indirectly, it would benefit Gingrich. But it wasn't just the end of Jim Crow that was transforming the South. The promise of plentiful employment, inexpensive living, and warmer weather was fueling a wave of migration from the North. Developments and suburbs were popping up and a New South was emerging.

Gingrich settled on a theme he'd turn to over and over in the years to come. He would run as a reformer and stoke a revolt against establishment power. He painted Flynt as an entrenched insider, a cog in a political machine that took care of its own at the expense of everyone else. The pitch sidestepped race and wasn't ideological. It was new versus old, stale versus fresh, the stagnation of the status quo against the excitement of blowing it all up. Flynt chaired the House Ethics Committee, which Gingrich seized as a weapon: If Flynt was doing his job, he'd ask, why was there so much corruption in Washington?

Of course, that was just the problem for Newt. The corruption that was on everyone's mind in 1974 involved Watergate. Nixon's scandalized presidency was in collapse, and by August he was forced to resign. When his successor, Gerald Ford, pardoned him a month later, the public's revulsion escalated further. As the November midterm approached, Watergate was dragging down Republicans everywhere. Democrats were on their way to gaining forty-nine House seats, a giant number by any measure, and particularly amazing since they already had 242 to start with. In northwest Georgia, it was too much for Gingrich, but only barely. He finished with 49 percent of the vote, nearly unseating a long-tenured Democrat in one of the worst Republican years in history.

At first, it looked like his big breakthrough was merely being delayed by two years, until the next election in 1976. But then Jimmy

Carter got in the way. Selling himself as a humble peanut farmer from Plains, the former Georgia governor engineered an upset for the ages and won the Democratic presidential nomination. A surge of home state pride gave Carter 67 percent of Georgia's vote in November, his best in any state and just enough to carry his fellow Democrat Flynt to another narrow win over Gingrich. Even in victory, Flynt sensed he was out of miracles. At sixty-four years old, he declined to seek reelection in 1978. There'd be no third race against the relentless college professor.

Instead, Gingrich drew as an opponent Virginia Shapard, a local state legislator. It was no cakewalk. Gingrich liked to run against the idea of entrenchment, which made Flynt an ideal foil, but the forty-two-year-old Shapard had the appeal of an outsider. She, too, was running on a reform platform, and her gender—only three women had ever won congressional races in Georgia—further marked her as a different kind of candidate. There was still plenty of residual Democratic loyalty in the district, too, and Shapard was poised to profit from it.

For Newt, it was make or break. With a third consecutive loss, he'd be branded a perennial candidate, maybe written off for good. The race also coincided with a foundational event for the modern conservative movement—and, as it would turn out, Gingrich's own career.

It was in June 1978 that California voters weighed in on Proposition 13, a ballot initiative that called for steep property tax cuts and severe limits on the ability of lawmakers to raise state tax rates in the future. Critics called it radical, leaders from both parties denounced it, and experts assumed it would fail. Instead, it passed in a landslide, and the shock waves rippled across the country. "A revolt of middle-income and upper-income taxpayers," the *Washington Post* declared it.

Suburban America had exploded in size since World War II, and now its denizens were growing hostile to a government that could seem distant and detached from their lives. Never before had the subject of tax cuts divided the parties, but now a band of conservative Republicans moved to make it their signature cause. Nationally, Jack Kemp led the way. A congressman from western New York, where he'd played football for the Buffalo Bills, Kemp latched on to a counterintuitive idea: that by stimulating economic growth, cutting tax rates could actually produce *more* government revenue. Skeptics belittled it as hokum, but supply-siders like Kemp bought in hard. At a time when the top marginal tax rate stood at 70 percent, Kemp called for a 33 percent cut and challenged his party to join him.

Back in western Georgia, Gingrich heard the call. It was an idea that matched his conception of politics. It was big and bold and had disruptive power. It was also ridiculed by the elites of both parties, while connecting with the grass roots. Republican orthodoxy had long regarded balanced budgets and restrained spending as sacred. But now there was a populist tide rising, one that could wash away all that old thinking, and Newt Gingrich was determined to ride with it. It wasn't just taxes. With a gathering conservative backlash to the Supreme Court's 1973 *Roe v. Wade* decision, Gingrich placed a new emphasis on abortion. Feminists were ascendant nationally, pushing for an Equal Rights Amendment to the Constitution, and now they were meeting resistance, especially in places like western Georgia. Moral traditionalism had always been prevalent with the district's old guard, but even some newcomers were hesitating in the face of so much change. Ads of mysterious origin showed up in local newspapers telling voters that "Newt will take his family to Washington and keep them together. Virginia will go to Washington and leave her husband and children in the care of a nanny."

The margin was close, but on his third try, Newt Gingrich was elected to the House of Representatives. He was just the fourth Georgia Republican since Reconstruction to win a seat. At thirty-five years old, he had made it to the national political stage. Of course, as a freshman in the minority party in the lower chamber, he'd be a bit player in Washington—unless he could find a way to reinvent the game. He already had a few ideas.

They called it the Permanent Democratic Congress. It stretched back to 1954, Eisenhower's midterm, and a quarter century later it was stronger than ever. The House that Gingrich walked into in January 1979 had only 158 Republicans in it, and 277 Democrats. And this was *after* a strong midterm election for the GOP, their best in more than a decade. No one in Washington even talked about the possibility of Republicans someday winning it back. They hadn't even crossed the two-hundred-seat mark since the 1950s.

The freshman from Georgia's Sixth District told the press, told his new colleagues, told anyone who would listen: it didn't have to be this way. If Nixon just a few years earlier could get forty-nine states to vote Republican in a presidential election, why couldn't they convince 218 House districts to do the same? The key, he believed, was in nationalizing congressional politics through confrontations with the ruling Democrats that would clarify the differences between the two parties.

A liberal consensus had dominated Washington since the Depression, when FDR combated economic crisis with a massive expansion of the federal safety net. In the sixties, LBJ added to it with Medicare and his "war on poverty," but his presidency coincided with explosive civil strife, including riots in Detroit, Newark, Los Angeles, and other cities; waves of antiwar protests; and the rise of a counterculture. A backlash was building in blue-collar and middle-class white America, which Nixon nodded to with his talk of "the great silent majority." There was space, Gingrich said, to redefine American politics, if only Republicans would look for it. A few of his colleagues thought he was on to something. Most took him for a blowhard.

He would have to show them. There was a Democrat from

Michigan, Charlie Diggs, who'd just been convicted on federal corruption charges. It involved his office payroll and a kickback scheme,
and he was still casting votes in the House while he exhausted the
appeals process—and no one seemed to care. Tip O'Neill, the House
Speaker, said it was up to Diggs what to do. The way O'Neill saw it,
this was just the gentlemanly way to handle things. It was also the
House's tradition, and Republican leaders raised no objection.

But Newt did. When Diggs voted for a routine bill to raise the debt
ceiling, Gingrich took to the floor to object. It's an outrage, he told
colleagues, for the House leadership to allow "for an indefinite period
a convicted felon to vote on the laws of the United States."

He demanded a vote to expel Diggs. No one had seen anything
like this before. Expulsion was for only the rarest of circumstances.
Three members had been booted for treason during the Civil War,
and none since. Colleagues just didn't do this to each other.

He made enough noise to compel the Republican leader, John
Rhodes, to offer a motion to expel. The vote wasn't close, 322–77
against expulsion. Not even half of the Republican conference went
along with Gingrich, but seventy-six of them did. It was a start. Gingrich was recruiting fellow travelers, restless, hungry Republicans
who believed they could topple the Permanent Democratic Congress.
"What are you doing for the next ten years of your life?" he'd ask
would-be lieutenants.

Rhodes soon retired and was succeeded by Robert Michel, a
courtly pol from Peoria, Illinois. He was a consummate inside player,
chummy with everyone—including Tip O'Neill, one of his favorite
golfing buddies—and a throwback to the days before politicians had
to worry about talking on camera. "Mr. Nice Guy" was his nickname,
"Golly!" was among his favorite expressions, and "a compromising
legislative technician" is how the *Washington Post* described him. "He
doesn't have a vindictive bone in his body," one of his many friends
gushed.

To Gingrich, Michel was exactly the kind of leader that permanent minority status creates. He'd been in the House since 1956 and had spent every day since at the mercy of Democrats. He was too used to it, too resigned to it, maybe too comfortable with it. The Gingrich crew was small but growing, every election depositing in the House a new batch of Republican freshmen receptive to the pitch.

Officially, they called themselves the Conservative Opportunity Society. Others labeled them the Young Turks. Their older colleagues preferred terms unfit for print. They were Republicans in the Reagan mode, in a party that still had plenty of moderates and liberals of the Rockefeller sort. Ideologically, they were moving the GOP to the right, closer to Gingrich and further from Michel. Tactically, they shared Gingrich's sense of urgency. They were becoming a guerrilla army, looking to provoke the Democrats and jar their contented colleagues into becoming fellow warriors. They held their own strategy sessions, separate from the whole Republican conference, and plotted to exploit the chamber's rules for maximum partisan definition and contrast. To find their most potent weapon, it turned out, all they had to do was stand on the House floor and look up.

The Cable Satellite Public Affairs Network was a private entity bankrolled by the fledgling cable television industry. It was the brainchild of Brian Lamb, who'd dabbled in politics and journalism in D.C. and who grasped the niche potential of the emerging cable universe. Lamb's idea was to place cameras in the House chamber and to broadcast its proceedings, gavel-to-gavel and without commentary or interruption. It was a democratic vision for the television age, bringing the people's house into people's living rooms in every corner of the country. C-SPAN was chartered in 1977 and the cameras went live in 1979, Gingrich's first year in Congress. The content was bone dry, save for the occasionally colorful floor speech or tense roll call vote. By 1984, there were nearly twenty million homes with C-SPAN as part of their cable package, and the number was only growing as cable became a must-have item for Americans.

When no one was paying attention, Gingrich saw an opportunity.

A House rule dictated that any member who desired to could speak on the floor about any topic for up to an hour after the conclusion of daily business. Special Order speeches, as they were called, were typically delivered by members who wanted to read into the official *Congressional Record* the name of someone in their district—to congratulate a local Little League team, for instance, or commemorate a recently deceased community icon. Often, members didn't even deliver these speeches and instead just had their comments entered directly into the record.

But what if, through the magic of television, this formulaic tradition could be made to look like something completely different? When Gingrich and two of his allies, Robert Walker of Pennsylvania and Vin Weber of Minnesota, began claiming Special Order time, no one initially thought anything of it. A few backbenchers from the permanent minority party were asking to speak to an empty chamber after hours with absolutely no power to shape, advance, or block legislation in any way. *You want to waste your time on that? Knock yourselves out.* Their colleagues didn't get it: They weren't the audience Gingrich was thinking about. By rule, the C-SPAN camera was trained on whoever controlled the floor. For all the casual viewer knew, the chamber was packed and there was serious action taking place.

That's exactly how the Gingrich gang treated their Special Orders. Night after night, they'd step forward and air their grievances against the majority party and its tyrannical rule, tear into the corruption of the modern Democratic Party, and sing the virtues of small government and personal freedom—all without the audience hearing a whimper of protest from the other side. There was no Fox News Channel in 1984, no MSNBC, and only a skeletal version of CNN; nor was there social media, YouTube, or any online world to speak of. Opportunities to reach national audiences directly were scarce, but Gingrich was inventing a brand-new one. He was the producer of a nightly television show, and also one of its stars. He and his buddies, sniffed liberal columnist Mary McGrory, were making themselves "the idols of people who do not have much to do."

They were pushing the limits, very intentionally, and on May 8, 1984, finally hit the nerve they'd been looking for. After the close of business, Gingrich and Walker claimed their usual Special Order time. They took turns reading from a polemic that accused the modern Democratic Party of reflexively treating the United States as a guilty party in the global arena. Then Gingrich brandished a letter that a group of Democratic congressmen had recently sent to Daniel Ortega, the leftist militant who'd seized power in Nicaragua a few years earlier. Its official purpose was to encourage democratic elections, and its signatories included Majority Leader Jim Wright and Edward Boland, a Massachusetts Democrat known as Tip O'Neill's best friend. Gingrich seized on its greeting—"Dear Commandante"—and accused the Democrats of violating the separation of powers and undermining President Reagan's ability to set foreign policy.

When word of what had happened reached O'Neill, he was enraged. To his ears, Gingrich and Walker weren't merely attacking Democratic foreign policy. They were challenging the patriotism of colleagues—by name and without giving them a chance to defend their good names. Two nights later, with another Gingrich Special Order show under way, O'Neill invoked his authority as Speaker and ordered the C-SPAN camera to pull away from the podium and show the entire chamber. The operator complied, and an aide relayed the news to Walker, who then announced on the air that "it is my understanding that the cameras are panning the chamber, demonstrating that no one is here in the chamber to listen to the remarks."

O'Neill was satisfied, believing he'd shamed the renegades. He had no idea he'd just taken the bait. The Speaker's action, Gingrich charged, wasn't just a violation of the rules established when C-SPAN came to the House; it was an egregious example of the abuse that an arrogant Democratic machine had been inflicting on Republicans for far too long. The chamber was often just as empty, he noted, when Democrats ran through official business during the day, but no one would ever think to pan the chamber for that. Now the press was perking up. So were Republicans.

When the House came to order on the morning of May 15, Gingrich rose and requested a point of personal privilege. Translation: he had a bone to pick with another member—in this case, the Speaker of the House. Presiding over the chamber, O'Neill granted him an hour of time and announced that "the chair wants to listen with keen interest to the gentleman from Georgia." He handed over the gavel, stepped down from the Speaker's rostrum, and joined his colleagues on the floor.

Gingrich positioned himself in his normal late-night spot, but this was no Special Order speech. The chamber was packed. Reporters from national outlets were watching. He claimed he'd given fair warning to the Democrats he'd called out by name. He'd sent them a letter, he said, giving them a heads-up and challenging them to show up and debate. When he began to read from it, there were objections from the Democratic side of the chamber. That letter had never reached their offices, the Democrats began to say. Gingrich cut them off. This was his time, and he would be yielding only for questions, not lectures.

Confidently, he continued for several minutes, insisting he'd questioned no one's patriotism and accusing Democrats of fomenting "a McCarthyism of the left." This was too much for O'Neill, who stood up behind a microphone. "Will the gentleman yield?" he asked. Gingrich was in his third term. He had no seniority, no major legislation to his name, no clout in any traditional sense. And here was the Speaker of the House, standing on the floor with all of their colleagues and requesting a dialogue with him.

"I'm always delighted to yield to our distinguished Speaker," Gingrich replied.

The seventy-one-year-old O'Neill was an old-school pol, burly and white haired, who'd come of age long before the television era. As a public performer, he was no match for the poised and glib Georgian. "It's very interesting the way the gentleman talks," O'Neill said, "as an apologist for the remarks that he made the other day. Let's look at the truth of the thing. Let's look at the truth . . ." Gingrich interrupted. "I reclaim my time!"

"I thought you said that you would give to the Speaker the courtesy . . ."

"I am not in any sense apologizing," Gingrich said. "I'm explaining, I'm putting in context, I'm not apologizing."

He let O'Neill continue. The Speaker pointed out that the House's postal service had recently informed members that due to high volume it would take four to five days to deliver mail to their offices—meaning, O'Neill said, that Gingrich's heads-up letter had been sent too late for them to get it in time. He thought he'd landed a nice blow. Gingrich wasn't fazed.

"Let me make the point to the Speaker. . . . My point to the Speaker is, as you have pointed out to us again yesterday, you are in charge of this House. If you are saying to me that your postmaster is incapable of delivering from one member to another the mail in the system you run, then I think you may want to appoint a commission to look into the post office."

It was a trivial point, but that hardly mattered. It was Gingrich's tone, his dismissiveness toward such an august figure, his defiance. Here was a backbencher from the minority party baiting the House Speaker into a fight and flinging his words right back in his face. No one had ever seen anything like this, least of all the Speaker himself. O'Neill continued, his rage now boiling over. He mentioned his friend Ed Boland, "a gentleman whom I have the greatest respect for."

"My personal opinion is this," the Speaker said. "You deliberately stood in that well before an empty House and challenged the people and challenged their Americanism. And it's the lowest thing that I've seen in my thirty-two years in Congress."

There were cheers. And gasps. Trent Lott, the Republican whip, spoke up. House rules forbade derogatory comments about members and the Speaker had just violated them, he said. Lott asked the presiding officer, Joe Moakley, an O'Neill ally from South Boston, that

O'Neill's comments be taken down, or removed from the record. This was heavy stuff.

Moakley stepped away to consult with the parliamentarian. For several minutes, the House was at a standstill. A beaming Walker strode over to Gingrich. They were used to performing for C-SPAN diehards, but now they had the whole political world's attention. The parliamentarian told Moakley that O'Neill's use of "lowest" was over the line. Moakley had no choice but to rule in Lott's favor. "The chair feels that that type of characterization should not be used in debate." For the first time in history, the words of a House Speaker had been taken down.

Gingrich moved to rub it in. "In many ways," he said after the ruling, "it is *my* patriotism that is being questioned."

"I'm not questioning the gentleman's patriotism," O'Neill bellowed. "I'm questioning his judgment. I also question the judgment of the chair. I was expressing my opinion. As a matter of fact, I was expressing my opinion very mildly, because I think much worse than what I said."

It went on for a few more minutes until Gingrich's time finally expired. He gathered his papers and marched off the floor. There were cheers, loud cheers, and not just from his merry band of fellow guerrillas. Almost all of the Republican members were clapping, and many were on their feet. Gingrich had been pleading with them to embrace confrontation as a strategy. Now he'd shown them what it looked like, and maybe more importantly, what it could *feel* like.

"What a great show!" Illinois Republican Henry Hyde exclaimed afterward. "Here you have Tip sitting up there like an emperor on his throne, lord of all he surveys, and there's a mosquito buzzing around his ear. And for all his might and power, he can't get rid of the mosquito—and it's driving him crazy!"

Lott acknowledged that "there were hurt feelings and tempers involved here, and both sides have some legitimacy," but now even the skeptics on the Republican side were starting to see potential in his tactics.

"I am just not the kind of person who seeks confrontation," Hamilton Fish, a liberal Republican from New York, said. "But as I watch what has happened this year, I can see that confrontation is effective. Those of us who were not inclined to confrontation have now discovered that pressure, and tough pressure, is the way to get results."

Gingrich was a gadfly no more. The rebuke of O'Neill was a national story. Word spread in conservative circles everywhere: Who is that guy Gingrich? Why don't we have more like him?

The Republican applause that May morning wasn't quite unanimous. As Gingrich soaked in the cheers, the Republican leader sat on his hands. Bob Michel, "Mr. Nice Guy," wasn't feeling an exhilarating burst of partisan pride. The currency of the House he knew and loved was cordiality and respect. This was not how business was done. Michel and Gingrich, two decades apart in age, represented two radically different paths forward for their party, Steven Roberts wrote in the *New York Times,* and there was growing tension in the ranks.

"Veteran Republicans like Mr. Michel . . . express frustration at their minority status and at the power of the Speaker, but they have generally chosen to influence events by dealing with the Democratic leadership, not by denouncing them.

"That basic concept is rejected by Representative Gingrich and his allies. They view the Speaker as a tyrant and compromise as the art of surrender. . . . Through visible public clashes, they feel that they can spotlight their agenda and draw a clear distinction between the parties. Their aim is not to pass bills but to influence the viewing and voting public at home. With most of their careers still in front of them, they are impatient and ambitious; before they leave public life, they want the Republicans to become the majority party in the House.

"Representative Michel and other Republican leaders share their aims but are troubled by their tactics, and the leaders feel that something precious is in danger."

The Speaker as a tyrant; the Democratic majority as a dissent-crushing machine; the House as an institution corrupted by a generation of single-party rule: Gingrich was planting seeds with his fellow Republicans, prodding them to renounce coexistence and take up arms against their oppressors. The O'Neill showdown certified him as a force. Then came the 1984 election, which would hand the Gingrich army the ultimate recruiting tool.

They called it the Bloody Eighth, the congressional district taking in the southwest corner of Indiana, from the college town of Bloomington on one end to the Kentucky border on the other. Politically competitive turf, it was known to incubate and then almost immediately snuff out one promising political career after another. Between 1966 and 1982, the voters of the Eighth District threw their member of Congress out of office six times, a level of volatility unmatched by any of the nation's other 434 districts.

The endangered incumbent in 1984 was a Democrat, Frank McCloskey, aiming to defy history and the Reagan reelection tide and win a second term. His challenger was a rising Republican star in Indiana, twenty-eight-year-old Richard McIntyre, a state legislator and military lawyer. It was one of the most closely watched House races in America, and when all the precincts had reported on Election Night, it looked like McCloskey had survived by just seventy-two votes, out of more than 230,000 cast.

McIntyre wouldn't concede, though. There were tabulation issues in several counties. A December recount put him ahead by thirty-four votes, and Indiana's Republican governor certified him as the winner. Now it was McCloskey's turn to dig in. "They are trying to steal it!" he cried. In Washington, his fellow Democrats agreed, and they had the power to do something about it. With their mammoth majority, they refused to seat McIntyre with the new Congress, then voted to turn the matter over to a task force of three House members to investigate. Two were Democrats and one was a Republican. The Republicans claimed the fix was in. "On any point of substance, I'm going to lose two-to-one," said Bill Thomas, the GOP member. "Why bother?"

The process wound along, slowly, torturously. In February, a re-count left McIntyre with an even bigger lead, but the Democrats wouldn't relent. There were ballots in Evansville, one of the dis-trict's anchor cities, that had been declared invalid that they wanted counted. Auditors from Washington were called in to examine the ballots, nearly five thousand of them, one by one. There was no clear standard, just chaos and cries of unfairness. In late April, the task force called a public hearing in Evansville to settle the matter. It ran for five hours, but every key vote broke the same way, until McCloskey was declared the victor by four ballots. "What happened . . . can be characterized as nothing short of a rape," Thomas said.

The action returned to Washington, where Republicans had been watching with building rage. From the backbenches, Gingrich had been priming his colleagues for a moment like this, and this time he wouldn't have to plead. Democrats had the numbers to seat McCloskey, but now the minority party was determined to make a statement. "I think we ought to go to war," said Wyoming's Dick Cheney.

The vote to seat McCloskey was called on May 1, 1985. There were 236 yeses and 190 nos. Speaker O'Neill called for McCloskey to enter so that he could take the oath, and as he did, Republicans, ev-ery last one of them, rose from their seats and exited. Even Mr. Nice Guy. O'Neill called out to Bob Michel, his golfing buddy. "Would the gentleman remain within until I have had an opportunity to ad-minister the oath?"

"No," Michel replied.

Moments later, Frank McCloskey officially began his second term representing the Bloody Eighth. Democrats cheered as their majority climbed to 269–166. From the Republican side, there was no reaction. They were long gone.

Michel called it "the most notorious example of raw political power." But he still had his limits. On his way out of the chamber, he offered a handshake to McCloskey. Even in this moment, Michel remained intent on preserving the culture of the House he'd always known. It was a gesture that irked his fellow Republicans, and not just the Gingrich wing. "I try to tell people," Michel protested, "that we've got to have at least one person left on our side to go over there and talk. We can't sever communication, the business of government goes on." His friends on the other side of the aisle, and he had many of them, sympathized. "We trust each other," Jim Wright, the majority leader, said, "but he cannot control the red-hots in his own party."

Gingrich knew a win when he saw one. The Michel-led walkout, he crowed, would never have happened even a year ago. "I think you're going to see a much tougher and a much more militant Republican Party," he predicted. "I think it changes permanently the nature of the Republican Party of the House."

THREE

Not long after Reagan's second romp, a group of Democrats declared war on their own party. They were mainly from the South, but some came from out West, and there were even a few northerners mixed in. They were political moderates, and almost all of them were white and male. Apprehension and dread united them.

By now, the Democrats' futility in presidential elections was well established. They'd lost four of the last five, and twice in that time carried just a single state. The fear now was of further erosion. The defections from the white South were trouble enough, but the party of FDR was also shedding its traditional base in the blue-collar Northeast and Rust Belt, where "Reagan Democrats" had flipped state after state into the Republican column. These were trends the breakaway Democrats resolved to reverse.

The chairman of the Democratic National Committee had lobbied them to stay put and to work for change within the party's structure, but they'd concluded that only pressure from the outside would work. They would call themselves the Democratic Leadership

Council, but most would come to know them just as the DLC, and they introduced themselves on the last day of February 1985.

"The perception is that the party has moved away from mainstream America," said Georgia senator Sam Nunn, one of the organizers. "We want to move it back to the mainstream."

Their immediate focus was on 1988. They wanted their party to stop putting up liberals like Mondale, and slowly a plan took shape. With their power in their home states, they would create a massive, single-day regional presidential primary, a wall of southern and border states voting on the same early date. It was a carrot meant to entice one of their own into the race, but it was also a stick, a blunt message to the liberals dreaming of a Cuomo-led revival: you're going to have to get through us.

Bill Clinton put his name on the list, and certainly he fit the mold of a DLC Democrat. He was a centrist who'd cultivated his state's rising black electorate while minimizing white resentment. Confronted with a divisive issue, he'd find a middle ground less artful politicians couldn't even see. As governor, for instance, he increased taxes to pay for more aid to schools, but only while forcing competency testing on teachers and their enraged union.

"People are saying they are willing to take the shirts right off their backs if it will help improve education," Clinton explained. "But they demand accountability. They want us to get rid of the teachers who are really bad."

This was his approach. He'd praise integration as a noble feat while still decrying "busing and breaking up neighborhoods." He'd voice approval of the Equal Rights Amendment "in principle," but would expend no political capital pursuing its ratification. Somehow, it seemed, he could convince liberals he shared their goals even as he made conservatives believe he was protecting them from the excesses of the left.

His mind was on 1988, but there was a pecking order in southern politics, and Clinton was far from the top of it. If anyone had the right of first refusal, it was Charles Robb, the pedigreed Virginia governor

THE RED AND THE BLUE

who, despite astronomical popularity, was term limited in 1985. He had been a Marine Corps captain during Vietnam, and his wedding in 1967 had been a national event; he tied the knot with President Johnson's daughter Lynda Bird in the East Room of the White House. Robb, observed David Broder, seemed to be the complete package— "movie-star handsome, comfortable on TV, middle-road but modern in his thinking, well-connected, southern and dazzlingly successful as a politician." He also, after some flirtation, decided to stay out of the '88 race, opting to run for a Senate seat instead.

Robb urged that Nunn, his fellow DLC founder, make the race. Nunn had learned politics through his great-uncle, Georgia congressman Carl Vinson, a segregationist and the long-tenured chairman of the House Armed Services Committee, on whose staff a young Nunn served before returning to Georgia to launch his own political career. Running for the Senate in 1972, he found himself linked in his Republican opponent's campaign posters to George McGovern. In response, Nunn secured an endorsement from a different George— Wallace, the Alabama governor who had once tried to block federal troops from integrating his state's flagship university. He won that election and had stayed away from such overtly racial politicking since. Nunn was a conservative Democrat, voting with the Reagan White House more than almost any other Democrat, and with his strong ties to the military and hawkish sensibilities he loomed as a one-man rebuttal to the Republican charge that Democrats were a bunch of McGovern-ish softies. But he could surprise you, too. He'd made headlines opposing Reagan's push for new missiles, and now, as the '88 race came into focus, he was one of the president's chief critics over the Iran-contra affair. When Robb said no, the Nunn boomlet was on.

"When he strode into the crowded conference room in the Capitol," the *New York Times* reported in early 1987, "several reporters began humming 'Hail to the Chief,' musical fanfare that has officially graced the entrances of only one Georgian, Jimmy Carter."

But Democrats had just won back the Senate in 1986, and now

Nunn was the chairman of the Senate Armed Services Committee. It was a perch he'd long coveted, so he, too, passed on the presidential race. Even then, with the South's two brightest Democratic stars saying no, it still wasn't Clinton's turn. There he was in Little Rock, itching to get started, but powerless to make a move until a different Arkansas Democrat made up his mind.

Dale Bumpers was known as "the northerner's favorite southerner." He had defeated two iconic segregationists in his famed career—Orval Faubus, in the 1970 gubernatorial primary, and J. William Fulbright for the Senate in 1974. He was something of an exception to the Democratic template of the New South, more willing to align himself with national liberal causes. He was, for instance, the only southern senator to oppose a bill to prevent federal judges from ordering busing. But he was still to the right of his national party, particularly when it came to trade and organized labor.

With Nunn and Robb out, Bumpers moved toward a candidacy, only to pull the plug in late March of 1987. He was sixty-one and had just undergone knee surgery, forced to walk with a cane as he recovered. The physical and personal toll of a national campaign, he said, would just be too much. "It would mean a total disruption of the closeness my family has cherished," Bumpers said.

Suddenly, finally, it was there in front of him: the opening Bill Clinton had been angling for all these years. He was forty years old. This was his fourth term as governor. The Democratic nomination was open, and now there was an obvious lane for him to run. Yes, there was one other Democrat from Dixie lining up to run: Albert Gore, a freshman senator from Tennessee and another DLCer, just a year younger than Clinton. He had no more automatic claim to the South's loyalty than Clinton. There was room for both of them to jump in and find out who was the better candidate.

Gore, thirty-nine, formally entered the race in late June. He'd previously vowed not to be a candidate in '88, but that was before Robb and the others said no. Now Gore was hearing from some deep-pocketed donors. Most of them weren't from the South, but they agreed that their party's best path to reclaiming the White House ran through Dixie. Gore hailed from a political family, the son of a three-term Tennessee senator, Albert Gore Sr., who'd lost his job— and his presidential aspirations—by moving too far to the left of his constituents, especially on issues of race and war. The younger Gore, elected to the Senate in 1984, would run as the politically moderate face of a new South and a new generation. At his announcement, he brought up the mega-primary the DLC's leaders had put together, vowing to capitalize on "the new and powerful role the South will play in selecting our presidential nominees next year."

In Little Rock, meanwhile, it was full steam ahead.

It wasn't by choice, but Clinton was getting a late start. He hit the rubber chicken circuit, party dinners and picnics in Iowa, New Hampshire, anywhere he could find a roomful of Democrats willing to hear him out—almost three dozen trips out of Arkansas in the first half of '87. The message was moderate and his touch was soft, a nod to the party's liberal values and traditions joined to a suggestion that, perhaps, they could be better packaged. At a gathering of the nation's mayors in Nashville, he talked of "the great dilemma for Democrats: how to speak to the possessed and the dispossessed at the same time." The reviews were encouraging. Pledges for more than one million dollars in donations materialized, this in an era when just a few million bucks could go a long way.

In addition to Gore, the Democratic field now included Jesse Jackson, the civil rights leader who'd run in '84; Delaware senator Joseph R. Biden, who was acquiring a reputation as a talented orator; Missouri congressman Richard A. Gephardt, who'd backed Reagan's tax program earlier in the decade but was now migrating to the left; Governor Michael Dukakis, a technocratic liberal who touted his "Massachusetts Miracle" as a model for a national economic boom; Senator Paul Simon of Illinois, a prolific writer and former academic who in addition to being the oldest candidate in the race was also considered its most cerebral; and former Arizona governor Bruce Babbitt, a long-shot candidate whose embrace of politically unpopular tax hikes and means testing for Social Security was earning him fans in the media class, if not among actual voters.

It was a competent and talented group, but far from imposing. There were liabilities, some of them glaring, with each candidate, and none had a built-in national following. A nickname for the Democratic field had taken hold: the Seven Dwarfs. (For a brief time, when Colorado congresswoman Patricia Schroeder was exploring the race, the nickname was amended to Snow White and the Seven Dwarfs.) When the first debate was held in early July of 1987, Clinton was missing. He was still officially undecided, considering the race, trying to make up his mind. Those who'd followed his career were puzzled: Why was this super-ambitious politician suddenly dawdling? Then a date was set. The national press corps was summoned to Little Rock for July 15. It had taken a while, but Governor Clinton would at last be making his big announcement—and everyone knew what it would be.

I t turned out there was a tangible reason for his hesitation. Just a few months earlier, the rules of American politics had in shocking fashion been rewritten. Reporters with the *Miami Herald* had received a tip that Gary Hart, the Colorado senator who had nearly knocked off Mondale in the '84 primaries and who was the overwhelming front-runner heading into '88, was having an extramarital affair. They tailed him and reported on what appeared to be a Saturday night rendezvous between Hart and the young lady, Donna Rice. A few days later, another reporter asked Hart point-blank if he had committed adultery. (He refused to answer.) It was the first presidential campaign sex scandal of the modern era. Within a week, an overwhelmed Hart was out of the race.

With that, the unspoken bargain between the press and politicians disintegrated. Questions of marital fidelity had, except in rare instances, been off-limits, but now a new precedent had been set. Not only was a candidate's private life fair game, the Hart episode had shown just how politically devastating any embarrassing revelations could be. Among those who'd been protected by this bargain was Clinton. The gossip in Arkansas about the governor's wandering eye was legendary, but never reported. As he prepared to seek the presidency in the new, post-Hart environment, Clinton now paused. Would the Arkansas press begin reporting what it had previously suppressed? What would the national media discover when they started sniffing around? What would he say when he was asked about the state of his own marriage?

He was still determined to try. This was the opening he'd been salivating over. How could he pass it up? He sat down with his chief of staff, Betsey Wright, for a final gut check just before July 15. She had a list of names with her, all the women linked by rumor to the governor, each a potential campaign-killing bomb waiting to explode. The Hart scandal was just a few months old, and everyone in the media was looking for the next one. If anything—*anything*—were to come out in a climate like this, Clinton would be done, and not just for 1988. For good.

The *Arkansas Democrat-Gazette* called it "a stunner, the one in a hundred chance that political writers, the press and many of the governor's close supporters had said would not occur." Instead of announcing his candidacy on July 15, Bill Clinton told the world he was staying in Little Rock. He sold it as a personal decision, the father of a seven-year-old girl grappling with the demands of a national campaign and opting to put his family first. "Frankly," he said, "to be perfectly selfish, I thought of missing all those softball games and soccer games, plays at school, consultations with teachers, and it mortified me." His wife joined him at the press conference. "I want to go to supper with my husband," Hillary Clinton said. "I want to go to the movies. I want to go on vacation with my family. I want my husband back."

Even those who knew Clinton were floored. "I know that as of eight this morning, they were getting ready to go," the *Democrat-Gazette* quoted one insider saying. "It makes no sense. Whatever happened had to be unexpected. I'm just shocked." There was, Clinton insisted, "no external reason" for his decision, his way of heading off chatter that Hart-like skeletons might be in his closet. He said he was at peace, but he could only pretend so much. "For what it's worth," he added, "I'd like to be president. Perhaps there will be another day. I believe there will."

M ario Cuomo had the opposite problem. It took no begging on his part, or any effort at all, to force his name into the '88 mix. But was he even interested?

At the end of 1985, Ted Kennedy announced he wouldn't be running in the next presidential election. This is how every campaign cycle had played out since 1972, with all eyes looking first to the last son of Camelot. Kennedy was still only in his mid-fifties, but as it turned out this would be his final no; never again would he seriously contemplate a White House bid. Kennedy's early exit solidified Cuomo's new status as his party's Great Liberal Hope. Cuomo bristled at the label. This was the eighties, the Reagan era. The "L" word was now a potent epithet in Republican messaging. "Pragmatic progressive," Cuomo called himself.

"Words and phrases like 'conservative,' 'liberal' may be unavoidable habits of our expression," he said, "but they are not useful as meaningful descriptions of the complex forces with which our policies must contend."

He had run for a second term in New York in 1986, hinting this would preclude a presidential bid two years later. But when he nearly cracked 70 percent on Election Day, he relented and said he'd give it some thought, then set out on a very un-Cuomo-like national speaking tour. Polls showed him trailing only Hart, who was again pitching himself as the candidate of "new ideas," the face of a new centrist breed of Democrat that could sell to independents in the suburbs. The contrast with Cuomo, the eloquent and unreconstructed New Dealer, suggested a lively battle to come.

On a Thursday night in February 1987, Cuomo arrived at the studios of WCBS radio in New York City. He did this every month, sitting with host Art Athens to talk state politics and take listener calls. This appearance went like all the others, until the final minutes of the broadcast, when Cuomo took out a prepared statement and asked to read it on-air. He had something he wanted to say about the presidential race.

"In my opinion," he said, "the Democratic Party offers a number of candidates who can prove themselves capable of leading this nation toward a more sane, a more progressive, and a more humane future. I will not add my name to that list. I will not be a candidate."

No one had been expecting Cuomo to reveal his decision, not on this night, not in this venue. Word reached Hart's campaign office in Washington. Could it be true? A call was placed to WCBS and the station played back Cuomo's comments over the phone. "Cheering could be heard breaking out in the background on the other end of the phone," as the Associated Press reported it. Cuomo would call it a hard decision and cite, as always, the extraordinary demands of his job governing New York State. "It's silly to say you can't win," he said. "You have to do what you think is right."

Why wasn't he going for it? With that one speech in San Francisco, he had made himself one of the most famous names in American politics and a hero to millions of Democrats. It had been nearly three years, but the appeals had never stopped coming into Albany from all over the country: *Please,* run for president. Who wouldn't cash all of that in? What kind of politician, what kind of person, could be immune to that kind of flattery? Well, Mario Cuomo, apparently. Nor did he budge when Hart later imploded and the pleas to run returned, now with even more intensity, even more urgency. A poll was taken, but Cuomo's name wasn't included, since he wasn't a candidate. But enough respondents volunteered his name anyway that he came in second. Maybe that would get him to reconsider?

"No, no, no," Cuomo said. "Why won't people simply believe me?"
Eventually they did, even if it mystified them. Was he afraid

of losing? Maybe there was some dark secret holding him back? It couldn't possibly be that he just didn't want to be president—could it? And the irony of it all: his refusal to run only enhanced his stature. The race proceeded without him only for the candidates to find themselves continually measured against the giant on the sidelines.

Biden, the bright light of the summer months, was gone by the autumn, felled by a plagiarism controversy. The early contests then revealed a fractured party. Gephardt won the Iowa caucuses, while Dukakis claimed next-door New Hampshire. Then the race moved south for Super Tuesday, the mega-primary the DLC's founders had conceived as both a check on the party's liberal wing and a catapult for one of their own. What they got instead was their worst nightmare.

The Reverend Jesse Louis Jackson was already a historic figure. Four years earlier, he'd mounted the first-ever full-fledged national campaign by a black candidate, powered by the same trend that so alarmed the DLC. The enactment of the Voting Rights Act in 1965 had erased most of the old Jim Crow barriers to poll access, and since then blacks in the South had been registering in droves and aligning themselves—overwhelmingly—with the Democratic Party. At the same time, the migration of southern whites away from the Democrats was only accelerating. By the 1980s, black voters made up a larger share of the Democratic Party than ever before, and Jackson believed they were getting a raw deal.

"The Democratic Party is accepting integrated votes," he declared, "but running segregated slates."

His 1984 candidacy revealed a gulf within the black political community. The existing leadership class kept him at arm's length. There were doubts about his stature—he was an activist who'd never held elected office—and the willingness of white voters to take him seriously. There were rivalries and resentments, too. Jackson was a captivating orator with an appetite for media attention—a publicity hound, the critics said. There were also charges of embellishments in his life story. Coretta Scott King, the widow of Martin Luther King, refused to endorse him. From King's circle there were hard feelings toward Jackson, a belief that he'd exaggerated his closeness to the slain leader and used King's death to build his own profile. More black leaders lined up with Mondale, the front-runner, than Jackson.

With black voters, though, he was a hit. Jackson contested every primary and caucus in 1984 and earned 3.1 million votes, the vast majority from African Americans. It was a transformative moment in black politics. The old guard had shunned him, but eight out of every ten black voters went for Jackson anyway. A poll asked African Americans to name "spokesmen for blacks who are well-known nationally." Jackson was the runaway top choice. When Coretta Scott King and Andrew Young, MLK's top aide and a revered civil rights leader in his own right, walked into a room of Jackson delegates at the convention that summer, they were serenaded with boos.

"I respect him for what he did twenty years ago," one of the delegates said of Young, "but twenty years later, things are not so good."

The footprint of black voters was expanding, and Jackson had made himself their broker. For party leaders, it created a dilemma. How much should they accommodate Jackson and his demands?

On the one hand, blacks were now pivotal to any winning Democratic coalition. No Democrat in the last generation had earned less than 80 percent of the black vote in a presidential election. To alienate Jackson might endanger that loyalty. But Jackson's new prominence was also turning away other voters, and his campaign established him as one of America's most polarizing political figures.

Racism clearly played a role. There were also his politics, far to the left of even the most liberal leader of the Democratic establishment. Jackson made no bows to Reaganism, demanding higher taxes, more domestic spending, and, even with the Cold War raging, a slashed military budget.

His critics delighted in depicting him as a friend to America's enemies. It stemmed from a tour of the Middle East a few years earlier, when he'd met with Yasser Arafat, the head of the Palestine

Liberation Organization. Arafat would later in his life claim a more respectable role on the world stage, but when Jackson met him he was known mainly as a militant whose group's fingerprints were all over the slaughter of eleven Israeli athletes and coaches at the 1972 Munich Olympics. A photo showed Jackson hugging Arafat. The condemnation back home was withering.

The Jackson campaign had exposed delicate cultural tensions between African Americans and Jews. Alarmed by his Middle East politics, national Jewish groups mobilized against Jackson, who bristled at the attacks. He defended his sensitivity to the Arab world ("I'm a third world person. I grew up in an occupied zone—Greenville, S.C.") and charged that "the most attempts to disrupt this campaign have come from Jewish people."

There were charges of anti-Semitism, and Jackson swore he was getting a bad rap. Jews had been integral to the civil rights movement in the sixties and he had long counted some of them as close friends and confidants. At his best, he could speak passionately about a shared morality binding the two groups together.

At his worst, though, he gave the accusation more life. He was introduced at a rally by Louis Farrakhan, the leader of the Nation of Islam. Before handing over the microphone, Farrakhan issued a message to the Jewish leaders criticizing Jackson. "If you harm this brother," he said, "I warn you in the name of Allah, this will be the last one you harm." When it was his turn to speak, Jackson ignored the remark, then later defended it, saying Farrakhan had been expressing the anger many blacks felt toward the murders of political leaders like Martin Luther King. "Jews went to the chambers silently," he said. "They should have gone fighting if they had to go at all."

What would haunt Jackson for the rest of his public career, though, was a word he had a penchant for using when his guard was down. "In private conversations with reporters," the *Washington Post* reported, "Jackson has referred to Jews as 'Hymie' and to New York as 'Hymietown.'"

That it was a dated expression didn't make it any less of a slur.

"'Hymie,'" explained the *New York Times*, "is a shortened form of Hyman, a name once relatively common among Jews; using the nickname to refer to Jews in general might be likened to referring to an Irish person as 'Mick.'"

Jackson denied it at first, then fessed up: "I was shocked and astonished that this ethnic characterization made in a private conversation apparently was overheard by a reporter. I am dismayed that a subject so small has become so large that it threatens relationships long in the making, and those relationships must be protected."

"However innocent and unintentional, the remark was insensitive and it was wrong," he said.

Jackson came to the '84 convention with a list of demands— a prime-time speaking slot, a formal role in Mondale's campaign, money for a Jackson-led voter registration drive, new rules for delegate selection the next time around. When Mondale balked, Jackson called his supporters together for a fiery pep talk. "You ain't got nothing!" Jackson thundered.

Mondale was paralyzed. Without the black vote, whatever slim chance he had in November would be lost. But linking up with Jackson risked its own blowback, and many white Democrats wanted him as far from the stage as possible.

"Jesse Jackson is destroying the Democratic Party in the South," one congressman told the *Washington Post*. "I took a poll in my district just three weeks ago. It showed Reagan getting 63 percent of the vote, Mondale 21 percent, and the rest undecided.

"And I'll tell you something else. If you break it down along racial lines, Reagan gets about 90 percent of the white vote. My God, with that kind of split we could stand to lose a lot of House seats in the fall. They could even knock me off."

Finally, Jackson gave his endorsement and got his speaking slot. It was a powerful address, with a call for unity that moved even some of his detractors in the hall. What brought the loudest response from Jackson's delegates, though, was when he told them his campaign had only been the warm-up act.

"I am not a perfect servant," Jackson told the delegates. "I am a public servant, doing my best against the odds. As I develop and serve, be patient: God is not finished with me yet!"

After Mondale's drubbing, Stanley Greenberg, a Democratic pollster, set out to study the mass exodus of blue-collar whites from the party. He keyed in on Macomb County, Michigan, a haven for the assembly-line workers who built cars in nearby Detroit. Macomb had once been the definition of a Democratic bastion, handing John F. Kennedy a larger majority than any other suburban county in the country. But by '84, it had swung hard the other way, siding with Reagan by two to one. Race, Greenberg reported, was a major factor. The working-class whites of Macomb, he wrote, "express a profound distaste for blacks, a sentiment that pervades almost everything they think about government and politics."

"Blacks constitute the explanation for their vulnerability and for almost everything else that has gone wrong in their lives," Greenberg concluded. "Not being black is what constitutes being middle class; not living with blacks is what makes a place a decent place to live."

To the Democratic Party of the 1980s, this constituted a political crisis. The electorate was still 85 percent white, and to compete for the presidency, Democrats would have to unite two groups that were separated by an ugly chasm: African Americans, who were now rallying to Jackson and his insistence that they'd been settling for too little for too long; and Reagan Democrats, who were showing zero interest in supporting the party of Jesse Jackson.

The 1988 version of Jackson was a better candidate. He was more mindful of ethnic sensitivities, more restrained in his rhetoric—more mature. The goal this time was more than symbolism. He tailored his message of "economic justice" to a wider range of audiences— including the blue-collar whites who'd reacted to him with such hostility the last time around. "The overwhelming difference is his welcome into the community," one union organizer said.

In heavily white Iowa, Jackson took 9 percent, after barely registering four years earlier. In New Hampshire, he scored 8 percent, another big jump. He had attracted almost no white support in 1984; now he was getting at least some. Then, at last, came Super Tuesday. The *New York Times* billed it as "the day designed to restore the South to the dominant role it once played in the Democratic Party"—fifteen southern and border states all voting on the same day. Gore, the lone white southerner in the race, was predicting it would launch him to the front of the pack for good.

Gore did do well, but Jackson did better. Across all the 1984 primaries, Jackson had won a total of two states. But on this Super Tuesday alone, he carried five and finished a close second in many more. Overall, he won more votes than Gore and gained more delegates. This was supposed to be the DLC's coming-out party. Instead, Super Tuesday had made Jesse Jackson a more credible and powerful political leader than ever before. "Things are not going to work out the way the Southerners hoped," R. W. Apple wrote in the *New York Times*.

Black enthusiasm for Jackson was even higher this time around; one estimate had him carrying 96 percent of the black vote. But he also doubled his share of the white vote from '84, cracking 10 percent. "My message is transcending ancient barriers," Jackson said. "Whites all over the country have opened their hearts to me. They know there has been this separation, and now they have been willing today to say, 'Let's call it even.'"

Super Tuesday left Democrats with a mess. Jackson edged close to first place in the delegate race, neck and neck with Dukakis, with Gore now in third and Gephardt far back and sputtering. The picture grew more muddled a week later in Illinois, where Jackson—who'd built his activist career in Chicago—far outpaced Dukakis. (First place actually went to Paul Simon, whose national effort was already dead; but Illinois Democrats, sensing a potential convention stalemate, were equipping their favorite son candidate with leverage.)

Then, on March 26, the earthquake: Jackson went north to Michigan and knocked off Dukakis and Gephardt—crushed them, actually. He won an outright majority, 54 percent, doubling Dukakis's share. Here was the breakthrough that was never supposed to be possible for Jackson. Michigan was a northern industrial state with a large white population, home to the famous Reagan Democrats of Macomb County. Yes, Jackson was rolling up a massive margin in predominantly black Detroit, but he was also winning in majority-white areas. This was a statewide victory, and a smashing one.

Gephardt dropped out on the spot. Dukakis was thrown on the defensive. Jackson now had received more votes nationally than any other candidate, and he had the most delegates, too. In Washington, the Democratic firmament was shaken to its core, confronted with the possibility—utterly inconceivable before now—that Jackson just might emerge as the party's presidential nominee.

Yet again, all eyes turned to the reluctant man in Albany. If no one emerged from the primaries with a delegate majority, power would shift to the superdelegates, the hundreds of party leaders who'd be free to vote for any candidate they wanted, no matter the primary results. Might Mario swoop in and save them? He seemed to rule it out, then he seemed not to. What did this guy want?

It came just as the race reached New York, which would be the test—was Michigan a fluke or a portent? Jackson was already laying the groundwork for a delegate stalemate, arguing the nomination should go to the candidate with the most total votes in the primaries. If he could keep winning big states, he'd make it awfully hard for the party to say no.

But he was still Jesse Jackson. He was still running on the most liberal platform since at least George McGovern, he still had the résumé of a professional activist, and he was still haunted by the charges of anti-Semitism, which were now thrust back into the spotlight by New York City mayor Ed Koch, an expert provocateur with a deep distaste for Jackson.

Koch, a hawkish advocate for Israel, was appalled by Jackson's support for an independent Palestinian state, along with his friendly relations with Arab leaders. To win New York, Jackson needed at least some support from Jewish voters, who'd make up a quarter of the electorate. The mere possibility offended Koch. His fellow Jews, he said, "would be crazy" to support Jackson. Koch reminded New Yorkers of the "Hymietown" controversy and revived the old accusation that Jackson had used MLK's murder to promote himself.

Jackson resisted the bait, but his supporters fought back hard, and with New York's tabloids egging both sides on, the tensions exploded. "For Ed Koch to proclaim himself the king of Jews is obscene," declared Charlie Rangel, the Harlem congressman who was backing Jackson. Koch fired back that if Jackson became president, he'd "bankrupt this country in three weeks and leave it defenseless in six."

Officially, Koch was backing Gore, but that hardly mattered. Gore was stuck in single digits and going nowhere. This was all about stopping Jackson. And it worked. Jackson won 97 percent of the black vote but only 17 percent of the white vote, which added up to a double-digit win for Dukakis. The question was answered: Michigan had been an anomaly. Order was restored to the race, Gore dropped out, and Dukakis began drubbing Jackson in one-on-one contests.

Jackson had grown exponentially as a public figure between his two national campaigns, but there was just too much separating him from the kind of sustained white support he needed to gain the nomination. He had the highest negative poll rating of any candidate in either party. To Democrats, he was still the same double-edged sword: the ambassador to one of their largest and most loyal constituencies, but a net liability with the rest of America. Still, he finished the primaries with over seven million votes, doubling what he'd gotten the last time.

Once again, Jackson made public demands of the presumptive nominee. He launched a seven-hundred-mile bus tour, dubbed the Jackson Action Rainbow Express, in the days leading up to the convention. "What do we want?" he asked his supporters. "We want to share." What he really wanted was the number-two spot on the ticket. It was only fair, he said, since he'd won the second-most votes in the primaries. Dukakis never had any intention of picking Jackson, but pretended to contemplate it, then neglected to give Jackson a heads-up before his choice of Lloyd Bentsen was revealed.

Dukakis tried to apologize and there was peace, at least in public. Jackson endorsed the ticket and even got a plane to travel the country in the fall. Otherwise, the Dukakis team labored to keep him hidden. A poll divided the white voters of America into two camps. Those who liked Jesse Jackson were breaking for Dukakis over Bush by seventeen points. Those who didn't like Jackson were against Dukakis by fifty-seven.

Preventing a Jackson eruption wasn't Dukakis's only challenge in Atlanta. There was also the question of the nominating speech. It was a crucial slot, right before the roll call of the states on the convention's third night, one of the few speeches that every network would air live. Cuomo was the obvious choice, but a dangerous one. Democrats already feared that, with Dukakis, they weren't putting their best foot forward. Another stemwinder from Cuomo would only remind them what they were missing. A reporter compared the nominee Democrats were getting with the one they dreamed of. "[Cuomo's] mere presence on a public platform makes Michael Dukakis . . . look short, cold and repressed." So Dukakis decided to offer Cuomo a smaller role. He wanted him to introduce the man who would deliver the nominating speech: Bill Clinton.

Clinton had withheld his endorsement until the nomination was settled, but Dukakis had taken a shine to him anyway. They'd worked together through the National Governors Association and Clinton had been cultivating him personally. Now Dukakis was looking to put a southern face on the Atlanta convention, so the governor of Arkansas was his man. In conventions past, it had been customary for a series of speakers to deliver nominating speeches. But this time there would only be one, Dukakis said, to be delivered by "an eloquent spokesman for our party." Any risk that Clinton might be upstaged was removed when Cuomo politely declined to be the warm-up act and decided he wouldn't speak in Atlanta at all.

Sitting out this presidential race had for Clinton been a bitter frustration, but here was his chance to make it all better. He'd seen what Cuomo did with the spotlight of a major convention speech, and now it would be his turn. That might have been the problem.

The delegates weren't hostile to him when he started, but there was a marked lack of electricity in the Omni Coliseum. For Clinton, this was the opportunity of a political lifetime; for the crowd, it was just another speech from just another politician. Cuomo had faced the same dynamic in San Francisco, but easily overcame it. When Clinton began, there was the same din of idle conversation that had greeted Cuomo, but this time it never went away.

The speech lacked coherence, one paragraph of dense statistics and platitudes piled on top of another, with no organizing theme. The delivery lacked artistry. It was supposed to run fifteen minutes, and if it had, it would have been considered a dud, but at least a forgettable one. But Clinton was moving through the words slowly, and as he plowed on, the boos started. Network television cameras scanned the crowd, revealing bored delegates talking with each other, reading their programs, staring around the hall, the polar opposite of the scene they'd shown with Cuomo four years earlier. Now the speech was becoming a different kind of dud, one that would be memorable, defining, maybe even catastrophic.

"You're listening to the *lengthy* nomination speech of Governor William Clinton of Arkansas," Tom Brokaw told viewers as NBC pulled away from the speech. Clinton, Brokaw said, "now is seriously in oversight. He only is about halfway through his prepared text and he should have been done about five minutes ago."

He threw to reporter Chris Wallace, with the New Jersey delegation. In the background, Clinton was still speaking, but now the commotion on the floor was drowning him out. "It seems," Wallace reported, "that Bill Clinton has overstayed his welcome in this hall."

Clinton did land one genuine applause line—when he uttered the words "In closing" and the delegates erupted in sarcastic glee. Quipped one Democrat: "It was either the longest nominating speech or the shortest presidential campaign in history. I've never seen a speech do so much damage." After all those years of trying, all the decades, really, Bill Clinton had finally gotten everyone's attention. By becoming a laughingstock. He played good sport a few nights later with an appearance on Johnny Carson's *Tonight Show*, at least earning a few points for self-deprecation. But the real opportunity was lost.

"It could have been the speech that propelled Gov. Bill Clinton into the upper tier of national politics," the *Arkansas Democrat-Gazette* wrote, "as the unchallenged rising star of the Democratic Party, one of its towering hopes for the future, especially should Mike Dukakis somehow falter at the ballot box in November. Alas, it was not to be so, and we feel a certain sadness for him."

November brought yet another presidential massacre for the Democrats. Dukakis lost forty states. Maybe there really was a Republican lock on the White House. Now Democrats were more desperate than ever for a savior. The convention had been Clinton's audition for the role, but all he'd succeeded in doing was making Mario Cuomo seem even more magical.

FOUR

By the fall of 1986, Tip O'Neill was ready to hang it up. He'd steered his party through the heart of the Reagan years, defying the dire early predictions that a rising national conservative tide would end Democratic hegemony in the House. A storyteller of a different era who could spin yarns about campaigning in Boston with the legendary "rascal king" James Michael Curley, O'Neill was regarded with universal affection by his party. But there was a growing sense that it was time for a passing of the torch. The postwar decades had radically shifted the center of American political gravity into the suburbs, where college-educated professionals were far more skeptical of government than their urban machine forebears. The lingering presence of the seventy-three-year-old O'Neill threatened to trap them in their past.

"Physically, he is the very model of the political boss," the *Los Angeles Times* wrote, "a cartoonist's dream: 6-foot-3, a hefty 280 pounds, with thatched, Olympian white hair, busted-plum nose, heavy lids drooping over warm, blue-gray eyes. And a friendly mug with something like a road map of County Cork etched from jowl to jowl."

O'Neill's House floor confrontation with Gingrich contributed to the perception that he was out of step with the times. Democrats wondered if the man they found so charming in private was the man Americans perceived in this new media age. "The Speaker kept many of his promises over the years," wrote columnist Mary McGrory, an admirer, "but never the one to diet. In the television era, none of that may matter more." (Actually, they were selling his appeal short: in retirement, O'Neill would author a widely hailed memoir, *Man of the House,* and star in several popular television ads.)

To replace him, Democrats adhered to tradition and went with the next man in line, James C. Wright Jr. of Texas. Wright had served as majority leader for the last ten years, and O'Neill said he felt a duty not to snuff out his lieutenant's aspirations by staying too long. At sixty-three, Wright was only ten years younger than O'Neill. He was also a product of the same political era, elected to local office in Weatherford, Texas, not far from Fort Worth, in the 1940s and then to the House in 1954, just two years after O'Neill. He was taken under the wing of his fellow Texan Sam Rayburn, a Washington icon who served a total of seventeen years as Speaker before his death in 1961, learning the ways of the chamber from one of its all-time masters. Just like Tip O'Neill, Jim Wright was very much a creature of the old House.

But in personality, he was anything but Tip. Wright was far more polished as an orator and far more reserved in private. The *New York Times Magazine* called his ascent a "triumph of the minutiae of politics, of things as basic as fund raising and memorizing a freshman congressman's name." The major D.C. social event of the fall of '86 was a bipartisan gala honoring O'Neill. President Reagan was there, and so was every congressional leader from both parties. "I love this guy," Michel, the House Republican leader, said. It was the kind of tribute Wright was unlikely ever to inspire. He strived for cordiality but could be thin-skinned. "I think he wants to become what the person who he's with thinks he is," his first wife had once said.

He was also more tactical than O'Neill, who favored leadership by consensus and a wide berth for committee chairs. Wright was determined to centralize power and muscle through an aggressive agenda. Reagan still had two years left, and Wright wanted confrontation. "He'll discipline us," Democratic representative Mary Rose Oakar said.

What Jim Wright really was, though, was the best thing to ever happen to Newt Gingrich.

At first, he implemented his vision with a force and effectiveness that stunned Washington. Republicans knew they didn't have the votes to pass their own legislation, but they did have tools to slow down and frustrate the Democrats—tools that Gingrich and his allies had been exploiting in ways never before seen. But Wright ran straight over them. He thrust himself into the peace process in Nicaragua, again arousing cries from Republicans that he was overstepping his constitutional bounds. "Stripped of their political leverage and outgunned by the Democrats 258 to 177," the *Washington Post* wrote as 1987 came to an end, "House Republicans have been no match this year for Wright and his troops during skirmishes over spending, taxes, trade, arms control and Central America policy."

It was like Wright was mocking their impotence. Republicans were now thirty-three years deep in minority party status. They had just 177 seats out of 435 in the chamber. They were seething at Wright, but really, what could they do about any of it?

Newt Gingrich had an idea. There was a story in the *Washington Post* in the fall of 1987 that noted the unusual book deal the Speaker had struck with a small-time publisher back home in Fort Worth. Carlos Moore, who was a longtime ally of Wright, had published a collection of Wright's essays and speeches under the title *Reflections of a Public Man*. Copies were sold for $5.95, with $3.25 from every sale going straight to Wright—a royalty of 55 percent per book.

The terms were unheard of in the publishing world, where royalties typically ran in the 10 to 15 percent range. So far, Wright had made fifty-five thousand dollars on a flimsy 117-page book that more resembled a pamphlet and that almost no one in Washington even knew existed.

It looked fishy, but the story fell off the radar quickly. Michel and other top Republicans didn't see the story as something to weaponize. That sort of thing just wasn't done.

But now Newt stepped forward to ask: Why not? In November, he went public, calling the book deal a case of "money laundering" and branding the Speaker of the House "a genuinely corrupt man" and "a genuinely bad man."

"Wright is so consumed by his power that he is like Mussolini, believing he can redefine the game to suit his own needs," Gingrich said.

By December, he was demanding a special committee to investigate the "corrupt and blatantly unethical behavior" of Wright. His joust with Tip O'Neill had trampled old boundaries, but this was venturing into far touchier territory. He was calling into question a fellow member's honor, calling the Speaker corrupt. Wright was indignant. "It's totally without foundation," he said. "My total career speaks for itself. Those who know me best, constituents, neighbors, friends, and colleagues, have repeatedly paid me the very great honor of electing me to positions of great responsibility."

With every insult, Gingrich was keeping the *Post*'s story alive, and now the media was getting hungry. Wright was asserting that he was a man of integrity; now reporters were asking him to prove it. Editorial boards were digging in. Republicans weren't lining up to join Gingrich, but pointedly Michel refused to rein him in. Then, in May 1988, Common Cause came aboard. Fred Wertheimer, its president, wrote a letter to the Democratic chairman of the House Ethics Committee asking for an investigation of Wright.

Now it really was a scandal. Common Cause was a good-government group that had become a bedrock institution of liberal, post-Watergate Washington. The Democratic line had been to dismiss Gingrich's charges as a transparently political sideshow. That would be a lot harder now. Within days, Gingrich had filed a formal complaint with the Ethics Committee.

Republicans were rushing off the sidelines. Seventy-two of them added their names to the Gingrich complaint. They included moderates who had never been aligned with him before and old-timers who'd never before imagined they'd join up with a mission like this. All of them were accustomed to being the minority party, but their treatment under Wright felt different, dismissive, peremptory. A desire to fight back was building, nudging them toward the troublemaker from Georgia and his dark arts. Newt Gingrich's one-man war against Jim Wright was now a project of the Republican Party.

The Ethics Committee voted to open a formal investigation of Wright. The committee, unlike others in the House, was evenly divided—six Democrats and six Republicans. The vote to investigate was 12–0. Wright's lawyers put out a twenty-three-page statement attempting to refute the accusations. The Speaker would concede only "mistakes of the head and not of the heart," insisting he'd broken no rules of the House. Asked his personal feelings toward Gingrich, Wright pronounced them "similar to those of a fire hydrant to a dog."

New details were emerging fast now. The *New York Times* sent a reporter to Fort Worth, who found wealthy Wright allies talking about purchasing bulk copies of *Reflections of a Public Man* as a way of helping their buddy make a few bucks. "I was just trying to make a contribution to Jim's income," a developer who'd bought about a thousand copies said. It turned out the Teamsters had bought bulk copies, too, through their political action committee. This suggested another possibility. House rules put strict limits on the income members could receive from outside speeches—30 percent of their congressional salary, a little more than twenty-five thousand dollars. Had Wright been using bulk sales of the book to skirt around the limit—leaning on political groups to buy the book in bulk in lieu of an actual speaking fee?

The *New York Times* wrote of Gingrich and his transformation from "political guerilla to Republican folk-hero."

"I'm so deeply frightened by the nature of the corrupt left-wing machine in the House that it would have been worse to do nothing," Gingrich said. "Jim Wright has reached a point psychologically, in his ego, where there are no boundaries left."

The Ethics Committee deputized an outside counsel, Richard J. Phelan, to pursue the investigation. Meanwhile, the Permanent Democratic Congress was reelected in 1988 and voted Wright in for another term as Speaker. He was under a cloud, though, and in April 1989 it opened up. Calling the book deal "an overall scheme" to subvert caps on outside income, the Ethics Committee accused the Speaker of nearly six dozen violations of House rules.

The feeding frenzy was on. The *Washington Post* reported a story about Wright's top aide, John Mack. As a nineteen-year-old, it turned out, he'd attacked a woman, Pamela Small, with a hammer, stabbed her, and left her for dead in 1973. She survived, somehow, while Mack went to prison and then—thanks to a family connection—landed in Wright's office as part of a work-release program. Now, as the country learned about all of this for the first time, he was the embattled Speaker's right-hand man. Feebly, Wright's fellow Democratic leaders offered a defense of Mack, which was quickly overshadowed by the news that one of them—Majority Whip Tony Coelho—was himself facing a federal investigation over illicit profits from junk bonds.

Gingrich was always talking about a corrupt Democratic machine bent on protecting its own at any cost. Now they were making his point for him. Coelho resigned, then on Wednesday, May 31, Wright sent word that he'd speak from the well of the House at 4 P.M. The chamber was overflowing when he took his place.

"Mr. Speaker," he began, "for thirty-four years I have had the great privilege to be a member of this institution, the people's house, and I shall forever be grateful for that wondrous privilege." He thanked his Texas constituents and noted a poll showing that many of them wanted him to stay as Speaker. Then he turned to the charges lodged against him.

He talked about the book deal, said his staff was overexuberant in pushing bulk sales to political groups, and granted that "I have to accept some responsibility for that if it was wrong." Then he pleaded with his colleagues to take a step back. The royalty income from all of the bulk sales, Wright noted, added up to about seventy-seven hundred dollars. "If monetary gain had been my primary interest, don't you think I would have gone to one of the big Madison Avenue publications, the houses there that give you a big advance?"

He brought up John Mack. He said he'd only learned two years earlier about exactly what Mack had done to Pamela Small, and by that point Mack had matured into a dedicated family man and accomplished legislative aide.

Finally, he got to the heart of it, to the words that Washington would think back to for the next generation. Without ever mentioning the name of his chief tormenter, the gentleman from Georgia, Wright offered up his own indictment of Gingrich.

"It is grievously hurtful to our society when vilification becomes an accepted form of political debate, when negative campaigning becomes a full-time occupation, when members of each party become self-appointed vigilantes carrying out personal vendettas against members of the other party. In God's name, that's not what this institution is supposed to be all about."

It wasn't meant as a compliment, but Wright was admitting that Gingrich had succeeded in changing the House of Representatives. He continued.

"All of us in both political parties must resolve to bring this period of mindless cannibalism to an end!"

At that, Democrats rocketed to their feet. Some Republicans did, too. Gingrich was seated next to Michel. The men whispered to each other, then both rose to join the ovation. The line was meant to cripple him; Gingrich was standing in defiance of it.

Wright still had to explain why he was resigning. He insisted he wasn't guilty, that he hadn't dishonored the House. His exit, he said, was an act of sacrifice.

"Have I contributed unwittingly to this manic idea of a frenzy of feeding on other people's reputation?" he asked. "Have I—have I caused a lot of this stuff? Maybe I have. God, I hope I haven't, but maybe I have. Have I been too partisan? Too insistent? Too abrasive? Too determined to have my way? Perhaps. Maybe so.

"If I've offended anybody in the other party, I'm sorry. I never meant to—I would not have done so intentionally. I've always tried to treat all of our colleagues, Democrats and Republicans, with respect. Are there things I'd do differently if I had them to do over again? Oh, boy! How many may I name for you!

"Well, I'll tell you what," Wright continued. "I'm going to make you a proposition: Let me give you back this job you gave to me as a propitiation"—defined as the action of appeasing a god, spirit, or person—"for all of this season of bad will that has grown up among us. Give it back to you. I will resign as Speaker of the House effective upon the election of my successor."

Wright's speech ran for exactly one hour. "God bless this institution," he concluded. "God bless the United States." The acclaim from his colleagues was thunderous. No one ever said Wright couldn't give a good speech. Emotions were raw. Democrat Jack Brooks, the Speaker's fellow Texan and staunchest ally, said his friend had been "lynched by leaks."

"There's an evil wind blowing in the halls of Congress today that's reminiscent of the Spanish Inquisition," Brooks said.

Wright's son called Gingrich "another Joe McCarthy" and predicted that "as soon as he becomes an embarrassment to [the Republicans], they'll back away from him and let him sink on his own." From retirement, Tip O'Neill warned that soon "you'll have nothing but the imbecilic sons and daughters of wealthy families who want to send their children to the Congress." Even Gingrich's allies seemed moved by Wright's words. "I think most of us are weary of inhabiting the congressional Beirut we've been living in," Henry Hyde told reporters.

Gingrich waited a few days, then responded. "Jim Wright was not the innocent sacrifice to some tribal ritual," he declared. "Wright and [Tony] Coelho pulled off this wonderful scam where they are the innocent victims of a mood in Washington. That is baloney. The guy was guilty."

Why, he demanded, were Democrats and the press now pointing fingers in his direction? "It would not have occurred to me that the man who filed charges against the Speaker of the House and turned out to be right, that I would be the devil."

There would be no public soul-searching, no self-flagellating on Gingrich's part. And why should there be?

It had been ten years since he'd come to the House a nobody professor from a nowhere college. He'd arrived with no powerful friends looking out for him, a pompous gadfly ridiculed from both sides of the aisle. "A professional pest," Wright branded him early on. "A pain in the fanny," Dick Cheney pronounced. Hill staffers took to circulating his speeches and writing as a joke.

So many who'd mocked him were there in the chamber on that final day of May, and even if they still despised him, they had to respect him now. Newt Gingrich had toppled the Speaker.

Gingrich's dream was to lead the Republicans of the House of Representatives out of the political wilderness and to majority status for the first time since 1954. Even after the Wright drama, though, the notion of a Republican Congress felt as far-fetched as ever. He was convinced it could be within reach, and sooner than anyone imagined.

When that day finally came, he needed to be something other than a glorified backbencher. It would be a historic occasion, the fall of the Permanent Democratic Congress and the birth of a new Republican order. If it was his destiny to lead that new Republican majority, he needed to position himself for it, and soon, and that would mean finding a way into the party's leadership.

It was a tall order, maybe an impossible one. His Republican critics were quieter now, but he still had enemies with influence, and even his allies weren't sure he was leadership material. On top of that, the path looked clogged for years to come.

The first obstacle was Trent Lott, a courtly Mississippian who'd come to the House in 1972, six years before Gingrich. Lott embodied the conservative economic and cultural tides of his region, but he was also a consummate backslapper, quick to ingratiate himself with the right power players. He moved up the ranks quickly, and when Michel was elected minority leader after the 1980 election, Lott slid into the number-two post behind him, minority whip. Though he was not even forty years old, his good manners and collegiality appealed to the old guard even as his ideology matched the demands of the newer arrivals. Now, here was leadership material. Lott, all of Washington knew, would eventually succeed Michel, then lead the House GOP far into the future, probably well into the 1990s.

But in 1988, Lott was confronted with an alluring possibility. Senator John C. Stennis, a conservative Democrat and stalwart of Mississippi politics for more than forty years, was retiring. The race would by no means be a slam dunk for Lott; while it had sided with Goldwater, Nixon, and Reagan at the presidential level, the state remained predominantly Democratic. He'd be risking everything— sixteen years of seniority in the House and a near-certain promotion to party leader sometime soon—in exchange for freshman status in the Senate, with the chance he'd lose the election and his future.

What swayed him was the possibility of finally serving in the majority. Republicans had controlled the Senate for the first six years of Reagan's presidency, but in the House they were stuck in the mud as always. The Senate just looked like a better bet, so he made the jump and won by eight points.

Now Lott was gone, but that still didn't do much for Gingrich— not with Dick Cheney waiting in the wings. At forty-seven, Cheney was already a Washington institution, having served as Gerald Ford's White House chief of staff before winning Wyoming's at-large House seat in 1978.

Cheney wasn't flashy, and while his voting record was quite conservative, he had little interest in public crusading. His strength was the Washington power game itself, and his mastery of its nuances made him invaluable to Michel and the party. When Lott left at the end of 1988, the vote was unanimous: Dick Cheney would be the new Republican whip—and the new heir apparent to Bob Michel.

At the same time, the incoming president, George H. W. Bush, was filling out his cabinet. To run the Pentagon, he picked an old Texas friend. John G. Tower had represented the state in the Senate for twenty-four years, before leaving after the 1984 election. He was the first Republican since Reconstruction to win a statewide race in Texas.

The sixty-three-year-old Tower was qualified; he'd chaired the Armed Services Committee in the Senate. He also enjoyed a reputation in Washington as an aging bachelor with a taste for carousing. He and his second wife had divorced a few years earlier and the filings had surfaced. Now there were suggestions of heavy drinking by Tower and carrying on with multiple women outside his marriage.

In an earlier time, none of this would have been reported, nor would the other party have thought to seize it as a political weapon. But Tower's nomination came after the Hart scandal, and just as Gingrich's campaign against Wright was reaching its peak. Nothing seemed off-limits anymore.

Concerns were raised about Tower's fitness. An FBI investigation was launched. Details started leaking. Now the media had a juicy drama to track, and Democrats had a weapon. Tower denied he was an alcoholic. "I'm a single man," he pointed out. "I do date women." Eventually, he offered to take an oath not to have a sip of alcohol if he was confirmed. It was a soap opera that endured for months. The Senate vote was finally called in March 1989. Tower was rejected, 53–47.

It was a big blow to Bush, who'd invested early capital in the nomination fight. Now, two months into his presidency, he still didn't have a defense secretary. He needed a new name, one that would be confirmed quickly and easily. He looked to Capitol Hill.

Cheney had allies in Bush's inner circle, including one of the president's closest friends, National Security Adviser Brent Scowcroft, and deep experience navigating the federal bureaucracy. And as a veteran of the Ford White House, he'd already undergone two rigorous FBI screenings. There weren't going to be any John Tower issues with this one. Bush called Cheney "the best and proper choice" and Democrats signaled their assent. He was as good as confirmed.

For Gingrich, the timing was exquisite, almost impossibly so. He couldn't have matched Lott or Cheney in a leadership race, but now they were both gone, and the number-two slot in the House GOP leadership—minority whip—was wide open. What's more, the election to fill it would take place against the backdrop of the Wright scandal. At any other moment during his ten years in Congress, under any other combination of circumstances, a campaign for a Republican leadership position by Newt Gingrich would likely have been a joke. Now, though, he had a fighting chance. Bush was still introducing Cheney at the White House when Gingrich started spreading the word: he was in.

The race would be the political equivalent of a hundred-meter sprint. Cheney was picked on March 10 and the House Republican Conference would vote on his replacement less than two weeks later, on March 22. Gingrich called a press conference and acknowledged the trepidation many in his own party felt toward him: "Everyone was stunned when Dick left. Their second thought was, My god, here comes Gingrich." The job of whip entailed overseeing the party's floor strategy and wrangling members on close votes, but Gingrich was pitching his candidacy as something bigger. "I'm a national leader who serves in the House," he said. "I am deeply committed to ending the thirty-five-year Democratic monopoly in the House."

Michel and the old guard didn't want Gingrich anywhere near the job. The question was how to stop him. They had some options.

California's Jerry Lewis wanted it. A pragmatist with a knack for navigating the GOP's moderate/conservative divide, Lewis had just months earlier leapfrogged into the number-three leadership slot, narrowly ousting Lynn Martin to become the conference chairman. That gave Michel and his allies pause, though. If Lewis won the whip's post, it would just open up his spot as conference chair, with Gingrich positioned to claim it as a consolation prize. Henry Hyde, best known nationally for his legislation banning public funding for abortions, was also interested, but then a survey suggested he was unlikely to peel votes from Gingrich.

That left a man who possessed an abundance of what had traditionally been the most important asset in a leadership race: friends.

In temperament, ideology, and ambition, fifty-three-year-old Edward R. Madigan was Gingrich's opposite—"an intellectual who has quietly become one of the GOP's most effective issues-oriented legislators," as the *Washington Post* described him. He hated press conferences, never popped up on television, and was no reporter's go-to for a quote on anything.

Madigan, everyone knew, was Michel's candidate. They were friends and fellow Illinoisans, each man a believer in consensus who could count more than a few friends across the aisle. If Mr. Nice Guy had a natural successor, it was unquestionably Ed Madigan.

It wasn't just Gingrich who'd been waiting for this moment. Conservative activists and pundits mobilized on his behalf. William F. Buckley called Gingrich's candidacy "a historic opportunity" and endorsed him. William Safire used his column to disparage Madigan as part of "the go-along, get-along group headed by Bob Michel."

"This is not a moderate vs. conservative split," Safire wrote. "The struggle is about the basic approach to how the Republicans do business in the House: dickering for crumbs from the table of Speaker Wright or sharing fairly in power with Speaker-to-be [Thomas] Foley."

The crusade against Wright was now an asset for Gingrich. "We are chafing under Jim Wright far more than we ever did under Tip O'Neill," said Vin Weber, who was running Gingrich's campaign.

Madigan thought he had more than a hundred votes, a solid majority, and he probably did have that many commitments. It was hard to say no to Ed Madigan—or, for that matter, to his campaign manager, a hard-charging third-termer from Texas named Tom DeLay. But this was a secret ballot, and when those were tallied the score came out 87–85 for Gingrich.

What put him over the top, incredibly, were moderates, including three women from New England: Olympia Snowe, Nancy Johnson, and Claudine Schneider. It was his fight with Wright that won them over. They were tired of feeling pushed around by the Democrats.

That was the theme. The vanquished Madigan talked of encountering a long-serving Republican member just before the vote. "He came up to me this morning with his lip trembling, saying he was going with Newt." Even Henry Hyde, regarded as a solid Madigan vote, sounded like a Gingrich acolyte afterward. "There is a fear that animosity and polarization will be heightened. Newt is a hard charger. But to that I say, 'tough cookies.' We have nice-guyed ourselves off the map."

A dirty secret was being revealed. It wasn't just hard-core right-wingers who wanted what Newt was offering. There were moderates and pragmatists and old-guard traditionalists who were looking around and realizing they didn't quite mind it either. Newt Gingrich was now the second-ranking Republican in the House of Representatives. He joined Michel for a press conference, which started awkwardly when Michel referred to his new number two as "Nit."

"Newt," he quickly corrected himself. "It will take a while."

Michel had never wanted this, and it showed. He found himself talking about the similarities between his own leadership style and Madigan's, then added, "Obviously, the House Republican Conference in this case, given a choice, opted by a narrow margin for a different personality, and we accept that."

The result was, in a way, a rebuke of Michel, and defensiveness crept into his words. "I've been one who has been flexible all through his political life," he said. "I've been torn from the right, torn from the left and the middle. And from time to time you have to go with the flow."

On this morning, Gingrich was conciliatory and accommodating, heaping praise on Michel ("If this election had been a test of Bob Michel's leadership, I wouldn't have gotten forty votes") and downplaying talk he'd be a thorn in the Republican White House's side.

"I think you will find this to be a team effort. It's not a conservative activist victory. It is the entire Republican team," he said.

But no one saw anything routine at work here. The *Los Angeles Times'* story on Gingrich's win carried the headline: "HOUSE ELECTS MILITANT MINORITY WHIP BY TWO VOTES." The *Washington Post* went with "GINGRICH ELECTED HOUSE GOP WHIP; INCREASED PARTISAN POLARIZATION SEEN." The *New York Times:* "'AGGRESSIVE' REPUBLICANS CHOOSE HOUSE WHIP."

"Suffering through their 35th year as the minority party in the lower chamber," congressional scholar Ross Baker wrote the next day, "Republican members evidently decided that their powerlessness could at least be enlivened by making a major party spokesman out of the great thunder lizard of the GOP conference, a member so diabolically nettlesome to Democratic leaders that he has lured them into violations of House rules in their eagerness to strike out at him."

Forget Trent Lott and Dick Cheney. There was a new next-in-line to Bob Michel—who, by the way, was now sixty-six years old and probably nearing retirement. While making sure to stipulate that it was a "highly unlikely" scenario, the *Boston Globe,* Tip O'Neill's hometown paper, dared to note that if Gingrich were to supplant Michel and the GOP were then to win a House majority, "then Gingrich would be in line to be Speaker of the House."

FIVE

T he year 1991 began with President George H. W. Bush leading the nation to war with Iraq. He did so over the objections
of Democrats, who conjured fears of a years-long quagmire,
the jungles of Vietnam replaced by the sands of the Middle East. But
Operation Desert Storm produced the almost immediate expulsion
of Saddam Hussein's forces from Kuwait, a quick Iraqi surrender, and
scant American casualties. Who knew war could be so easy? Bush's
leadership stood validated on the grandest of scales. Weeks of parades
and patriotic celebrations ensued. Not since World War II more than
forty years earlier had Americans been able to bask in such a decisive
military triumph. "By God," Bush declared, "we've kicked the Vietnam syndrome once and for all!" His approval rating soared to stunning heights, north of 90 percent. Since the advent of public opinion
polling, no American president had ever been so popular.

Bush's political deification coincided with what was supposed to
be the early stage of the presidential race. But not this year. His postwar sheen reinforced what by now even Democrats had come to see
as a Republican lock on the presidency. Bush's defeat of Dukakis in

'88 had marked not just the third straight national victory for the GOP but the third straight landslide, with Bush taking forty states. It was now five out of the last six presidential races that Republicans had won, and with the switch of a small number of votes in Jimmy Carter's 1976 squeaker over Gerald Ford, they would have gone six for six. Since 1968, Republican presidential candidates were now, on average, winning 417 out of a possible 538 electoral votes per contest.

There were other ways of expressing the GOP's dominance and the Democrats' futility. A total of nineteen states had now voted for the Republican candidate in every one of the previous six presidential races. Democrats couldn't claim a single one. Besides the redoubt of the District of Columbia, there were no electoral votes that Democrats were assured of winning. Factor in Bush's war triumph and the savage effectiveness of his campaign team, which had methodically dismantled Dukakis, and, well, 1992 wasn't shaping up as the year for Democrats to break the curse.

"The view of Washington," reported the *New York Times*, "is that President Bush is all but unbeatable."

When Bush appeared before a joint session of Congress at the war's conclusion, Republicans in the chamber greeted him—and taunted the loyal opposition—with buttons announcing, "I voted with the president." They pinched themselves at how thoroughly the politics of the Gulf War, a conflict Americans were initially hesitant to wage, had rebounded in their favor. Now they had new ammunition for a time-honored strategy: painting the other party as soft on defense. The label had haunted Democrats since the nomination of the antiwar McGovern in 1972, and surely it would do so again in 1992. The chairman of the Republican National Committee pointed with anticipation to the prominent Democrats who'd tried to warn Bush off the war: "I see many of those quotes emerging once again in 1992."

Democrats faced a puzzling predicament. Over the past generation, each party had dramatically opened up its nominating process, empowering rank-and-file voters in primaries and caucuses and ending the tradition of brokered conventions. No longer could candidates bide their time and play coy, refraining from campaigning while hoping their party might turn to them come convention time. This new system depended on ambitious politicians declaring early, building organizations, and taking their case straight to the people. There'd never been a shortage of interest, but now there was. The next election was a year away, the Democratic nomination was wide open—and no one wanted it.

One after another, the party's weightiest figures were running away from the starting line. Bill Bradley, a onetime NBA star marked as a presidential prospect from the minute New Jersey elected him as a senator in 1978, took his name out of consideration. Everyone had assumed Dick Gephardt would make a follow-up bid after his overachieving run in '88; but now he was the number-two Democrat in the House and opted to stay put there. Gore, another '88 veteran penciled in for a second go-round, was a scratch, too. Senator Lloyd Bentsen of Texas had been a breakout star as Dukakis's running mate, delivering a devastating putdown of Dan Quayle in their vice presidential debate ("You're no Jack Kennedy"); now there was a clamoring for him to run, but he said no as well. Ditto for Senator Jay Rockefeller, the well-heeled West Virginia liberal, and George Mitchell of Maine, the top Senate Democrat. Sam Nunn of Georgia still made electoral sense as a moderate southerner, but his vote against Bush's war now threatened his reputation on national defense; he took a pass.

The *Washington Post* declared that "the Democratic presidential campaign industry is in a full-blown depression." When Mitchell appeared on *Meet the Press* at the end of March and said he'd like to be president someday, but not in 1992, David Broder called it the unofficial start of the 1996 campaign. Since "the odds now strongly suggest" a Bush win in '92, Broder argued, the smart play for Democrats was to sit out the cycle and wait for an open seat four years later. Soon enough it was fodder for *Saturday Night Live*, which aired a send-up of a Democratic presidential debate. Instead of spoofing the actual candidates running, the sketch—titled "The Race to Avoid Being the

Guy Who Loses to Bush"—featured all the noncandidates pleading with their party to choose someone, anyone other than them.

Mario Cuomo won a third term in 1990 and raised eyebrows by refusing during the campaign to rule out a White House bid. But that was before the Gulf War, back when Bush had a mortal's poll numbers. Now there was a survey, conducted at the height of the war, putting Cuomo sixty-one points behind the president, 77 to 16 percent. Cuomo also had his hands full in Albany, where a budget standoff with the Republican-led state senate was grinding government to a halt. Those who asked about 1992 now got his stock answer: "I have no plans to run. And I have no plans to make plans." Pressed for a firm disavowal, he sneered and reminded reporters of what happened four years earlier, when he'd announced early that he wouldn't run. "Everybody ignored it and, in effect, they called me a liar for six months. I'm not going to do it again. I wouldn't insult myself or lose my dignity by playing the media's game."

It sounded like the same old Cuomo. He hadn't been temped in '88, so why on earth would the prospect of facing Bush now be any more enticing? As always, Cuomo's reluctance only fed the legend.

Bill Clinton, on the other hand, wanted it badly, but he'd taken himself out of the mix, too. The Gulf War wasn't the reason. His survival in Arkansas was. In 1990, he faced a midcareer crisis. In a way, he remained a star of the future, just forty-three years old and still one of the youngest governors in America. But within Arkansas, Clinton was in danger of going stale. He'd been governor for ten years and was accumulating powerful enemies. Polls showed him exhausting his welcome with the electorate. Gubernatorial terms were now four years in Arkansas, after the law was changed in 1986. If Clinton won another in 1990 and served it out, he would become the longest-serving governor in state history.

Running for reelection would be risky. If he lost, his days as a national prospect were done. There was even a chance he wouldn't survive the Democratic primary, not with the state teachers' union and the AFL-CIO both vowing to oppose him. He thought about quitting, declaring his tenure a success and spending the next two years preparing for a presidential campaign. He floated the possibility of a different Clinton running in 1990. Hillary, he said, would make an "unbelievably good" governor.

But, really, he had to run again, even if it meant going bust. His national prospects were indelibly linked to his appeal in the South— a Democrat who could bring home the party's lost voters. To back away from a race in Arkansas would be an admission that he'd fallen out of favor. He had to prove he could still pull it off. He recognized, too, that no one is more quickly forgotten in politics than a former anything. Days before the deadline, he announced he'd run for a fifth term, even though "the fire of an election no longer burns in me."

"The joy I once took at putting on an ad that answered somebody else's ad, that won some clever little argument of the moment, is long since gone," Clinton said. "But what I believe in is what you and I and we can do together."

The primary was a slog. He hung on with 54 percent of the vote. He knew the electorate was restive, that it wouldn't take much to stir up a revolt in November. He could see one of the strongest cards his Republican opponent was holding, the sense that this famously ambitious governor was already looking past his job and toward the direction of Washington. So he made a promise. If the voters reelected him, he'd serve his full term in Little Rock. The only job he wanted in 1992, he said, was the one he had now.

He regretted it right away. Just before Election Day, when it was clear he was going to survive, he even opened up to a reporter: "I may have blown my timing." As soon as the election was over, the political world would be moving on to 1992, but now he was saddled with a public statement—a promise—that would take time to wriggle free of. His head was already there, though. "People will let you change your mind," he told the reporter, "if you give them a good explanation for why you're changing."

For Clinton, the Gulf War was a godsend. His campaign promise prevented him from being out on the hustings in the early months of 1991, but now everyone else was sidelined, too. They were already thinking of '96, following Broder's advice, but his focus was trained on the present. His time was running out back home. If he was ever going to test all that potential, it had to be now—or never.

He needed to create some national buzz, without reneging on his pledge yet. That's where the Democratic Leadership Conference came in. In its early years, he'd stood back like a silent partner, letting Robb and Nunn and the others blast away at powerful liberal interest groups and the national Democratic Party. He wanted the DLC label, not the enemies that came with it.

The DLC was maturing, though, and the dialogue within the party was turning pragmatic. Whites in the South, Reagan Democrats in the North, the suburban middle class everywhere: What could be done to get them voting Democratic again? The DLC was determined to provide the blueprint. Political and policy pros drafted an official platform. They had a plan to transform the organization into a full-fledged national network, with state chapters and grassroots members. Now Clinton wanted in.

He became the DLC's chairman in 1990, and a national conference was called for that March, the group's first ever. The platform would be ratified, a statement of principles Clinton would then challenge the entire Democratic Party to adopt. The New Orleans Declaration, as it became known, talked of "expanding opportunity, not government" and argued that "the promise of America is equal opportunity, not equal outcomes"; it struck a hard line on crime, chastising attempts at "explaining away" the behavior of lawbreakers; it was hawkish on national security, advocated linking financial aid for college students to national service, and embraced more "choice" for public school parents.

"While we favor tax fairness and think the middle class is overtaxed," Clinton said, "we don't think the Democratic Party should lead with class warfare."

The DLC had been conceived in animus toward the Democratic National Committee, but it wasn't the surrender of the party's national leadership that Clinton was seeking. He was looking for acceptance and alliance. He was asking them to give a little, to acknowledge that there were valid grievances with modern liberalism, that some old orthodoxies had to be abandoned. But he was offering himself as a partner, someone who in his core shared their basic values—and knew how to sell them.

Clinton cultivated Ron Brown, who in 1989 had become the DNC's first black chairman. A high-powered lawyer and lobbyist in Washington, Brown had firm roots in the party's traditional liberal wing. He'd worked for Ted Kennedy's 1980 presidential campaign and pitched in for Jesse Jackson at the 1988 convention. As a young law student at St. John's University, he'd been taught by Mario Cuomo himself. On the left, hostility toward the DLC remained the rule, and Brown had been one of the critics. For Clinton, then, it was a coup when Brown came to New Orleans and told the DLC: "I applaud your work. I applaud the contribution you've made to the party."

It was a start, but the resistance would be stiff. Liberals tended to regard the DLC label as the sign of a self-loathing Democrat, an opportunist who would sell out the party's most vulnerable constituencies. When the DLC held its next national gathering, in May 1991, a group of liberal leaders organized a rival event and the tensions exploded into public view.

It was held the same weekend in Des Moines. The Coalition for Democratic Values had been launched by Ohio senator Howard Metzenbaum as an emphatically liberal counterweight to the DLC. "I don't know whether they consider themselves another party," Metzenbaum said. "We sure as the devil do not." To a fired-up audience of three hundred, speaker after speaker preached the gospel of FDR and LBJ, excoriating Republican leaders and the Democrats who would appease them. "Other Democrats say we ought to be more accommodating," Iowa senator Tom Harkin thundered. "We're here to fire a shot across the bow of those Democrats."

Harkin, a second-term senator with a limited national profile, was preparing to run for president. With so many others intimidated by Bush, the Democratic race lacked an unapologetically liberal voice. Harkin was determined to give it one.

The dueling conventions underscored how precarious Clinton's challenge would be.

I n August, Clinton resigned as the DLC's chairman and formed a presidential exploratory committee. If the election were held that day, he said, Bush would win—"but nobody has made the alternative case in a long time." On October 3, 1991, he summoned the national press to Little Rock for an announcement. This was the moment in which he'd blinked four years earlier, but not this time. With a vow to fight for "the forgotten middle class," Clinton became an official candidate for president.

He joined what was regarded as a field of leftovers. Paul Tsongas had been first in. A former one-term Massachusetts senator who'd left politics after a cancer diagnosis in 1984, he had no name recognition and little charisma, and spoke with a slight lisp. Raw memories of Dukakis made Tsongas an even tougher sell. "Just what we need," one Democrat told columnist Mary McGrory. "Another sanctimonious Greek from Massachusetts."

Tsongas announced his candidacy in his native Lowell, Massachusetts, at the end of April. For a while, he had the field to himself, a lonely long-shot candidacy that symbolized how bleak his party's prospects were. Eventually, he was joined by Virginia's Doug Wilder, who two years earlier had become the nation's first popularly elected black governor; Nebraska senator Bob Kerrey, a decorated combat veteran who'd lost a limb in Vietnam; and Iowa's Harkin, who at least knew how to stoke the base's passions.

What these candidates lacked was stature. To a party coming off three straight national drubbings that was now tasked with challenging a president who'd won a war, an average, serviceable candidate just wouldn't do. To win a presidential election, or to even seriously compete in one, Democrats had come to believe, required nothing short of a political all-star. And all of their all-stars were sitting this one out.

Unless you counted the sixth candidate to enter the race. Jerry Brown at least had been a political all-star, for a few months back in 1976, when, as the young, iconoclastic governor of California, he'd made a late entry into the Democratic presidential race and nearly derailed Jimmy Carter. It was supposed to be the start of something big, but it didn't work out that way. He ran again in 1980 but went nowhere, then whiffed on a Senate campaign in 1982, left the governorship, and renounced politics. He spent time studying Buddhism in Japan and working at Mother Teresa's Calcutta mission. His critics, and they were many, took it as proof that Governor Moonbeam, the nickname columnist Mike Royko had hung on him, really had cracked up.

He made a return to politics in 1989, nabbing the chairmanship of California's Democratic Party, and in the summer of '91 was readying to run for the U.S. Senate. There was grumbling. Brown commanded residual loyalty from Democratic primary voters, but polls foretold big trouble in the general election. Voices in the party were speaking up, trying to talk him out of it. It was around this time that he was struck with an epiphany. It was the kind of thing that only seemed to happen to Jerry Brown. The economy was down, corruption scandals were eating up Washington, and discontent was in the air. A political revolution was brewing, Brown decided, and it was his calling to lead it. So he shut down the Senate campaign and drafted a five-thousand-word letter to his political network, telling them he would instead run for president, with one simple mission: to blow up the political system.

"There is no national figure," he wrote, "who is prepared to enter the presidential race and take on not only George Bush but the entire corrupt system, including those entrenched Democratic politicians who have turned our party from a voice of opposition into a party of complicity."

Brown was a name, all right—the name of a has-been who seemed to be completing his transformation into an unhinged radical. This wasn't the kind of star power his party had in mind.

Clinton had cause for optimism. None of these rivals was unbeatable. Then again, he was carrying his own baggage into this race. Many Democrats knew him only as the windbag who'd bombed in Atlanta. Others couldn't get past his DLC credentials. There were also the rumors, rampant in political circles, of marital infidelity. The post-Hart consensus held that a revelation of adultery would be sufficient to destroy any national candidacy. There was no reason to think that Clinton, if any of the chatter about him was true, would be the exception to this rule.

Clinton and his fellow candidates were "starting to look like Bush's ultimate insurance policy," wrote Adam Nagourney in *USA Today:* "What once was shrugged off by Democrats as a passing gust of the political winds—the absence of heavyweight challengers to President Bush—is now looming as a major political embarrassment."

I t was almost Thanksgiving and the field remained formless. No candidate was gobbling up endorsements or raking in campaign cash or leaping ahead in the polls. The first primary, in New Hampshire, was less than three months away now, and no one could remember a campaign when so many were so undecided so late in the calendar.

That's what made Chicago so important. Technically, it was the annual meeting of a group of state Democratic leaders, about a hundred of them in total, in a ballroom at the Palmer House Hilton. They'd invited each presidential candidate to speak and take questions, but this was no ordinary cattle call. With the race so unsettled and the calendar so tight, this was a casting call for a front-runner. The audience included not just the party chiefs in the room but also the national press corps and Democratic donors, activists, and influencers around the country. For the candidates, this was as good a chance as they were going to get to transform their image with the folks who mattered most.

The opening day, a Friday, featured four contenders with four familiar messages. Harkin played up the prairie populism ("If I can get in this ring, I can flatten George Bush") and warned against nominating a "warmed-over Republican," a shot at Clinton and the DLC set. Kerrey focused on universal health care, the latest—and not the last—Big Theme he'd settled on. Tsongas called himself a "pro-business liberal," in favor of cutting the capital gains tax to goose the economy but against reducing middle-class rates for fear of exploding the deficit. Jerry Brown showed up, too, but the party establishment was hardly his renegade campaign's target audience. There was nothing to surprise anyone.

Saturday morning was for the other two candidates. One was Wilder, whose bid was already on life support and would be over in a few weeks. The other was Clinton, and expectations for him were minimal. After all, this was the same guy who'd been hooted off the stage in Atlanta.

All the better to wow 'em, it turned out. Clinton's pitch was crisp, passionate, and personal, everything he hadn't been in that convention speech. He grabbed his audience straightaway. "I believe this is the eleventh hour. Our country is going in the wrong direction and we must struggle to keep the American dream alive." He talked of tax cuts, education reform, overhauling welfare—DLC issues all of them, the stuff Harkin was railing against.

The packaging was delicate, and Clinton knew it. If he sounded like a fed-up southerner lecturing the liberals on how out of touch they were, they'd tune him out. When the floor was opened to questions, the DLC stigma came up immediately. What about the accusation that he was a warmed-over Republican? Clinton was indignant. He was "a Democrat by heritage, instinct, and conviction," he said. "When my grandfather died, he thought he was going to see Franklin Roosevelt." He told them to look at his record in Arkansas—more money for schools, more money for infrastructure. "These people call me a Republican because I want to change and push the party into the future, not pull it to the right or the left," he said. He kept going.

He could think of two other Democrats, he said, who had been hit with the same charge: "Their names were John Kennedy and Jimmy Carter. And if you want history to repeat itself, that means you ought to support me instead of the people throwing the rhetoric out."

The verdict was close to unanimous.

"It was rather startling," the Arizona state chairman pronounced. "He seemed much further along than the others." Ohio's Democratic chairman had publicly written off Clinton's candidacy. Now he reconsidered: "I think he successfully beat the rap that he's a warmed-over Republican." To the Texas chairman, it was "a big-time speech." With a Chicago dateline, the *Washington Post*'s Dan Balz reported that "Arkansas Gov. Bill Clinton jumped out of the pack of Democratic presidential candidates with a performance here today that party leaders from around the country agreed had outdistanced his announced rivals for the nomination."

Clinton, Balz wrote, seemed ready for the question he'd aced. And he had been. It turned out his campaign had planted it. But this was a small detail. Clinton, the *Chicago Tribune* wrote, "left a crowd of Democratic Party leaders with their mouths agape Saturday in Chicago as he transformed a two-day forum for presidential candidates into a seminar on his ideas and a pep rally for his struggling candidacy."

It had been forty months since the Atlanta debacle, but this time Bill Clinton had met the moment. In this field of B-listers, it should have made him the front-runner on the spot. And it would have, if it weren't for Mario Cuomo.

America's Gulf War high was cooling faster than anyone had anticipated. By the fall of '91, unemployment was nearing 7 percent, up almost two points from just a year before. The recession, officially declared in April, felt like it was dragging on with no end in sight. Economic anxiety was on the rise and Bush's approval rating was at last exiting the stratosphere, back into the sixties by September, down into the fifties in October.

The psychology of the opposition party was shifting. Some Democrats now dared to wonder: What if the economy kept worsening? What if Bush's numbers kept falling? Could he actually be vulnerable in 1992 after all? The clamoring intensified for a candidate capable of capitalizing on this.

This was the moment that Mario Cuomo chose to say something he had never before said about running for president. He said it in private first. The occasion was a closed-door breakfast with his political fund-raising team at the Regency Hotel in Manhattan. The date was October 11, the Friday before Columbus Day. It was supposed to be a routine gathering, but as soon as the session broke attendees raced to nearby phones to share the news. Governor Cuomo, they said, had told them he was thinking about getting in the presidential race.

Communication was slower in those days, but by lunchtime everyone in the game of politics was talking about Cuomo. He called a press conference. "They said, 'Will you think about it?' I said, 'Sure, I'll think about it. I'm always thinking about it.' I said I'd have to be mindless not to think about it. I don't talk about it, but I think about it. Of course I do."

He made it sound so simple, so obvious—*Of course I do*. But make no mistake, this was a real shift. For years, he'd been contemptuous of this subject, treating any talk of presidential politics, any speculation about his own ambition, as cheap and tawdry, even harmful to democracy. "I cannot get two minutes to talk about the economy," he would say. "But I can get a half hour to talk about a candidacy." For Cuomo to go even this far in public signaled a receptiveness that had never existed before.

The possibility of making a late entry had actually been on Cuomo's mind since at least a month earlier, when he'd made a rare out-of-state pilgrimage to California. The official purpose was to address the National Association of Broadcasters, an event that returned Cuomo to the same Moscone Center stage from which he'd delivered his 1984 keynote address.

He was also in San Francisco to headline a fund-raiser for the state party, a favor for the most powerful local Democrat, Representative Nancy Pelosi. They'd first met more than a decade earlier in the Italian province of Avellino, dispatched by President Carter to lead the U.S. delegation surveying the damage from the catastrophic Irpinia earthquake of 1980. Pelosi back then was climbing the ranks of the California Democratic machine, a Baltimore transplant and the daughter of a mayor. Cuomo, who was New York's lieutenant governor then, was still nursing the wounds from his failed 1977 bid for New York City mayor. He and Pelosi bonded over their shared heritage, Catholic faith, and urban political backgrounds. When Pelosi sought the chairmanship of the DNC in 1985, Cuomo, by now his state's governor, threw the support of New York's delegation behind her. She fell short, but two years later claimed a seat in Congress, where her insider skills and fund-raising savvy now marked her as a rising star. To no avail, she'd been urging Cuomo to run for president, and nothing he told her during his September visit suggested his resistance might be loosening.

"I don't see any signs that he actually is going to run," Pelosi said after the fund-raiser. "I say that with discouragement, frankly, but this is a subject that he just won't talk about when you try to bring it up." The quote became the basis for a *New York Times* headline the next day: "California Democrats Take Cuomo at Word." Even those who dearly wanted him to run, the paper was reporting, were finally giving up. The story even had Cuomo saying, "Am I pleased to hear that people believe me? Very pleased."

Cuomo was back in New York when the *Times* story ran. It was Saturday, September 14. Upon reading it, he telephoned his chief political lieutenant, John Marino, with an order: call Pelosi and tell her he was, in fact, considering running for president, and keep her from endorsing anyone else. Marino had never heard his boss talk this way, with such focus, such determination, about the presidency. There was more. Cuomo asked Marino to research logistics—primary dates, filing deadlines, delegate rules, fund-raising restrictions—and to draw up a blueprint for a campaign. "I want you to plan this as if I'm running," Cuomo said. "That is your job."

Something about this particular challenge appealed to the romantic in him—fighting the good fight against a supposedly invincible president. There was a class component, too. Cuomo, the grocer's kid from Queens, carried a chip on his shoulder, and Bush, product of the New England aristocracy, son of a senator and inheritor of wealth, status, and privilege, made for a dream foil. And now, of course, it was beginning to look like the campaign might actually be winnable.

"What makes you think President Bush is going to be impossible to handle in an intellectual disputation?" Cuomo now asked. "What's he going to say? 'I won the war'? 'The other guy's a bum'?"

Instantly, Cuomo's move froze the Democratic race in place—just like he wanted. Almost no major party leader, elected official, or fund-raiser was about to side up with one of the mites when there was a chance Cuomo the Giant was about to come in and crush them. "He'd certainly be the favorite," acknowledged Kerrey's campaign chief. A former chairman of the Democratic National Committee, John C. White, had been leading an effort to coax a major name into the race. Cuomo had been his top target, and Cuomo's sudden interest, White now said, "changes the game totally. It changes the projections. And if he does run, Bush will be in for the political fight of his life." A week after his breakfast session, Cuomo was back on Art Athens's radio show in New York City, the same forum where he'd announced he wouldn't be a candidate in 1988. Speculation swirled: this time he was going to say he was in. The *Boston Globe* reported on its front page that "New York Gov. Mario Cuomo is on the verge of deciding to run for the Democratic presidential nomination, according to party activists who say that at least 20 fund-raisers and organizers have been asked not to make commitments to other candidates so they can enlist in his campaign."

Instead, all Cuomo did was talk up his responsibilities in Albany. "It's difficult to imagine a time when the state needed a governor more than it does now," he said. In another interview, he hinted he might have to take the dramatic step of resigning to seek the White House. "Let's say I wanted to run for president. I'd have to figure out how to handle this deficit, next year's budget deficit and still go to Iowa, New Hampshire, all these other places. How do you do it? Well, then you'd have to quit being governor."

On and on it went like this: conflicting ruminations, philosophical circumlocutions that only confused everyone further. Was he executing some long-game strategy to maximize the suspense and drama? Was it all one big tease? Or was he genuinely torn, publicly airing a tormented internal monologue?

It was driving the other Democrats nuts, especially Clinton. He'd become a hostage to Cuomo's indecision, and was staring at two radically different paths forward.

If Cuomo stayed out, the party leaders he'd dazzled in Chicago and elsewhere would migrate to him and Clinton would become the new Democratic front-runner. Instead of running against Cuomo from the right, his chief rival in New Hampshire would be Tsongas, who was promising to be Wall Street's best friend as president. It would be the perfect setup for Clinton's soft takeover strategy. He could retain his New Democrat credentials and still have space to go after Tsongas from the left. He wouldn't have to fight the party establishment. In fact, he might be the candidate of the party establishment, the old Democratic coalition's best hope to stop Tsongas and his Wall Street takeover. He could succeed in winning the nomination and repositioning the party toward the middle without inciting a knockdown fight with the left. He just needed Cuomo to make up his damn mind.

December came and Cuomo's intention remained a mystery. Even with his inner circle, he was sparing with what he shared. Marino, who spoke with his boss every morning, believed he was more likely to run than not. He was carrying out Cuomo's orders, laying the groundwork for a late-starting bid, organizing a loose network of Cuomo devotees ready to swing into action when the moment came—*if* it came.

Beyond whatever inner angst he was feeling, there was something now complicating his choice. With the economy foundering and tax receipts drying up, New York's fiscal condition was in peril. All year, Cuomo had been at odds with the Republican-controlled State Senate over how to close an $875 million gap in a state budget totaling $30 billion.

Cuomo now told one interviewer, "If the negotiations aborted or we don't have a current year plan, under those circumstances I couldn't in good conscience declare a candidacy."

Beneath all of this was a serious political concern. The Democratic indictment of Bush leaned on the enormous budget deficits that had become routine in the last twelve years of Republican rule. How would it look for Cuomo to present himself as a national candidate with his home state drowning in red ink? Could a governor just leave behind an unresolved fiscal crisis and run for president? Already, Clinton was sensing a weakness, branding Cuomo "a powerful spokesman for the Northeast liberal base of the party" and pointing out that there were other Democratic governors out there who "were able to balance budgets."

Clinton was threading a tricky political needle. His dismissal of Cuomo sounded very much like something from the DLC's more strident faction, the ones who delighted in picking fights with the national party and wore the scorn of liberal activists as a badge of honor. Very deliberately, Clinton had shunned that approach, knowing it would pigeonhole him as a regional candidate. But Clinton knew he needed to bring Cuomo back down to earth. That meant attacks from the right on the liberal hero of much of the party. With every punch, Clinton would risk making himself exactly the kind of DLC candidate he didn't want to be.

The Friday before Christmas, December 20, was the drop-dead date, the last day possible to file for New Hampshire. Even Cuomo acknowledged that this was it, and the whole week felt like one steady march to his emergence—at last—as a candidate for president.

On Monday, the top Republican in New York politics, U.S. Senator Alfonse D'Amato, showed up in Albany. He was there to score points for the Bush team, which now saw Cuomo as its most likely opponent. Cuomo's negotiations with Republican legislators remained stalled and D'Amato bellowed: "If the governor can't take care of the budget of New York—which is totally out of control—how's he going to undertake the economic problems of the nation?"

Cuomo sensed that the budget stalemate was being used as a Republican ploy to keep him on the sidelines in '92. If this was the case, he said, there was one obvious response: "If you made up your mind that they were doing this only to prevent you from running, then the solution would be to run. Then they would no longer have the motive to slow you up."

"The more I hear their attack," he added, "the more enticing it becomes."

A poll came out that day. For the first time since the war, Bush's approval rating was below 50 percent. He was leading Cuomo by just five points. On Wednesday, Cuomo moved to jump-start the budget talks, granting new concessions to Republican legislators.

The uncertainty was getting to Clinton, whose irritation spilled over in public. "I always thought he'd run," he said, "and I thought he'd wait until the last minute. . . . He waited long enough to see which way the wind was blowing." That night, Cuomo told Marino he had an idea: a sneak attack in Iowa. The other candidates were ceding the leadoff caucus state to Harkin, but Cuomo had decided he was weak. What better way to demonstrate his own strength than to go to Iowa and knock off the favorite son? Cuomo was thinking like a candidate now. Marino went to sleep that night convinced his boss was about to do it.

Thursday brought no progress on the budget front. Republicans weren't even nibbling. Negotiations went until nearly midnight, the participants emerging with nothing to report. Cuomo was asked again if Albany's Republicans were trying to sabotage his candidacy. "The evidence," he said, "piles up in one direction."

The filing deadline in New Hampshire was the next day. If he was going to run for president, it was now apparent, Cuomo would have to do it without a budget in place. The plan for a campaign announcement was in place and ready to execute. Marino had arranged for a private plane to stand by the next morning at Albany's airport. When he was ready, Cuomo would board it for the 120-mile trip to Concord, New Hampshire. He would then be driven to the state capitol a few miles away. There'd be cheering supporters outside to greet him. Cuomo would enter the statehouse, walk to the office of the secretary of state, and hand in the signed form and a check for the one-thousand-dollar filing fee. Then he'd head back outside for an immediate show of force: scores of Democrats lined up would endorse him on the spot.

Early Friday, the major papers were reporting that Cuomo was going to enter the race—"It looks and smells like a go," the *New York Post* quoted a top Cuomo aide saying—and Marino expected the governor to arrive in New Hampshire around noon. Meanwhile, on television, one of the first campaign dramas of the cable news era was playing out. CNN, the sole twenty-four-hour news channel, had a camera trained on the plane in Albany. All morning, the political world sat riveted, waiting for New York's governor to step into the screen and onto the plane for his appointment with history.

Typically, Cuomo and Marino chatted first thing each morning. But not on this day. Marino thought Cuomo was heading to New Hampshire, but he still wasn't sure. Given the magnitude of the day, Marino wasn't too concerned by the lack of communication, at least not at first. When he still hadn't made contact with his boss by mid-morning, he started doing the math—how long to get to the airport, how long for the flight, how long to drive to the statehouse. It added up to worry. The deadline was 5 p.m., and now they'd be really cutting it close. By now, a crowd of 150 had gathered outside the statehouse. The suspense was all encompassing.

It was during all of this that Cuomo reached out to his old right-hand man, a former staffer he'd loved like a son who was now a big-time player in Washington. Tim Russert was the Washington bureau chief for NBC News and had just that year been appointed to host *Meet the Press*. Before his media career, though, Russert had been a New York political operative, starting out with Senator Daniel Patrick Moynihan before joining Cuomo's team. Russert had played a key role in Cuomo's upset win in the 1982 gubernatorial primary over Ed Koch, the victory that made possible everything that had happened in the last decade. At first, he'd stuck with Cuomo in Albany, hoping to go national with his boss. When Cuomo proved ambivalent, Russert decamped to D.C. and switched to the world of journalism.

Cuomo felt particular warmth and affection for Russert, but in this moment he was to play a different role. For all his national fame, Cuomo had kept his distance from the Washington world. Russert was the exception. Here was the one elite D.C. figure Cuomo truly trusted, someone to confide in, someone who would offer candid counsel in return. This was the man Cuomo sought out as he grappled with the biggest decision of his life. What Russert said to him this Friday morning reinforced what Cuomo saw as the biggest barrier to his candidacy. "If you go up to New Hampshire without a budget," Russert told him, "Bill Clinton will beat you to death with it." Perhaps Russert really believed this. Or maybe he just had a sense that, deep down, it's what his anguished mentor wanted—and needed—to hear from him. When the call ended, Cuomo dialed Marino and told him he'd made up his mind. He wasn't running.

The press conference was around three in the afternoon. Cuomo thanked the people who'd urged him to run. "My responsibility as governor is to deal with this extraordinarily severe problem," Cuomo said. "Were it not, I would travel to New Hampshire and file my name as candidate in the presidential primary.

"That was my hope, and I prepared for it. It seems to me I cannot turn my attention to New Hampshire while this threat hangs over the head of New Yorkers I was sworn to put first."

He was blaming it on the budget, and maybe that's all it was. Then again, who would turn down a clean shot at the presidency just because of some obstinate state legislators from the other party? Cynics said it was a cover story for an indecisive politician who just didn't have the nerve to go through with it—Hamlet on the Hudson. Others saw a man who took the measure of himself and realized he didn't want—or need—to be president. Whatever the rationale, it was over. Mario Cuomo was out. And Bill Clinton was on his way.

SIX

The rebel commander was now one of the highest-ranking members of the Republican Party's congressional leadership. There was nothing irreconcilable about these two roles, Newt Gingrich liked to say. Then came the fall of 1990.

It was two years after George H. W. Bush looked into the eyes of tens of millions of Americans and swore over and over that he would never, ever raise their taxes. Now, he wanted to do just that.

The deal was a classic Washington compromise, billed as the biggest single collection of spending cuts and tax increases ever. It was also an epic reversal. Bush was breaking his word, and the grief he'd take would be monumental—and also, he was convinced, worth it.

The fiscal case was straightforward. Massive deficits had been piling up for a decade. From the country's founding until the start of Ronald Reagan's presidency in 1981, the national debt had never exceeded one trillion dollars. But the Reagan years produced a steep drop in tax rates—from 70 percent for the highest earners all the way down to 28 percent by the time he left office—and a tidal wave of red ink. The debt cleared one trillion dollars in 1982, two trillion by 1985, and now, in 1990, it was well over three trillion, with no sign of slowing down. Democrats pointed to the tax cuts and defense buildup of the Reagan years. Republicans said the Permanent Democratic Congress was addicted to spending.

The blame game was old, but there was new urgency now, as the boom years of the eighties faded away. The economy was slowing, interest rates were up, joblessness was rising—unmistakable signs of an approaching recession. Running up big deficits when things were humming was one thing; could the country handle the strain it was about to face? Amplifying the potential of crisis was something called the Gramm-Rudman-Hollings Act, a five-year-old law that represented Congress's only response to the red ink of the eighties.

Named for the senators (two Republicans and one Democrat) who crafted it, Gramm-Rudman-Hollings mandated specific annual reductions in the size of the deficit. If Washington failed to meet the target, the law decreed, spending would be indiscriminately slashed across the board. Already, the number for 1990 was big, and that was before the spring announcement from Bush's budget director that even less revenue was coming in, thanks to the sputtering economy. The hole was getting larger.

"The prospect of a huge 'sequester'—automatic budget cuts—looms for the fall," the *Washington Post* reported, "and while the administration last year was willing to endure such automatic cuts, they would be so large this time . . . as to be unacceptable to both the White House and Congress."

Bush decided he needed a deal, which meant confronting the most famous promise in presidential campaign history. When he'd run in 1988, Bush had been adamant about how he'd approach Congress in an impasse like this. The Democrats would push him to raise taxes, "and I'll say, 'No.' And they'll push, and I'll say, 'No.' And they'll push again, and I'll say to them, 'Read my lips: no new taxes.'"

It was the line of the campaign. But by 1990, Bush was thinking pragmatically. With their huge congressional majorities, Democrats had proven adept at ignoring his demands at no apparent political cost. Since his inauguration, Bush had been crusading for a cut in the capital gains tax rate without moving Democrats even an inch. Bush, the Connecticut-raised Yankee, had never really been the Dirty Harry type to begin with, and the kind of drama he'd acted out in his campaign wasn't about to jar this Congress.

Democrats told Bush they'd meet him at the table, on one condition: he would have to publicly back off the "Read my lips" promise. In early May, the president complied, dispatching his press secretary, Marlin Fitzwater, to announce he'd invited both parties' congressional leaders to a new round of talks on the deficit. The negotiations, Fitzwater added, would be "unfettered" by anything said or done in the past. It wasn't a formal assent to a tax increase, but everyone could read between the lines: Dirty Harry was blinking. At the end of June, Bush was explicit. Any deal struck, he said in a White House statement, must include "tax revenue increases." The flip-flop was complete.

It took nearly five months, but with the government on the verge of a shutdown and the automatic cuts about to trigger, the deal was finally cut on the morning of Sunday, September 30. Bush would sign off on tax hikes and give in on his treasured capital gains cut while Democrats would swallow hard on Medicare and back off their insistence on a tax increase for the rich—and on any income tax hike at all. Instead, it was agreed, taxes on gasoline, alcohol, and cigarettes would all rise. Add it up and it would supposedly lop about a half trillion dollars off the deficit over the next five years. The media was called to the Rose Garden for an afternoon press conference. Congressional leaders from both parties joined Bush.

"Sometimes," Bush said, "you don't get it just the way you want, and this is such a time for me, and I expect it's such a time for everybody standing here."

Those standing with him constituted a bipartisan who's who of Washington power. Senate Majority Leader George J. Mitchell, House Speaker Tom Foley, and House Majority Leader Richard A.

Gephardt headed the Democratic contingent. Senate Minority Leader Bob Dole, Senate Minority Whip Alan K. Simpson, and House Minority Leader Bob Michel led the Republicans in attendance. None of them were joyful, but all of them were relieved that months of grinding negotiations had at least produced something. They would avoid a shutdown, stave off the automatic cuts, and take a real whack at the deficit.

"The naysayers, the nitpickers may have a field day because the easy vote in this case is to find something you don't like and vote no," Dole said. "But in my view, we owe more to the American people than finding fault with what I consider to be a good agreement."

They also hoped it might come with a political payoff. For Democrats, here was proof they weren't the obstinate obstructionists Bush had spent the last two years railing against. He'd invited them to the table and they'd taken him up on it, and now they'd have something to show for it.

Bush was thinking longer term. Reelection was still two years away, but the ominous economic signs already had him nervous. Right now, his poll numbers were actually quite strong. His approval rating had been well north of 50 percent—and often above 60 or even 70 percent—since his inauguration. There was public exhilaration at the sudden weakening of the Soviet Union; the Berlin Wall had fallen in 1989 and democratic revolutions were sweeping across Eastern Europe. Plus, voters tended to like Bush personally.

But there was the threat: if the economy was just starting to tank now, Americans wouldn't really start to feel it until the 1992 campaign was under way—when it would be too late to do anything about it. The play, then, seemed obvious to Bush: break the taxes pledge now, with his approval rating still high, take the hits, endure the jokes, pay the political price, then reap the benefits in '92 of the stronger economy that this deal would ultimately produce.

Making deals just made sense to George H. W. Bush. His politics had shifted with the winds in the Republican Party—Goldwaterite in his first campaign back in 1964, moderate alternative to Reagan in the 1980 primaries, Reagan-style conservative since. Really, he was a pragmatist. If you didn't have the votes to impose your will, then you worked something out with the other side. It was practical. It was responsible. It was what you had to do. What was the alternative— digging in your heels and getting nothing? That wouldn't be just reckless and irresponsible. It would also be political suicide. After all, what the voters really wanted was leaders who got things done. Didn't they?

This was Bush's read on politics. And it was shared by the Republicans who joined him in the Rose Garden that Sunday afternoon in September 1990. They were used to being outnumbered in Washington. They'd seen their share of Republican presidents, but those Republican presidents had all faced hostile Democratic majorities on Capitol Hill.

The Permanent Democratic Congress: it was all every Republican in the Rose Garden that day had ever known. Bush, Dole, and Michel were all nearing seventy years of age, each with decades of Washington service under his belt. Not one had been in office the last time there was a Republican Speaker of the House. To get things done, they all knew only one way: work with the Democrats.

For their part, Democrats were delighted they'd made Bush budge on taxes, but they didn't want the deal passing with just their votes. There was an election in a few weeks, and they wanted Republicans on the hook just as much as themselves. So they told Bush: We've got the votes as long as more than half of your party is on board, too. A majority of the minority, or the deal would be off. Which seemed easy enough. The Republican president would be twisting Republican arms, and so would the Republican vice president, the Republican leader in the House, and the Republican leader in the Senate. Surely, that would be plenty of juice.

Although, come to think of it, there was one Republican member of the congressional leadership who wasn't onstage that Sunday afternoon. He was locked in a tough reelection fight back home, so maybe he was just preoccupied with that. But he was also notoriously volatile, a lover of political combat who'd previously taken down a House Speaker, won a formal rebuke of another, and maneuvered into his party's number-two spot in the House. As this deal was sealed, he'd kept quiet, uncharacteristically so. Now, every other top Republican in Washington was about to fight for this deal, fight for their president, fight alongside the Democratic leadership. But what about Newt?

Gingrich kept his distance from the months-long negotiations. At meetings, he'd pull out a newspaper and start reading it when the topic turned to taxes, demonstrating his contempt for the conversation. Starting with that embrace of Jack Kemp's rate-slashing platform all those years earlier, Gingrich had made cutting taxes one of his defining crusades. Now he was being told to rally the troops for a rate hike. Resisting was a matter of philosophy, but just as much it was strategic.

House Republicans held an emergency meeting after the White House ceremony. Reporters flocked to Gingrich. "Still looking at it, still looking at it," he told them. Now there was suspense.

The White House unleashed its full-court press on Monday. Vice President Dan Quayle was sent to Capitol Hill to make the sale to Republicans in the House, where scores of Gingrich acolytes would have to be mollified. Quayle was joined by the president's chief of staff, John Sununu, and his budget director, Richard Darman. "I feel like a friendly dentist," Quayle said, "applying Novocain and trying to extract a few votes." From New York, where he was to address the United Nations on the showdown with Saddam Hussein, Bush publicly urged lawmakers to "exercise leadership for the good of the country." The Bush team marketed it as good politics. The midterm elections were five weeks away. Every single House Republican would be facing the voters. Wouldn't it make sense to have something to show them? "I think it would be politically risky not to pass a good budget package," Quayle said. "If, in fact, a budget package is not passed, all are going to suffer, the Republicans as well as the Democrats." Sununu was blunt, warning that the president wouldn't be inclined to hit the trail for any Republican who wasn't with him on this vote.

They needed half the room. To get the support Democrats had promised, half of the 176 Republicans in the House would need to support the deal. Really, though, they just needed one. If Newt Gingrich, despite his obvious dislike of it, were to walk out of the meeting and announce he was nonetheless on board—that this was a moment to stand with the president—then there might not be much of a rebellion at all. The press stood waiting outside for an answer. Would the power of the presidency force Gingrich into line? Or would there be a civil war? The meeting wrapped, the doors opened, and Gingrich made his way to the reporters and photographers. He was flanked by several loyalists, the rabble-rousers he'd been cobbling together from the day he arrived. Now, finally, he broke his silence on the deal.

"It is my conclusion," Gingrich said, "that it will kill jobs, weaken the economy, and that the tax increase will be counterproductive. And it is not a package that I can support."

The war was on.

The Gingrich army was ready for battle. This was what he'd spent years priming them for, a betrayal by the party establishment on a sacred issue. Chuck Douglas of New Hampshire called it "the fiscal equivalent of Yalta." To Indiana's Dan Burton, it meant "guerrilla war, and you may rest assured that whatever it takes to scuttle this terrible budget package, we will do it." Dick Armey from Texas said it would guarantee a recession, that "the historians will look back and call this the pre-plummet summit."

Meanwhile, the White House was accusing Gingrich of duplicity. Sununu, the chief of staff, said he'd signed off privately then gone back on his word. "Newt Gingrich laid down some specific concerns," he said. "He asked that the tax rates not be raised. The tax rates have not been raised." Sununu was referring to income taxes, which indeed weren't part of the agreement. Reporters asked Gingrich if the charge was true. He insisted he'd told Bush his decision and "the answer was the president saying to me he was going to be very strongly for this package. I thought it was a clear conversation." As for the idea that he had a responsibility as a Republican leader to support his party's president, Gingrich said, "We are required to render independent judgment."

Bush was thinking of 1992. He wanted the benefits of a good economy for his reelection campaign and was satisfied this was how to ensure it. All of the dealmakers believed, too, that there might be a reward in the idea of bipartisanship itself. The compromise they'd struck, predicted Dole, the Senate Republican leader, will "demonstrate to the American people, who are sometimes somewhat cynical, that the Congress and the president of the United States can work together, and we can look ahead, and we can do the right thing for our country."

Already, though, there were signs that the deal was a loser with the public. An overnight poll by ABC News found 65 percent of Americans saying they were opposed to it. Members from both parties were reporting calls to their offices running decisively against it.

The problem was twofold. Since it was a grand compromise, the deal almost necessarily offended the true believers in each party. Charles Schumer, a Democratic House member from New York, summed it up: "Democrats think it's a Republican budget and Republicans think it's a Democratic budget."

There was the way supporters were talking about it. "Obviously, it is not a proposal that most members of Congress want to rush out screaming in the streets with joy about," Foley, the House Speaker, said. Virginia senator John Warner, a Republican, likened it to "drinking a gallon of castor oil, but we're going to do it." They were making it sound dirty and disreputable, exactly how Newt Gingrich was always describing Washington dealmaking.

With the midterms approaching, every Republican candidate in the country now faced a dilemma. "They should take on this agreement, label it a Democrat tax increase agreement that was basically stuffed down our president's throat," Vin Weber, Gingrich's lieutenant, advised. "They should oppose it out in the hinterlands."

Michel was asked about his whip's defection. "He has his reasons" was all he'd say. The GOP leader predicted he'd come up with the votes anyway, but now he was on edge. He urged Bush to get involved personally. The House was set to vote on the blueprint as early as Wednesday, two days after Gingrich's revolt, and the situation was volatile. The president complied. On Tuesday, dozens of House Republicans received phone calls from Bush. More were invited to the White House to meet him face-to-face. He was pressing hard for commitments and not getting nearly as many as he expected. Now the White House was alarmed. The television networks were alerted: the president wished to address the nation that night.

When the camera came on, Bush's message was ominous. "If we fail to enact this agreement," he told Americans, "our economy will falter, markets may tumble, recession will follow." Bush didn't name Gingrich, but did tackle his argument: "Those who dislike one part or another may pick our agreement apart. But if they do, believe me, the political reality is, no one can put a better one back together again."

By their custom, the television networks also offered time for a Democratic rebuttal, but this time it was used to echo the president's message and turn up the heat on the GOP. "We hope," Mitchell, the Senate majority leader, said, "that Republican members of Congress will also set aside partisan differences as we have done, and join us in doing what's right for our country." This was just more ammunition for Gingrich, who said it proved "this is essentially a Democratic package."

"The Democratic leadership should have the burden to pass it," he said. "The Democrats got what they wanted: tax increases, no growth incentives, cuts in the defense budget, and no reductions in discretionary spending."

The fractures inside the GOP were really starting to show.

Jerry Lewis, the Californian who'd wanted to run for whip the year before, branded Gingrich's actions "very unfortunate" and all but accused him of betraying the president. Both Gingrich and Lewis, who sat one rung beneath him on the leadership ladder, aspired to replace Michel whenever he stepped down. Now Lewis was wondering if there was a chance to shove his rival out of the way early. Others were rallying behind Gingrich. "We're in prison as the House's minority party," California's Bill Thomas said. "I'd rather plot the great escape with Newt than mark off the days with someone else."

By Wednesday, the White House was in full crisis mode. The plan for a House vote that night was put off another day. The votes weren't there yet, and there was only one way left to reel them in. "The president's political life is on the line," Michigan's Guy Vander Jagt, who ran the GOP House campaign committee, declared. Two Republican former presidents, Gerald Ford and Richard Nixon, weighed in, urging the party to close ranks. Bush called a press conference and warned of "economic catastrophe" if the plan went down.

Democrats were feeling apprehensive. They had a lot of skin in this game, too. Gephardt, the House majority leader, called it an "acid test" of Bush's leadership. "If a president can't get his own party members to follow him, he's in real trouble."

Gingrich said they were all being a bit too dramatic. "You know, we are not going to suspend the Constitution and eliminate the legislature the morning after if this fails."

"Mr. Gingrich's preference for the outsider's role was never more apparent than today," the *New York Times* wrote.

Hours before the vote Thursday night, Bush tried one last maneuver. Originally, Democrats said they wouldn't provide the votes for passage unless at least half of the Republicans were on board. Now it was very iffy whether Bush and his allies could reach that quota, so the president made a concession. If the blueprint passed the House that night, Foley's committee chairmen would be given "flexibility" to shape the final package, something the Democratic gavel wielders had been clamoring for.

These were the same entrenched committee chairs who, in the Gingrich lexicon, were cogs in the corrupt Democratic machine that had been trampling Republicans forever. In his frantic bid for votes, Bush was giving him one final gift. "Every committee chairman will be allowed to run amok!" Gingrich fumed.

The vote was taken after 1 A.M. on the morning of Friday, October 5. For a few moments, the tote board suggested a squeaker, but soon the bottom fell out. At the end, there were 179 ayes and 254 nays. "The conference report is rejected," Pennsylvania Democrat John P. Murtha, who was presiding over the chamber, announced. For five arduous months, the president and nearly two dozen congressional leaders from both parties had bargained their way, one tiny piece at a time, to what was supposed to be a grand deal to wipe red ink off the country's books. Now it was dead.

The opposition came from both parties, but it was clear who killed it. Only seventy-one Republicans stayed loyal to their president. The other 105 went with Gingrich, the renegade party whip who told them there were more important things than loyalty to the White House. No one had expected defections of this magnitude.

Democrats had been ready with their promised votes, but when they saw what was happening, they told their members to go ahead and vote no if they wanted. There was no point in supporting the deal if the Republicans were in revolt like this. An analysis showed that of the twenty-five House members—Democrat and Republican—facing competitive reelection races, every single one of them voted no. Gingrich had kicked up a revolt not just in his own party but in the whole Congress.

He took no victory lap. In fact, Gingrich didn't even deliver a floor speech before the vote. Killing the deal was something his fellow Republicans wanted to do, he wanted to show, not something he was compelling them to do. There was truth to this, but only so much. Here was a rebuke of a Republican president delivered by his own party, and Gingrich's fingerprints would always be on it.

"In past days," David Broder soon wrote, "when individual ambition burned less brightly in some politicians' breasts and party loyalty meant a whole lot more, what Newt Gingrich did this week would have been unthinkable."

Newt Gingrich had been in the House for a dozen years now and upended the nature of the institution. The lifers agreed: the People's House was feeling less and less like the place they'd always known and more and more like something very different.

SEVEN

ill Clinton came rolling into 1992. Endorsements and cash had been in short supply during Cuomo's tease, but now both came his way in abundance. His campaign rushed to release its new fund-raising total—north of two million dollars in the final days of 1991. No one else in the field was close. His performance in Chicago before Thanksgiving had piqued the party elite's imagination. Now, with Cuomo out of the way, Clinton was reaping his reward.

Polls had him gaining ground in New Hampshire, where next-door neighbor Tsongas had initially been penciled in as the post-Cuomo favorite. The state's neighbor-to-neighbor political culture fit Clinton well. His speeches and town hall meetings won positive reviews. By the middle of January, he passed Tsongas and took the lead. National media outlets declared him the front-runner for the nomination; their designation steered yet more money and more endorsements his way. Clinton, the *New York Times* wrote, "by virtue of his early strength in organizing and fund raising, not to mention message-making and pundit-stroking, has become the hot candidate of the post-Cuomo period."

There was the prospect of a virtual coronation. Iowa, the traditional leadoff state, would be off the board this time around. The rest of the field was ceding the caucuses there to Harkin, the home-state senator, who would get all of the delegates without a speck of momentum. That only amplified the importance of New Hampshire, with its primary eight days later. Clinton was supposed to be a tough sell in Yankee country, but now he was defying those early assumptions. Win in New Hampshire and he'd knock Tsongas out and be in position to finish off his weakened foes in his own backyard in the Super Tuesday mega-primary.

It would take a surprise to derail him—maybe like the one that landed in several major newspapers on January 17. The source was the *Star*, a supermarket tabloid that traded in rumors and gossip about celebrities. It was teeing up a story to run the next week that accused Clinton of carrying on at least five extramarital affairs in Arkansas. Names were named, but none of them had admitted to anything. The *Star* had gotten the information from a lawsuit filed a few years earlier against Clinton by a disgruntled state employee named Larry Nichols. For publicity's sake, *Star* shared its story with mainstream outlets before its issue hit the stands, a necessity in these pre-online-media days. Most of the big boys passed, but both New York City tabloids bit, and so did the *Boston Herald*, giving the story prominence in the same media market that served a big chunk of New Hampshire.

Clinton played it off as nothing. The *Star* confined its coverage to the realm of popular culture, but it sat on shelves with even less reputable tabloids like the *Weekly World News*, which touted wildly imaginative—and obviously fake—scoops about paranormal activity. For Clinton's purposes, they were all the same. "The *Star* says Martians walk on the Earth and people have cow's heads," he scoffed.

He could make a good case, though. The lawsuit that the *Star*'s story was based on was old news in Little Rock media circles. Larry Nichols, the man who'd filed it, had been fired as the marketing director for the Arkansas Development Finance Authority in 1988 for running up fourteen hundred dollars in long-distance bills on a state phone. (The calls were to Nicaragua, where Nichols was hoping to aid the contra rebels seeking to overthrow the country's Marxist leaders.) Nichols claimed it was a smear job—that he had damaging information about the governor and that Clinton wanted him discredited. He filed his suit during Clinton's reelection campaign in 1990, naming names of women supposedly involved romantically with the governor, but when reporters in Arkansas investigated his accusations, they found nothing and printed nothing.

Girding for an eruption like this, Clinton had already set about trying to manage the reaction of the press corps. Just before entering the race, he and his wife arranged a meeting with top national reporters, where they discussed their marriage. "We have been together for almost twenty years and we are committed to each other," he said. "It has not been perfect or free from problems, but we are committed to each other and that ought to be enough."

Confronted with the *Star* story, most outlets at first deemed it unworthy of coverage, but not all of them. The *New York Post*'s write-up branded Clinton "Wild Bill." The paper's editor, Jerry Nachman, alluded to the precedent set five years earlier with Gary Hart. "It's become part of what we do in campaigns, going over the character thing," he said. With the *Post* and a few other big-city tabloids playing it up, the outlets that passed were forced to follow up and put Clinton's denial on the record. It was embarrassing for Clinton, who called the story "an absolute, total lie," but the damage seemed contained. He was denying everything, there was no proof of anything, his wife was standing by him, and none of the women mentioned in the lawsuit were contradicting him.

Then, one of them did. Her name was Gennifer Flowers. She was forty-one years old and worked for the state unemployment agency in Little Rock, where she'd also scratched out income singing in nightclubs. She'd started her career in local television news, first crossing paths with Clinton just as he rose to power in the state in the late seventies. Flowers was among the women named in the Nichols suit. Now, she stepped forward to say it was true.

It was another *Star* exclusive. She and the governor, Flowers now claimed, had engaged in a sexual affair from 1977 to 1989, and his denials of the initial story had wounded her. She had tapes, too, parts of fifteen phone conversation from the past two years that seemed to establish at least some kind of relationship.

The *Star*'s editor, Richard Kaplan, previewed his story for the national press on January 23. Flowers herself wasn't present and Kaplan only played one small bit of the tapes. It was from a conversation that had occurred, supposedly, just before Clinton entered the race the previous summer. In it, a voice that sounded like Clinton's seemed to instruct Flowers to lie about the affair to the media: "If they ever hit you with it, just say, 'No,' and go on."

"I'm so tired of all the lying and hiding," Flowers was quoted saying. "For twelve years I was his girlfriend and now he tells me to deny it—to say it isn't true."

Clinton was in New Hampshire when the *Star* unveiled its scoop. He acknowledged only that Flowers was an acquaintance and said there was far more to their phone conversation than the *Star* was suggesting. "I read the story," Clinton said. "It isn't true. And I told her over and over, I said, 'Just tell the truth. Tell the truth.'"

There was murkiness. Back in Arkansas, Flowers had been linked by rumor to Clinton for several years, and a radio station had even broached the subject on the air a year earlier. Flowers had responded with a forceful denial and a lawsuit threat. The public airing of a bogus allegation, her lawyer said, was hurting her employment prospects. Now, she was taking it all back—and in exchange for money, too. The *Star* acknowledged paying Flowers for the story, but wouldn't say how much (around $150,000, it later turned out). Kaplan insisted it was standard practice. "She's obviously taken money to change her story," Clinton said.

For the media, it was a test of boundaries. That night, ABC and CBS didn't mention Flowers or the *Star* story on their newscasts, and NBC brought it up only in passing. This counted for something; their combined audience still exceeded thirty million. CNN, its influence growing after its wall-to-wall coverage of the Gulf War the year before, took a pass, too. But *Nightline*, ABC's late-night current affairs show, turned its broadcast into a story about the dilemma. The show opened with a voiceover from anchor Ted Koppel that recalled the Hart imbroglio: "Charges of sexual misconduct that could ruin the chances of a presidential candidate, this time Bill Clinton. Once again, it raises the same old question. When should the private life of a politician become fair game for the media, especially when the story has not been confirmed?"

Koppel's theme for the half hour amounted to: *Should we even be talking about this?* It was the meta-approach, addressing the *Star's* report without dwelling on its details. One of his guests was Mandy Grunwald, a media consultant working with Clinton's campaign. She called the setup "just a neat little way to get at the story" and accused Koppel of granting legitimacy to a "cash for trash" tabloid story. She also admitted that Clinton would have to address it in more detail— "not because he should and not because I think the American people care . . . but because you reporters continue to care. He's not going to be able to get his message out until he answers the questions in a fuller way than he has already."

Newspapers agonized over how to play it. In some form, though, they did cover it—all of them. A *Boston Globe* editorial laid out the rationale: "Checkbook journalism is abhorrent. But if Flowers had been paid $100,000 to write her memoirs by Simon & Schuster, and if those memoirs included an assertion that she had an affair with Clinton, no newspaper would have worried that money was involved when deciding whether to report her story." For all the media's agonizing, Clinton was now face-to-face with a sex scandal, and the only precedent for it was grim. Once the story about his rendezvous with Donna Rice hit in 1987, Hart had lasted just five days.

Just before the first *Star* story, a tracking poll had shown Clinton opening a lead of better than two-to-one over Tsongas in New Hampshire. Now, after the Flowers bombshell, Clinton was back in second, two points behind Tsongas and plummeting. Clinton was determined to speak out, and so was his wife. They went shopping for a high-profile venue, and here, the timing was helpful. The Super Bowl was coming up on Sunday, three days after the Flowers accusations. Eighty million Americans would watch the Washington Redskins play the Buffalo Bills, easily the largest audience for any broadcast of the year. CBS, the network carrying the game, offered the Clintons a chance to appear as soon as it was over. They would be interviewed by Steve Kroft of *60 Minutes*, the premier newsmagazine on television. Bill and Hillary would never have a better chance to bypass the media filter and tell their story directly to the public.

Inside the Minneapolis Metrodome, there wasn't much suspense as the game wound down. A pair of garbage-time touchdowns for the Bills made the score look more respectable at 37–24, but Washington's dominance was clear. Viewers at home watched as the Vince Lombardi Trophy was presented to the victors in their locker room, the Redskins' third title in a decade. Then, Pat Summerall signed off and the iconic ticking stopwatch filled the screen: "I'm Steve Kroft. And this is a special abbreviated edition of *60 Minutes*. Tonight, Democratic presidential hopeful Governor Bill Clinton and his wife Hillary talk about their life, their marriage, and the allegations that have all but stalled his presidential campaign."

The interview had been taped hours earlier in a Ritz-Carlton suite in Boston, then edited down to about ten minutes. The candidate and his spouse sat together on a couch with no space between them. Kroft faced them from a chair about five feet away. A coffee table was in between.

Kroft first asked about Flowers. Clinton said he'd known her on and off since the late seventies, that it was a "friendly but limited" relationship, and that he'd been talking to her on the phone the past few years because she was so upset about the rumors. Those calls, he added, had taken place with his wife's knowledge. He reiterated his charge that Flowers was fabricating a story for cash. "There's a recession on," Clinton said. "Times are tough. And I think you can expect more and more of these stories as long as they're handing out money."

"I'm assuming from your answer," Kroft said, "that you're categorically denying that you ever had an affair with Gennifer Flowers."

"I've said that before," Clinton replied. "And so has she."

His phrasing—placing his own response in the past tense—sounded peculiar. Was Clinton maintaining that there had never been an affair? Or was he playing some kind of word game and just saying that he had once said that? There were reasons to doubt Flowers's claim of a twelve-year romantic relationship, but there was also reason to suspect that at least *something* had happened between them. To an already confusing situation, Clinton was now adding more confusion. He was also acknowledging "difficulties" in his marriage. Kroft asked him to be more specific: "Does it mean adultery?"

"I think the American people—at least people that have been married for a long time—know what it means and know the whole range of things that it can mean," Clinton said. His voice was calm. His pace was measured. He seemed serious, but not panicked.

Kroft continued. "You've been saying all week that you've got to put this issue behind you. Are you prepared tonight to say that you've never had an extramarital affair?"

"I'm not prepared tonight to say that any married couple should ever discuss that with anyone but themselves. I'm not prepared to say that about anybody. I think that the issue . . ."

Kroft tried repeatedly to get Clinton to give a direct answer. Eventually, Hillary spoke up, amplifying her husband's message. "There isn't a person watching this who would feel comfortable sitting on this couch detailing everything that ever went on in their life or their marriage," she said.

"I couldn't agree with you more," Kroft replied, "and I think—and I agree that everybody wants to put this behind you." He was looking back at Bill Clinton now. "And the reason it hasn't gone away is that your answer is not a denial. Is it?"

Clinton took a breath. "But interesting. Let's assume."

"But it's not a denial."

"Of course it's not, and let's take it from your point of view."

Clinton's pace was quickening now. He almost seemed to be smiling. "If—that won't make it go away. You know that you can cut this round or cut this flat. I mean, if you deny, then you have a whole 'nother horde of people going down there and offering more money trying to prove that you lied. And if you say yes, you have just what I've already said by being open and telling you that we've had problems. You have, Oh good, now we can go play gotcha and find out who it is."

Five times, Kroft had sought a direct answer from Clinton on whether he'd ever cheated on his wife. And five times, Clinton had refused to give him one. He seemed to be gaining confidence, too, turning the tables on his inquisitor. Reporters like Kroft, Clinton was saying, had a choice about whether to play "gotcha" with his sex life or focus on "the real problems of this country." His body language was shifting; his right arm was now resting on his wife's, and she was nodding as he spoke.

Maybe in desperation, Kroft tried one final approach: "I think most Americans would agree that it's very admirable that you have stayed together, that you have worked your problems out—that you seem to have reached some sort of understanding and an arrangement . . ."

Bill Clinton recognized the danger of even seeming to grant that premise. "Wait a minute, wait a minute, wait a minute," he said. "You're looking at two people who love each other. This is not an arrangement or an understanding. This is a marriage. That's a very different thing."

Hillary voiced her agreement, an acquired southern drawl seeping into words. "You know, I'm not sittin' here some little woman standing by my man like Tammy Wynette. I'm sittin' here because I love him and I respect him and I honor what he's been through and what we've been through together." The camera shot caught Bill reaching for her hand and clasping it. She continued. "And you know, if that's not enough for people, then heck, don't vote for him."

The interview ended with Kroft asking Clinton if he thought he'd succeeded in putting the issue behind him. "That's up to the American people," he said, "and to some extent the press. This will test the character of the press. It is not only my character that has been tested."

That was it. No one was quite sure what to make of it, mainly because nothing quite like it had ever happened before. The *Los Angeles Times* put it this way: "Although many presidents have engaged in extramarital affairs—biographers have, for example, documented affairs on the part of Franklin D. Roosevelt, Dwight D. Eisenhower, John F. Kennedy and Lyndon B. Johnson—no serious presidential candidate ever has openly discussed such matters before a vast television audience."

The Clintons had wanted to reach the largest audience possible, and on this front they'd succeeded. Overnight ratings estimated that one out of every three television sets in use during the interview were tuned to *60 Minutes*. Clinton himself seemed buoyed. After the taping, he headed north to Portsmouth, New Hampshire, where he turned the scandal into the emotional apex of his stump speech.

"A lot of you have told me to hang in here the last couple of days, how rough all this is," he said. "Yeah, it's rough, but it's nothing compared to somebody going home at night and sitting down over a table and looking at their children and wondering if they're going to have another job again and feeling like they have failed their kids.

"You talk about rough, that's rough. And if I've taken a couple of tough days in the campaign to get a chance to resume the campaign fighting for those kinds of people, I gladly accept that burden."

The cheers were loud, but there was more trouble coming. On Monday, the day after the interview, Flowers held a press conference in New York and accused Clinton of lying. "The man on *60 Minutes* is not the man I fell in love with," she said. "I would have liked to think, after a twelve-year relationship, he would have had the guts to say, 'Yes, I had an affair with this woman. But it's over. And that's the truth.'"

Flowers spoke next to a poster-size image of the *Star*'s front-page. MY 12-YEAR AFFAIR WITH BILL CLINTON, it blared. This time, no major media organization had any qualms about being there. Reporters quizzed Flowers about inconsistencies in her story, and she tried to brush them off. Another caught the room off guard by asking, "Did Governor Clinton use a condom?" There was shock, and some chuckling too, especially from those who recognized the questioner as "Stuttering John" Melendez, a prankster from New York shock jock Howard Stern's top-rated morning radio show. Melendez had a follow-up question, too: "Do you plan to sleep with any other presidential candidates?" Even Flowers guffawed at that one. A supermarket tabloid; broadcast television networks and top national newspapers; and now Howard Stern's vulgar sidekick: worlds were colliding.

The headline from the press conference was that Flowers was releasing more from the conversations she'd taped. Again, it was murky. A voice sounding like Clinton's told Flowers the press wouldn't be able to run any stories if "all the people who are named" deny anything happened. But none of the conversations conclusively established that anything sexual had actually happened. So was this Clinton suggesting that Flowers lie, or was he explaining what would happen if she told the truth?

There was something else on the tapes, though, that created a brand-new crisis for Clinton, and it had nothing to do with the possibility of an affair. It was an exchange in which he and Flowers discussed Clinton's prospects as a presidential candidate—specifically, the possibility that Mario Cuomo might run. "I don't particularly care for Cuomo's, uh, demeanor," Flowers could be heard telling Clinton, who replied that Cuomo was "a mean son of a bitch." She then said that she "wouldn't be surprised if he didn't have some mafioso major connections."

"Well," Clinton said, "he acts like one."

Now Mario Cuomo was part of the story. It had been just over a month since the climax of his Hamlet act, but there'd been no slippage in the goodwill Democrats felt toward him. He and Clinton had eyed each other as rivals for years, sniping in public on occasion. The last thing Clinton needed was for those tensions to erupt now, but there was nothing that incensed Cuomo more than insinuations that he was somehow linked to organized crime. The whispers had percolated for years, without any substantiation, and he considered the topic a form of ethnic stereotyping. Quickly, Clinton moved to preempt a conflict. The conversation, he explained, had taken place at a moment when there'd been some "political give and take" between the two men, and he'd only meant to say that Cuomo was "a tough, worthy competitor."

"If the remarks on the tape left anyone with the impression that I was disrespectful to either Governor Cuomo or Italian Americans, then I deeply regret it." Back in Albany, Cuomo was in no mood for it. He scoffed at the phrasing of Clinton's apology. "What do you mean if? If you are not capable of understanding what was said, then don't try apologizing." He called the Clinton-Flowers conversation "part of an ugly syndrome that strikes Italian Americans, Jewish people, blacks, women, and all the different ethnic groups" and "an indication of snide condescension. Bigotry."

Clinton was trying to reach Cuomo to apologize personally. Through the press, Cuomo relayed a message back: "He ought to save himself his quarter."

The episode was crucial in two ways. First, it forced Clinton to admit that the voice on the tape was his. Until now, his campaign had been intentionally ambiguous on this question, suggesting he might have been the victim of tape doctoring. More significantly, it dramatized the contrast between him and Cuomo at the worst possible moment. Cuomo's appeal had always been as a white knight who might ride to the rescue of his party if it needed him badly enough. Now, with Clinton submerged in sleaze, Cuomo was claiming the high ground and all but urging Democrats to imagine the candidate they could have had. The old talk was starting up again. It was too late for Cuomo to run in New Hampshire, but if Clinton had to drop out, it still might not be too late for Cuomo to save the day.

For many Americans, the Gennifer Flowers scandal served as their introduction to Bill Clinton. Even fewer knew anything about his wife, whose comment about not being "some little woman standin' by my man like Tammy Wynette" now stirred controversy of its own. It was a reference to the biggest hit of Wynette's career and one of the biggest hits in country music history. First recorded in 1968, "Stand by Your Man" was, depending on whom you asked, either a rousing testament to the complexities of love and relationships or a backward anthem of patriarchy.

With her derisive remark, Clinton was airing the prevailing view of feminists, who were starting to see in her a kindred spirit. In style and résumé—Wellesley, Yale Law, Watergate committee lawyer— she was the embodiment of the modern career woman, and she was determined to upend traditional notions about the role of a political wife. In Little Rock, she'd been a force in her husband's administration, and now Bill Clinton was talking about putting her in a presidential cabinet. Months earlier, Clarence Thomas's Supreme Court nomination had nearly been derailed when a former staffer, Anita Hill, told senators—and a national television audience—that he'd subjected her to sexual harassment. Some believed her. Many didn't. Activists were now mobilizing to recruit female candidates and to pay back her doubters at the polls. With no women running for president, feminists were seeing in Hillary the closest thing to a candidate they could find. Her readiness to stand with Bill and fight in this moment was a powerful inducement for them to give him the benefit of the doubt.

At the same time, she risked running afoul of more traditional cultural sensibilities. Country music's fan base was heavily white, blue collar, and southern—a demographic Bill Clinton was promising to win back to the Democratic fold. Wynette, still a bankable star in her late forties, lashed out with an open letter to Hillary.

"I believe you have offended every person who has 'made it on their own' with no one to take them to a 'White House,'" Wynette wrote.

"Where in the song do you find my lyrics such that you have the audacity to refer to me as 'a little woman standing by my man'? With all that is in me I resent your caustic remark. After all, guilty or not, aren't you standing by your man?"

To some, Hillary's comment had come with a strong whiff of elite condescension. She'd been "looking right down her nose," columnist Jimmy Breslin wrote, "at all these grubby little Tammy Wynette tavern waitresses who live with a man who drives a butane truck." In her letter, Wynette inserted one more dig on their behalf: "I can assure you, in spite of your education, you will find me to be just as bright as yourself."

A week after the *60 Minutes* sit-down, Clinton was still standing. His New Hampshire nosedive had stopped, and maybe even reversed. Polls put him neck-and-neck with Tsongas. No other women had stepped forward, and Clinton declared the subject closed: "I'm through with it. I've said all I have to say."

Then, the next bombshell. It landed on February 6, twelve days before the New Hampshire primary, with a *Wall Street Journal* expose on Clinton's personal history with the Vietnam War draft. In it, the army officer who had arranged a military deferment for Clinton claimed he'd been duped.

The story dated back to the early months of 1969, when Clinton, who was studying at Oxford University in England, received an induction notice from his local draft board back in Arkansas. He was ordered to report at the end of July, one of hundreds of thousands of young men called up for military duty that year. Like many of them, Clinton was opposed to the war and didn't want to go. Some were openly defying their induction notices, destroying their draft cards, or even leaving the country. Those measures were out of the question for Clinton, who was already nursing big political dreams.

That's where Colonel Eugene Holmes, who was running the Reserve Officer Training Corps program at the University of Arkansas, came into the picture. The official aim of ROTC was to provide basic military training for college students, who upon graduating would enter the service as commissioned officers. Unofficially, it was the most respectable way for a young man to get out of Vietnam. Clinton had been accepted by the law school at Arkansas. If he could join the ROTC program there, he'd win a draft deferment for the length of his studies, buying him several years while the war raged. And if it ever then came to it and he was forced into active duty, as a commissioned officer he'd be much less likely to face front-line deployment in Vietnam.

Not just anyone with a draft notice could trade it in for a spot in ROTC, though. Clinton would need to convince Holmes, who would be more than a little suspicious about his intentions. Just getting a meeting was a challenge. Clinton tapped his burgeoning political network, far more extensive than the average twenty-two-year-old's, and found a back channel. An aide to Arkansas senator J. William Fulbright, on whose staff Clinton had briefly worked, knew Holmes and made the introduction. Clinton then convinced the colonel that his interest in ROTC was genuine. They struck a deal: Clinton would enroll at the law school the next fall, join ROTC, and fulfill all service obligations. In return, Holmes would see to it that Clinton's induction was deferred for four years. Clinton signed an ROTC letter of intent, told Holmes he'd be returning to Oxford that fall, and promised to be in touch monthly.

Then things broke down. Holmes didn't hear from Clinton again until December, months later than Clinton had promised. There was a reason. President Nixon and Congress had agreed on a new system for calling up able-bodied young men: a draft lottery—366 little blue capsules in a giant container, one for each day of the year. If you were eligible to serve and your date of birth came up early, you'd be getting a call. If it came up late, you'd be in the clear.

For Clinton, this changed the equation. It would be a gamble, but if he entered the lottery, there was a decent chance he'd get a low enough number to stay out of Vietnam. Holmes had saved him from induction over the summer, but now Clinton reneged on their agreement, backed out of his ROTC commitment, and entered the lottery, then watched as his date of birth was the 311th drawn. It was one of the lowest numbers possible. The gamble had paid off. He was now free and clear of the draft and he'd broken no laws to do it—only his word.

Two days later, he finally wrote to Holmes. In the letter, Clinton thanked Holmes "for saving me from the draft" and admitted he'd taken his chances on the lottery "to maintain my political viability within the system." He acknowledged to Holmes that "had you known a little more about me, about my political beliefs and activities," Holmes likely would have denied his ROTC application. He laid out his objections to the war and offered a defense of the young men who had taken more direct and explicit steps to avoid it. "To many of us, it is no longer clear what is service and what is disservice, or if it is clear, the conclusion is likely to be illegal," Clinton wrote.

To Holmes, it read as a taunt, a cunning young man scheming his way out of service and rubbing it in his face. There was nothing he could do, though, but sit back and watch as Clinton eventually returned home from Yale Law School—he never did attend Arkansas Law—and climbed the political ladder. Reporters would occasionally ask Holmes about Clinton, but he stayed quiet. Privately, though, he stewed, and now, at seventy-two years old and with Clinton seeking to become the commander in chief of the U.S. armed forces, he was ready to say something.

"Bill Clinton was able to manipulate things so that he didn't have to go in," Holmes told the *Journal*.

Clinton stood accused of draft dodging, a potentially fatal political charge. It had been less than twenty years since the fall of Saigon and feelings were still raw. "No one really wanted to go, but we felt it was our duty," one Vietnam veteran was quoted saying. "Even today, if I met somebody who avoided the draft I wouldn't associate with them and wouldn't vote for them."

That was the conviction of many of Clinton's fellow baby boomers, who'd answered the call to serve or had friends and loved ones

who did. Their emotions were intensified by America's belated embrace of its Vietnam veterans. Following the war, returning service-
men were often greeted with cool indifference, even hostility, by a
country that had trouble separating them from the contentious debate
over the war. Only in the 1980s, with the passage of time and the
construction of a memorial in Washington, did a widespread recognition take hold that those who'd served deserved public honor and
recognition, no matter the politics of Vietnam. Now, in 1992, they
were viewed with reverence. Feelings for the men who'd found a way
out were much more complicated.

The controversy took shape in two stages. At first, it was just
Holmes's allegation; the *Journal* didn't have Clinton's letter, and no
one knew if it even existed. Clinton swore there'd been no duplicity
on his part. "You've got to understand that if you look at the facts as
they then existed, instead of some twenty-three-year-later rewrite,
the facts are, I put myself into the draft, I didn't take it out," he said.

Then, a few days later, ABC got the letter, and Clinton agreed to
go on *Nightline* to address it. "Many of you will hear it and find in it a
reaffirmation of everything you like and admire about Bill Clinton,"
Ted Koppel told viewers. "Others among you will be angered by what
you hear."

He read it on the air and conducted a live interview with Clinton,
who again denied any scheming. He leaned on the short window of
time between when he'd withdrawn from the ROTC program in the
early fall of '69 and when the lottery was held in December. "I was
put back into the draft before the lottery came along, before I knew
my lottery number, and I was in the draft," Clinton said.

Koppel was incredulous. The lottery had been held on December 1. The immediate next day, Clinton had applied to Yale Law.
The day after that, he'd written his letter to Holmes. It sure looked
like young Bill Clinton had lined up all his ducks in a row. "But
what you're saying is . . . that's just coincidence of timing, I mean,
there's nothing to read into it?" Koppel asked.

"I say, I just don't remember, and there's nothing to read into it," Clinton replied.

There were some, especially fellow baby boomers, willing to give him slack anyway, recognizing in his words and actions the same attitude they'd felt at his age. Vermont governor Howard Dean, who had avoided the war with a medical deferment, called the letter "a telling reminder of what it was really like." Others were less sparing. "I've got some real misgivings as to whether or not he's leveling with the American people about what really happened," said Bob Kerrey, who'd lost a foot in Vietnam and was now running against Clinton.

For Clinton, the crisis was now far bigger than just the Vietnam controversy. It was the weight of scandal and the expectation of more. Not a single vote had been cast and already he was facing charges of womanizing and draft dodging. How much more was out there? How many dark secrets would reporters unearth? And what about Bush and the Republicans? Look at how they'd gutted Dukakis four years earlier; just think of what they'd have in store for Clinton if he ever made it to the fall. Electability was supposed to be his pitch. Now it was feeling like a joke.

Clinton's difficulties benefited Paul Tsongas, who emerged as his main rival. Tsongas opened a commanding lead in New Hampshire with the Vietnam accusations. In many ways, Tsongas was ideally suited to capitalize on the scandals. He lacked Clinton's effortless charisma and flair for the spotlight. He was dry and self-deprecating, speaking with a slight lisp and pronounced Massachusetts accent. Normally, this near-total lack of polish would doom a candidate, but now the contrast worked in Tsongas's favor. Clinton's character and integrity were in question, and his smooth style was starting to feel like a manufactured façade. It turned Tsongas's unvarnished attributes into the politically potent mark of an unpackaged candidate.

The strength of his image was underscored by his personal story.

Tsongas was running as a cancer survivor, diagnosed in 1983 with non-Hodgkin's lymphoma, a condition thought to be treatable for a while but ultimately terminal. He'd been a young political star back then, forty-two years old and already a senator. He chose to abandon his seat and return with his family to his hometown of Lowell, Massachusetts, terrified, he would later write, not so much of dying but of leaving the world before his two young daughters were old enough to know him. His health deteriorated, but then an experimental bone marrow transplant sent him into remission. Five years later, in 1991, he declared himself cancer-free and decided to run for president. He entered the longest of long shots, a forgotten one-term senator, but professed not to care. "I've faced death," Tsongas said, "the worst thing there is, so winning or losing maybe doesn't mean so much to me."

He called his campaign a "journey of purpose." His sickness, Tsongas said, had made him consider the world he would leave behind. This was the core of his campaign message. He offered himself as the candidate of hard truths, a man driven not by personal ambition but by a belief in "generational responsibility."

"Why I am here, I don't know," he told a rally. "Do I believe that I survived to become president of the United States? I don't believe that. I believe, however, that my survival carries with it an obligation."

It gave him an air of moral authority that was now strengthened by the contrast with the scandal-tarnished Clinton—and he was learning to wield it as a weapon. Clinton was promising a tax cut for the middle class, to be financed by higher rates on the rich. It polled well, but Tsongas turned it around on him, saying it just showed that Clinton was more interested in winning an election than being fiscally responsible. He even took to holding up a plush bear to represent Clinton—"a pander bear," Tsongas said, who was willing to say anything to get elected.

Somehow, though, his platform was actually to the right of Clinton's. Among the hard truths Tsongas was offering was his conviction that the Democratic Party had become hostile to the private sector, too focused on redistribution instead of growth. He wanted a capital gains tax cut, warned against "corporate-bashing," and talked of the federal deficit as a moral issue. "Nobody trusts us with the economy," he said, "nor should they." Organized labor's top priority in Congress was a bill to protect striking workers by banning companies from hiring replacements for them—scabs, to the union crowd. Every Democratic candidate supported the bill, except Tsongas.

Nor did he finesse his breaks with orthodoxy the way Clinton did. Massaging the sensitivities of party leaders would only undercut the image he'd crafted. "The problem with most Democrats," Tsongas said, "is that they care about employees but not about employers. They don't get the connection." Warm words were tossed his way from across the aisle. Senator Phil Gramm called him "the one Democrat who agrees with the president that we ought to be talking about creating wealth." If they had to vote in the Democratic primary, said Gingrich, conservatives "clearly would vote for Tsongas." Tom Selleck, one of Hollywood's few Republican stars, wrote him a check for one thousand dollars.

The Tsongas candidacy was catching on far more than anyone had imagined possible—at least with a certain type: white, college educated, economically upscale, politically independent. There were a lot of these voters in New Hampshire, in bedroom communities right over the Massachusetts line. With Clinton drowning in damning headlines and time running short, Tsongas's lead mushroomed to twenty points.

The Friday before the primary was Valentine's Day. It was also Tsongas's fifty-first birthday. A party doubling as a campaign event was held at the Sheraton Tara Hotel in South Nashua. The room was overflowing. The birthday, he said, was a milestone that only a few years earlier he'd never thought he would reach. "I will tell you that I'm grateful to be here, to be here physically," Tsongas said. "And I'm grateful for the opportunity of trying to change this country, so that we can go around to our children and say to them, 'This is what we're leaving to you, and we feel good about it.'"

Clinton, meanwhile, was trying to build a campaign-saving

message out of his own crisis. On the night his Vietnam letter surfaced, he was in Dover, an old mill town near the New Hampshire
coast. He was at the local Elks Club in front of a crowd of more than
two hundred. Times were tough everywhere, but especially in this
corner of the state. Clinton told the story of George H. W. Bush,
who'd come to New Hampshire four years earlier a wounded candidate, fresh off a humiliating third place in the Iowa caucuses. He was
facing the end of his campaign, the end of his dream of becoming
president, if he lost again.

"He came here," Clinton told the Elks lodge, "and asked you
to give him a second chance. And you did. You gave him a second
chance. And what did he do with it? He said, 'Give me an election
and I'll give you fifteen million jobs.' He's fourteen million seven
hundred thousand jobs short and fifty thousand in the hole in New
Hampshire."

He segued to his own plight. Clinton was facing the same stakes
now that Bush had in 1988. "Let me tell you something, folks. They
say I'm on the ropes now. Not because of anything I've done in my
public life. Not because of anything I've said to you."

His voice was lower than usual, his cadence a bit slower, as if he
was taking the audience into his confidence. The room was quiet.
"They say I'm on the ropes because other people have questions about
my life after years of public service. I'll tell you something, I'm going
to give you this election back, and if you give it to me I won't be like
George Bush. I'll never forget who gave me a second chance and I'll
be here for you until the last dog dies."

The intensity of the response told Clinton he'd found just the
right note. In its own way, his message was just as emotionally resonant as Tsongas's: *If you believe in second chances, give me one and I
promise we will bounce back together.* It wasn't just the words. It was his
presence, his defiance. No presidential candidate had ever been hit
with such a barrage of scandal, but it hadn't sent him scurrying back
to Little Rock. He was standing in there, refusing to bow his head in
shame. He was showing a toughness that could impress even his crit-

ics. His final television ad before the primary told voters: "He's been tested in ways other candidates haven't."

By primary day, February 18, Clinton's free fall had stabilized and he'd begun slicing into Tsongas's lead, but the expectations game lagged. News coverage still treated a Tsongas victory as a foregone conclusion. The only suspense was how dominating it would be. When the late-afternoon exit polls showed a single-digit race, Clinton and his team breathed a sigh of relief—and hatched a plan: there would be no congratulating Tsongas on his win. They would act like the victory was theirs and theirs alone.

It got better for them when the polls closed and the initial returns showed a dead-even contest. For a few minutes, it seemed possible that Clinton would win the primary outright. All the talk on television was about Clinton's surprisingly strong performance. As if on cue, Clinton left his hotel suite and went downstairs to address his supporters—and a national television audience. On-screen, the tally noted that only a quarter of the votes had been counted. As if that mattered to the beaming candidate who now appeared. He kissed his wife and then bellowed: "New Hampshire tonight has made Bill Clinton the comeback kid!" His supporters screamed with delight. "This has been a tough campaign," Clinton said, "but at least I've proven one thing—I can take a punch."

The close early count was a fluke. Steadily, Tsongas opened up a comfortable lead as the night progressed, and when the state's far-flung precincts finally reported, he was ahead by nearly ten points. At Tsongas headquarters, the mood was equally ebullient. "Well, New Hampshire," the winner said, "you did it again. You gave 'em hell."

By then, though, the narrative had been set. This was a good night for Tsongas, as everyone had expected. But it had been a great one for Bill Clinton. He was now, the *New York Times* wrote the morning after, "in a position to take his resilient candidacy on to the string of primaries in his native South."

For all practical purposes, it was a two-man race. Tsongas leaned on the contrast that had worked so far, his tough medicine versus

Clinton's free candy. "I'm not running to be Santa Claus," he said. "I'm running to be president of the United States, and there's a difference." Clinton seethed at his opponent's depiction of him and what he saw as the media's love affair with "St. Paul." In the dark days before New Hampshire, he'd held it in, but now he was brimming with swagger again. "No one can argue with you, Paul," he snapped in one debate. "You're always perfect."

"I'm not perfect," Tsongas shot back, "but I'm honest."

Tsongas had to prove he could win beyond his backyard, and two weeks later he did, taking first in Maryland's primary. But he was powered by the same demographic he'd rallied in New Hampshire— white, professional class, independent. To win the nomination, he'd need to expand his base, and fast, because there was demographic trouble approaching.

Here, at last, is where Clinton truly reaped the benefits of Tsongas's rise and Cuomo's absence. The Democratic race was shifting to larger, delegate-rich states dominated by the party's traditional constituencies. With Tsongas as his foe, Clinton no longer faced the challenge of gently convincing liberals not to view him as the enemy. Now, he was their last, best hope, the only thing standing between them and a nominee who was promising to be "the best friend Wall Street ever had."

Florida's March 10 primary was the key. This was the contest Michael Dukakis had used four years earlier to demonstrate that he had broad, national appeal. Tsongas hated the Dukakis comparison ("I'm a Greek from Massachusetts who fights back"), but his campaign now hinged on replicating the Duke's feat.

Tsongas stuck with his message, but this was a very different audience. His tough medicine included entitlement reform—Medicare and Social Security benefit cuts for those with higher incomes. In the white-collar suburbs of Boston and D.C., this had been seen as a principled and courageous stand. But to the retiree masses of South Florida, who'd been attached to the Democratic Party since the New Deal, it was heresy.

Clinton pounced. These were benefits they'd earned through a lifetime of work, he told them, and he wasn't about to take anything away from any of them. Tsongas, he charged, was protecting Wall Street and the wealthy at the expense of the middle class—just as Reagan and the Republicans had in the eighties. "We cannot put off fairness under the guise of promoting growth," he said. Clinton had a big edge in campaign cash and now he put it to work, bombarding the airwaves with attack ads. The Tsongas plan, one of them declared, "smacks of trickle-down economics."

This was the language liberals had been using to attack Reagan Republicans for more than a decade. It was also the language liberals had been using to attack the DLC—and the language that Clinton had once assumed would be used against him in this campaign. But now he was the one running from the left, and it became his weapon as he rallied the old guard to thwart a hostile takeover of their party. Tsongas accused him of distortion and scare tactics: "The American people are going to find out how cynical and unprincipled Bill Clinton is."

But he never had a chance. Exit polls showed Tsongas still strong with his core groups, but there just weren't enough of them anymore. Clinton won big with senior citizens and lower-income whites and ran up more than 80 percent of the vote among African Americans. Overall, Clinton took Florida by a 51 to 35 percent margin. It was the centerpiece contest on a day that featured ten other primaries and caucuses—Super Tuesday. The pattern held across the board: Clinton, the candidate of the DLC, harnessing the Old Left for landslide wins. When the night ended, the delegate race wasn't even close.

"I must say, it is only tonight that I fully understand why they call this Super Tuesday," Clinton said.

Even as he played to the left, Clinton retained the positions he believed would mark him as a different kind of Democrat in the fall. He still wanted to overhaul the welfare system. He was still in favor of the death penalty, still called for streamlining the federal bureaucracy. But was he winning over his party because of this, or in spite of it? Against Cuomo, or against any of the top-tier prospects who'd bailed in the face of Bush's post–Gulf War popularity, it might have been different, very different. But as long as Tsongas was his main rival, the limits of Clinton's centrism weren't going to be tested.

After Florida and Super Tuesday, the race moved to the Rust Belt—Illinois and Michigan. Organized labor loomed large, and the attitudes of union leaders toward Clinton, now the clear Democratic front-runner, ranged from deeply skeptical to downright hostile.

Arkansas was a right-to-work state, something Governor Clinton had never shown much interest in changing. He'd even used the state's low wages as a selling point to corporations interested in relocating. The Bush administration was now hammering out a major trade agreement with Canada and Mexico—the North American Free Trade Agreement, NAFTA, it was being called—and labor was desperate to nip it in the bud. Clinton, though, said he was for giving Bush fast-track authority to negotiate, a major step toward enactment. The president of the Arkansas United Auto Workers union was now warning his brothers and sisters about Clinton: "Every time he's claimed to be our friend, we've ended up with a knife in the back."

Labor leaders had wanted Cuomo to run. When he didn't, they rallied to Tom Harkin, the Iowa senator whose candidacy rendered his home-state caucuses a nonevent, tilting the battleground in Clinton's and Tsongas's favor. Harkin had proved a one-dimensional candidate, though, and was long gone. The only other active candidate now was Jerry Brown, who'd barely registered in the early contests and seemed far too erratic to win the nomination, let alone the presidency. In other words, by the time the primary campaign finally reached union country, only two truly viable candidates remained: Clinton, who was deeply distasteful to the union crowd, and Tsongas, who was even worse. Some union chieftains threw their support to Brown, hoping to stall the process and create a brokered convention, but more of them stood back as spectators. The paralysis only helped Clinton, who came in with all the momentum. In both Illinois and Michigan, he buried Tsongas, who the next day returned home to Lowell and called up the local paper with a scoop: he was getting out of the race.

As Clinton put Tsongas away, Bush's job approval rating hovered at 40 percent. His vulnerability was now obvious. But even as Clinton racked up the wins, a consensus was building within the party: they were about to nominate a sure loser.

The price of weathering so many scandals was becoming apparent. An image of Clinton was taking hold in popular culture, which lampooned him as a walking sleaze machine. "Every day there's a new revelation about Clinton," Johnny Carson said in a *Tonight Show* monologue. "They've got tapes, files, phone calls. Clinton could open up a presidential library before he gets elected."

It went deeper, though, beyond just the titillating accusations and obvious image problem, to a perception that underneath it all was an uncommonly calculating politician, a man uniquely skilled at making statements that could at the same time be both technically true and entirely deceptive.

Take the Gulf War, which had presented Clinton with a torturous dilemma just as he was preparing to run for president. In the early days of 1991, Bush was massing troops in the Persian Gulf and asking Congress to authorize war with Iraq. Most Democrats were against it. Memories of Vietnam ran deep, and polls showed widespread apprehension. Whatever happened, Clinton understood, it would be a defining issue in the presidential race. He was going to have to take a position.

Reporters in Little Rock sought comment from Governor Clinton. Did he support war with Iraq or oppose it? He managed to duck them, over and over, until finally, on January 14, 1991—the night before the deadline Bush had set for Hussein to withdraw from Kuwait or face war—a reporter cornered him in the statehouse and refused to relent. Now Clinton had to say something. He mentioned the debate in Congress, where a resolution authorizing force had just squeaked through. "I guess I would have voted with the majority if it was a close vote," he said. "But I agree with the arguments the minority made."

The wording was exquisite. Now, if the war went badly, he'd have room to maneuver. He'd agreed with the case against it, he'd be able to say, and he could always chalk up the rest as some sort of unifying sentiment on the eve of a war that was going to start no matter what he told that reporter. As it turned out, the war was deemed a smashing success, and a year later Clinton found himself running against a group of Democrats who'd all definitively opposed it. Only he, Clinton took to boasting, had shown the courage and foresight to support the war, and only he could neutralize the issue against Bush.

Put him in front of a crowd and Clinton's charm was undeniable. He wasn't a mesmerizing orator like Cuomo, but he was warmer. He could draw people in, make them feel a personal connection. What was he actually saying, though? That was the problem. There wasn't a pronouncement that didn't seem to come with its own set of fine print. The parsed phrases, tortured sentences, transparent craftiness—he could seem like more of a used-car salesman than a future world leader. A nickname was starting to stick: Slick Willie.

When Tsongas stepped out, Democrats began looking toward November. They shuddered at what they saw. Polls showed a generic, unnamed Democrat leading Bush. But use Clinton's name and the president led by nine. Clinton's favorable rating with Americans sat at 24 percent, compared to 41 percent unfavorable. By a more than two-to-one margin, one survey showed, voters believed Clinton was less honest than the average politician. Asked to say what bothered them most, 20 percent of respondents said his extramarital affairs, 12 percent said it was his dishonesty, 7 percent cited his morality, and another 7 percent said it was just the weight of so many scandals. "Too slick"; "arrogant"; "lacks character" were among the other responses. They remembered only too well what Bush's team had done to Dukakis in '88; now Democrats were bracing for something even worse.

"Just when the country realizes that Bush was, in fact, no bargain, the Democrats get ready to serve up damaged goods in Bill Clinton," wrote David Nyhan, a liberal *Boston Globe* columnist. "Why can you always depend on the Democrats to do the wrong thing when it counts?"

In the dying days of his own campaign, Bob Kerrey said of Clinton, "I think he's going to get opened up like a soft peanut in November of 1992." Kerrey had been a dismal failure as a candidate, but there weren't many Democrats quarreling with his assessment.

It was so bad, in fact, that Democratic voters staged a revolt. It came in Connecticut, the first state to hold a primary after Tsongas's exit. The contest was on no one's radar. Clinton's only official competition was Jerry Brown, whose guerrilla campaign seemed a particularly poor match for this upscale suburban state. Out of nowhere, though, Brown won by two points, 37 to 35 percent, with the sidelined Tsongas still drawing over 20 percent.

The jolt of an unexpected rejection threw Clinton's campaign straight back into crisis. The political world had just declared the Democratic race over; now, Democratic voters were responding by saying: *Not so fast.* In the exit poll, 58 percent of Democrats said they wanted a new candidate, and 46 percent said Clinton lacked the honesty and integrity needed of a president. Brown was the official winner, but no one was fooled. The result, one columnist wrote, represents "greater evidence of continued resistance to Bill Clinton than of a groundswell for Jerry Brown."

The next big test would be in New York, and it would be the ultimate referendum on Clinton. If he could bounce back and win, then that would be it. For better or worse, Democrats would be stuck with him. But if he lost again, then all bets were off.

Already Tsongas was flirting with jumping back in the race. Cuomo, smarting from Clinton's comments on the Flowers tapes, perhaps regretting not boarding that plane, stoked talk that he might ride to the rescue. "The presumption of ascendancy by Governor Clinton is now clearly rebuttable," he said the day after Connecticut. Democrats buzzed about the usual suspects, too: Bentsen, Gephardt,

Bradley—who would save them if the staggering front-runner now collapsed? "If he does have a setback in New York," Bob Matsui, a California congressman and deputy DNC chairman, said of Clinton, "everybody would have to reassess."

Just before the Connecticut earthquake, Clinton had been up fifteen points in New York. Now, a few days later, Brown was breathing down his neck—with Tsongas, his name still on the ballot, threatening to contend, too. Since entering the race, Brown had existed in what he called "a deep hole of media obscurity," but now he was back in the national spotlight for the first time in more than a decade, and he was loving it.

Even with his Connecticut victory, he still had only a fraction of Clinton's delegates and no plausible path to the nomination. Too many of his fellow Democrats believed that somewhere in his long exile from politics Brown had become unhinged. But in this moment, he was a major player again. Any Democrat who wanted to slam the brakes on Clinton now needed Brown to beat him again in New York. Only then would the door be open for a new candidate. If it was the only way back into the big time, then Brown didn't mind playing the stalking horse. He saw Clinton as a packaged candidate starting to come apart—and now he'd finish him off. He called Clinton "a right-to-work, union-busting, scab-inviting, wage-depressing, environmental-disaster governor" and "a Humpty-Dumpty candidate."

"There are so many scandals and so many problems, questions for Clinton that we're not going to put him back together again," Brown said. "And all the media and the insiders and one-thousand-dollar donors that created the Clinton campaign can't really sustain it in a way that will overcome George Bush and all his powers."

The national press corps was now joined in its daily pursuit of Clinton by the New York media and its famously in-your-face approach to journalism. New York City's two daily tabloid papers took it as a challenge to deliver the death blow that the big boys in the national media hadn't been able to produce.

Both took aim at "Slick Willie" with screaming front-page head-lines every morning. Clinton's New York hands told him this was just how it was done here—that he had to roll with it and take it as another test of toughness. So he sat down for an interview with the *Daily News* editorial board, where he was asked if he'd ever used illegal drugs. "I have never broken the laws of my country," he replied. The subject died there, but a reporter for WCBS television, Marcia Kramer, took note of what sounded like overly precise wording. When Clinton appeared on a show with her several days later, Kramer pointed out that he'd spent time living abroad while studying at Oxford.

"Have you ever broken a state law with regard to drug use?" she asked him. "Have you ever broken an international law—for example, when you were a student in England?"

"The answer to that question is I've never broken a state law," Clinton responded, "and when I was in England I experimented with marijuana a time or two. And I didn't like it, and I didn't inhale, and never tried it again."

I didn't inhale. With the boomer generation reaching middle age, past marijuana use was becoming a new issue in American politics. Many people Clinton's age had used it in their younger years, but now lots of them were raising families and didn't want their kids getting any ideas. The "Just say no!" campaign that Nancy Reagan had led in the eighties was still fresh in the air, and older Americans—the parents of the boomers—were even less forgiving on the subject. So any admission of past marijuana use came with political risk. Usually, those acknowledging it simply spoke of it as a part of their foolish, experimental college years, something they'd long since matured out of. But now here was Clinton adding a whole new layer of absurdity. Really, who else but Slick Willie would cop to using pot, then in the same breath try to exculpate himself by claiming he didn't actually inhale?

The comment exploded into the news on the same day as the Academy Awards. With more than forty million Americans watching on ABC, host Billy Crystal stared into the camera with a quizzical expression and, without mentioning Clinton's name, said: "Didn't inhale?" The audience roared. A newspaper cartoon depicted a television newscast, with the anchor saying, "In a related story, Clinton admitted he committed adultery but said he didn't like it."

Clinton started punching back. First, he targeted Brown, who'd left himself wide open with his proposal for a 13 percent flat tax. It was an idea that would with time be embraced by conservative Republicans, but in '92 it had no partisan home and Brown saw it as the perfect expression of the antiestablishment rage he was channeling. At rallies he would toss thick bound copies of the federal tax code into trash cans, promising to liberate Americans from the tyranny of the IRS. "A silver bullet solution for the 1990s," Brown called it.

But the flat tax, at least as presented by Brown, also amounted to a massive windfall for the rich, with far less benefit at the other end of the scale. For Clinton, it was political gold. Once again, the candidate of the DLC somehow found himself in position to attack his opponent from the left. "I'll tell you what," Clinton told a crowd, "if anybody can pull this proposal over the eyes of the voters of New York, they'll be the slickest politicians to ever come along in the history of American politics. We can grease the wheels of the Long Island Rail Road from now to kingdom come with the slickness it would take to shove this down the throats of the American people."

For Brown, that dark hole of media nonexistence had come with an upside, freeing him from scrutiny. Now the anonymity was all gone, and the flat tax was just the beginning of his trouble. There was also his promise, seemingly offered on a whim, to make Jesse Jackson his running mate. It made some sense. Weeks earlier, Clinton had been caught on tape making disparaging comments about Jackson, who had pointedly refused to endorse him. More recently, Clinton had also been embarrassed by revelations that he'd golfed at a whites-only country club in Arkansas. With black voters making up a quarter of New York's primary electorate, Brown sensed an opening.

With his Jackson gambit, Brown risked alienating as many voters as he was appealing to. Memories of the "Hymietown" slur ran deep. "You insult the Jewish community by picking Jesse Jackson!" a prominent Jewish Democrat shouted at Brown during a Brooklyn event. Another activist stood up to tell Brown that "the Jewish community is certainly not opposed to a black vice president. We don't think you have selected the best representative of the black community. We think you have not chosen wisely."

Brown tried to sell his choice as a profile in courage. "When I come to New York and say I'd like the Reverend Jackson to be a vice presidential candidate, I know it's controversial," he boasted. But in truth, it was a strategic move he just hadn't thought through. Brown would end up barely cracking 10 percent of the Jewish vote on primary day—in a state with the highest concentration of Jewish voters.

Clinton lashed back at the press, too, and landed some blows. He agreed to sit down with Phil Donahue, whose nationally syndicated television show taped in Manhattan. A pioneer of the daytime talk genre, the unabashedly liberal Donahue was known to mix discussions of weighty issues with trashy and exploitative elements. Now with Clinton in the guest chair, Donahue demanded the details that Clinton hadn't provided in his *60 Minutes* interview.

"Governor, you do acknowledge that your marriage hasn't been perfect. Throughout this imperfect part of your marriage, did you ever separate, you and Hillary?"

Clinton stopped him in his tracks.

"How would you like it if I spent a half million dollars looking into your life and asking you questions like this?"

It worked, on Donahue's turf that day, and at the ballot box on April 7. There were still all sorts of doubts about Clinton, but not quite enough for New York's Democrats to go for the nuclear option. Clinton took 41 percent and Brown ended up with just 26 percent—not even enough for second place. Tsongas, who'd barely lifted a finger, still got 27 percent. In the ultimate media cauldron, Clinton had passed his do-or-die test.

New York had sealed Clinton's claim to the Democratic nomination. It had also, plenty of Democrats were convinced, sealed their party's November fate.

"After a quarter-century of chaos," the *Washington Post*'s David Von Drehle wrote, "the conclusion is becoming inescapable that the party of Jefferson, Jackson and Roosevelt is no longer serious about competing at the presidential level."

And things were about to get worse.

EIGHT

N ewt Gingrich had once called George Bush an emblem of "the pre-Reagan party—the side of the party Reagan had to beat to win." It was true. In 1980, Reagan had run as the heir to the Goldwater insurgency, poised to topple the old eastern establishment once and for all, and it was Bush who emerged as his final obstacle.

Their clash was ideological. Reagan embraced the new theory of supply-side economics, with its emphasis on tax cuts and deregulation, and forged an alliance with Christian conservatives. Bush decried Reagan's "voodoo economic policy" as fiscally reckless and ran on a liberal social platform, including support for the Equal Rights Amendment and opposition to a ban on abortion. There was a cultural aspect, too. Bush was an aristocrat, the son of a financier turned senator, raised in Greenwich, schooled at Yale, endowed with impeccable old-money manners. He was, in other words, the embodiment of a pedigreed elite the conservative grass roots found repellent.

Reagan entered the heavy favorite, but Bush pulled off an upset in the Iowa caucuses. Reagan then steadied himself with a New

Hampshire win, but Bush stalked him well into the spring, taking half a dozen states before finally relenting. It was a surprisingly strong showing, one that convinced Reagan to make Bush his running mate. He announced his decision at the Republican convention in Detroit that July. The true believers were dismayed. "By a single stroke," the *New York Times* reported, "the party's Northeast Republican establishment was resurrected, placed on the national ticket by the man on whom they had pinned their hopes."

Bush could see where the party was moving, though. He pledged his loyalty to Reagan and his program and remade himself in his new boss's political image. Cultural moderation gave way to a vow to pursue a constitutional amendment outlawing abortion and to restore prayer in public schools. Doubts about the depth of his conversion lingered on the right, but eight years as the Gipper's number two was enough to carry him to the Republican nomination in 1988. The "Read my lips" pledge at that summer's convention was meant to capture the Reagan spirit and cement a bond.

To the right, then, it wasn't just betrayal when Bush cut his tax-hiking deal with the Democrats in 1990. It was confirmation of what they'd suspected all along. He just wasn't one of them. "Bush has had the great luxury of inheriting the conservative legacy of Ronald Reagan—without so much as lifting a finger to earn it," one columnist wrote.

The revolt in the House left the deal in smithereens. Unable to deliver his own party's votes, Bush lost leverage to the Democrats, who were waiting with new demands. There still had to be more tax revenue, but this time it was going to come from the wealthy. Democrats wanted a hike in the top marginal rate from 28 percent to 33 percent, plus a surcharge for the superaffluent. Bush settled for 31 percent and no surcharge. Democrats gave him the votes and Bush signed it just before the midterm election.

He'd warned the Gingrich crowd that rejecting the original deal would lead to a worse one, and now he had his proof, or so he thought. But Gingrich just said he should barnstorm the country attacking the

Democrats until they relented and agreed to a package without any new taxes. Bush was learning the hard way what Newt and his crew meant when they called taxes a foundational issue.

Just months earlier, with Bush's approval rating holding steady at around 70 percent, Republicans had crafted an ambitious plan to capitalize. They'd make a big dent in the Democrats' House majority in the fall of '90 and then, aided by redistricting and the president's reelection coattails, take back the chamber for good in 1992. Instead, when Bush finally put his pen to the budget at the end of October, his approval was down twenty points and the party's chief campaign strategist was saying Republicans were "looking into an abyss" on Election Day. They ended up losing eight House seats, dropping back to where they were after Reagan's midterm debacle in 1982.

The conservative wing was more potent than ever, and now it was training its sights on Bush. "We are in the middle of an argument which a lot of us thought ended in 1989," Gingrich said. "There really is a fundamental difference between the managerial/accommodationist approach to the welfare state and the reform/replacement approach." A group of disgruntled conservatives met to discuss the most radical recourse. "There is absolutely a mounting belief that Bush needs to be challenged in '92," one of them said. "The issue is, Bush calls himself a conservative, but is not."

Before it could explode into a full-blown insurrection, though, the country was at war. Bush had set a January 15 deadline for Saddam Hussein to withdraw his forces from Kuwait. When the date passed without the Iraqi strongman blinking, Bush ordered the launch of Operation Desert Storm. When the American-led coalition liberated Kuwait in lightning-fast time—and with a minimum of American casualties—the country rejoiced and the president's approval rating reached record heights.

In March, a triumphant Bush strode into the House chamber to deliver his State of the Union address. He was greeted with exuberance, from both sides of the aisle. The Republicans, so many of whom had defied Bush on his tax deal, now sported buttons that declared:

"I voted with the president." They were talking about the war vote, not the budget, which no one was talking about anymore. In this moment, with Bush seemingly on course for a landslide reelection, it was easy to reimagine the tax revolt as a meaningless little scuffle of no long-term consequence.

It didn't last. The patriotic celebrations died down and Americans returned their focus to the domestic state of affairs. Deteriorating economic conditions had brought Bush to the table in 1990 and convinced him that abandoning "Read my lips" would pay off with a vibrant economy. He didn't end up with the deal he wanted, but he still got the deficit reduction he believed was key—nearly half a trillion bucks. The fall of 1991 arrived, though, with no boom in sight. Joblessness, sitting at 5.9 percent when Bush signed the budget, was now 7 percent and moving upward. Now, everyone was talking about the deal again.

Bush tried to project confidence and talked of "staying the course." America was caught up in a global slowdown, he'd say, and his program would eventually produce dividends—especially if the Democratic Congress would give him more of what he was looking for. But conservatives were spinning a different story now, one that drew a straight line from Bush's tax increase to the worsening economy a year later.

The war had frozen the political game for a while, but with his poll numbers now dropping, the talk of a primary challenge stirred anew. Time was short. The New Hampshire primary would be in February, and for all his vulnerabilities, Bush would still bring vast— and almost certainly overpowering—strength to any nomination fight. He was still an incumbent president who'd won a war. He also benefited from the sudden collapse that summer of the Soviet Union, bringing to an end decades of harrowing tension between the enemy superpowers. Anyone stepping forward to oppose him would be doing so with the knowledge that victory was all but impossible.

For much of the fall, the only prospect to emerge was Ron Paul, a former three-term congressman from Texas who'd left the GOP in 1988 to run as the Libertarian Party's presidential candidate. The

fifty-six-year-old Paul was now back to practicing medicine and tend-
ing to his small political network through a monthly newsletter. In
mid-September, he announced that he was considering returning to
the GOP for the purpose of running against Bush. "The greatest frustration at the moment is among the old-line Reagan Republicans who
are disenchanted with George Bush," he said, "and the place to tap
into that is in the Republican primaries."

But Paul was a fringe player, and few even noticed when he spoke
up. Then, around Thanksgiving, a more substantial figure entered
the picture. He had experience in government, but it was television
that had made him a brand name in politics. He shared the right's
usual grievances with Bush, but brought some surprising ones to the
table, too, on war, trade, and immigration. Rather than intimidating
him, the scale of the challenge only seemed to give him motivation.
Patrick J. Buchanan was a born brawler who'd become America's most
famous pundit—and now he was about to run for president.

One of nine children in a devoutly Roman Catholic family, he'd
come of age in Washington, D.C., in the 1940s and '50s, back when
Mass was still given in Latin and women worked outside the home
only if they had to. His father, an accountant, demanded regular sessions with a punching bag, "in the prayerful hope," Buchanan would
later say, "one of his sons would be challenged to a neighborhood fight."

He was schooled by the Jesuits, first at Gonzaga High and then
at Georgetown, where his studies were interrupted when he picked a
fight with city cops. They tried to arrest him, he resisted, and then "I
put a size 10-and-a-half cordovan where I thought it might do some
good," Buchanan wrote in his memoir.

He liked to fight with words, too. With a flair for vivid and
punchy prose, he was lured to St. Louis at the age of twenty-three to
write editorials for the *Globe-Democrat*, which styled itself as the conservative counterweight to the dominant *Post-Dispatch*. Soon, he linked
up with Richard Nixon, who was plotting a comeback for the 1968
election, becoming a speechwriter and then a White House adviser.
He stayed loyal to the very end, and then some. Long after Nixon's

resignation, Buchanan would still refuse interviews with reporters from the *Washington Post*, scorning the paper for its role in unearthing the Watergate scandal.

Free from the White House, he turned to the vocation that would ultimately make his a recognized face far beyond the corridors of power: punditry. He started simply enough, penning a thrice-weekly column that nearly a hundred papers picked up. Then came radio.

In the pre-Limbaugh world of the 1970s, political talk wasn't seen as a moneymaker. But when one of D.C.'s most powerful AM signals, WRC, switched from music to an all-news format in 1978, the station decided to experiment. In the afternoons, Mort Sahl, the pioneering political satirist, was given the coveted four-to-seven spot. A morning opening went to Buchanan and Tom Braden, a liberal columnist.

The friction was infectious. Buchanan, a polished polemicist, and Braden, twenty years his senior and possessed of a more down-to-earth bearing, played off each other naturally. *Confrontation*, as their show was called, was a hit, and within months it replaced Sahl's in the afternoon drive slot, where it became one of the top shows in the market. There was nothing else like it. More than half of all members of Congress tuned in regularly to hear Buchanan and Braden, the *Post* reported, and "reliable estimates put their salaries into six figures."

"The program has furnished . . . to a greater extent than letters to the editor, an outlet for people who are worked up, upset, angry, scared and want to say something about it," Braden told the paper.

After Hours, a nightly television version, soon followed on a D.C. station—another instant hit. Then they went national. In June 1982, two years after its birth, CNN announced the creation of a new half-hour show that would air weeknights at eleven thirty. *Confrontation* became *Crossfire*, and now Buchanan and Braden had a coast-to-coast television audience. CNN's press release promised: "Their political expertise, sophistication and wit assure 'Crossfire' viewers of provocative, informative political debate." Within a year, it was moved to the plum 7:30 P.M. spot, where it would stay for the next two decades.

lection night, 1974. The numbers don't add up for Newt Gingrich in his bid to nseat Democratic Rep. John Flynt. After losing a rematch two years later, he finally ins a seat in Congress on his third try in 1978.

Too big for his britches. Thirty-three-year-old Arkansas governor Bill Clinton appears on NBC's *Today* with Tom Brokaw in July of 1980. With voters wondering if their ambitious governor was already looking beyond their state, Clinton would be defeated for reelection months later.

House Speaker Thomas P. "Tip" O'Neill, an old school charmer from North Cambridge, Massachusetts, pilloried by Gingrich as the face of the "Democrat machine"
that had run Congress since the 1950s.

"Mr. Nice Guy." An affable institutionalist, House Republican leader Robert M:
chel was troubled by the rise of Gingrich and his combative political style. Here th
two men confer as ABC's Brit Hume looks on.

"All of us . . . must resolve to bring this period of mindless cannibalism to an end!" Jim Wright in May 1989 becomes the first Speaker in history to resign due to an ethics scandal—and a prize trophy for Gingrich, who parlayed his crusade against Wright into a spot in the House Republican leadership.

"By God, we've kicked the Vietnam syndrome once and for all." President George H. W. Bush welcomes troops home from the Gulf War in March 1991. In the wake of Operation Desert Storm, his approval rating surges past 90 percent—a modern record.

Hamlet on the Hudson. As he agonizes over whether to make a last-minute en
trance into the 1992 presidential race, Mario Cuomo places a fateful phone call t
one of his favorite former aides, Tim Russert, who had since gone on to host NBC
Meet the Press.

Gennifer Flowers, a cabaret singer and former Arkansas state employee, rocks th
1992 campaign with the claim that she and Bill Clinton engaged in a twelve-yea
extramarital affair.

Bill and Hillary Clinton respond to Flowers's accusations in a special edition of *60 Minutes* that airs immediately after the Super Bowl on January 26, 1992.

There is no love lost between Bill Clinton and former Massachusetts senator Paul E. Tsongas in the 1992 primary race. To Clinton, Tsongas is "Saint Paul," a self-righteous scold. To Tsongas, Clinton is a dishonest "pander bear."

1-800-426-1112. Former California governor Edmund G. "Jerry" Brown turns his third presidential campaign into a guerilla effort in 1992, touting a revolutionary toll-free hotline for donors and refusing any contribution over $100.

A euphoric Pat Buchanan celebrates his stronger than expected showing in the February 1992 New Hampshire Republican primary, a result that signals just how vulnerable President George H. W. Bush has become.

Jesse Jackson listens as Bill Clinton addresses his Rainbow Coalition in Washington in June 1992. Jackson later calls Clinton's speech "a very well-planned sneak attack."

One of Texas billionaire Ross Perot's many appearances on CNN's *Larry King Live* in 1992, the venue he used that February to challenge Americans to place his name on the ballot for president in all fifty states.

Heeding Perot's call, volunteers march to the State House in Austin, Texas, on May 11, 1992, to submit petitions to place his name on the state's ballot. They needed 54,000 signatures. They hand in 225,000.

In a break from the traditional ticket-balancing formula, Clinton selects as his running mate a fellow southerner with a reputation as a moderate, Tennessee senator Al Gore.

Around the same time, one of Buchanan's closest friends was launching a television show of his own. John McLaughlin had originally made his name as a renegade Jesuit priest who defied the church to run for the Senate from Rhode Island in 1970. Calling himself a "progressive Republican," Father McLaughlin ran as an opponent of the Vietnam War and critic of the Nixon White House. Demolished in the Republican primary, he met up with Buchanan and asked for a White House job. Public opinion was moving against the war, but McLaughlin went the other way and recast himself as a champion of the Nixon policy. The conversion was enough to get him in. It was what he really wanted. He was a player in Washington now.

McLaughlin stayed in D.C. after Nixon's demise, renouncing the priesthood and marrying the woman who'd run his Senate campaign. Like Buchanan, he saw his future in media. With his booming voice, unwavering certainty, and rich Latinate vocabulary, McLaughlin had the makings of a unique television character, and in 1982, just as *Crossfire* was starting up, he found his own national platform. *The McLaughlin Group* was a panel show with an edge—"an unrehearsed weekly television program bringing you inside opinions and predictions on the major issues of the day," the voiceover would say.

The format would eventually become commonplace, but with cable news still in its infancy, it was a jarring break from the plodding Sunday shows viewers were accustomed to. Four commentators from across the spectrum would analyze a series of topics while McLaughlin, seated in the moderator's chair, listened impatiently, pouncing at any opinion he deemed ill supported. The pace was quick, the arguments intense, and the energy frenetic. McLaughlin was the star, but Buchanan was along for the ride, too, a fixture in the rightmost seat. Hundreds of stations began carrying the show, and by the end of the decade *Saturday Night Live* was parodying it, with Dana Carvey playing McLaughlin and Phil Hartman as Buchanan.

Without ever running for office, Buchanan had become one of the best-known conservatives in America. He looked at the Republican presidential race in 1988, but chose not to enter. The field was crowded and he would have had lots of company on the right. Sticking with punditry for a few more years was the smarter play. When the 1992 campaign came around, he decided the time was finally right. President Bush would be seeking a second term, and Buchanan would challenge him in the Republican primaries.

To the rest of the world, it looked like a gigantic miscalculation. Yes, Bush's tax hike had infuriated conservatives and confirmed their long-standing doubts about whether he was really one of them. But he'd followed up that betrayal with a military triumph, the rout of Saddam Hussein's forces, rallying his conservative critics right back around him. The dismal economy might jeopardize Bush's reelection chances against a Democrat, the thinking went, but there was little appetite among Republicans to turn on a commander in chief who'd just won a war.

That wasn't going to cow Buchanan. He'd been against the Gulf War, and even after all the ticker-tape parades, he still called it a mistake. In a Republican Party that now embraced internationalism, Buchanan identified as a nationalist. He regarded treaties and alliances with suspicion, fearful that they would erode American sovereignty.

His party celebrated free trade, but to Buchanan it was a sellout of the American worker. So was mass immigration, which Buchanan proposed to combat with an impenetrable wall along the Mexican border and a five-year moratorium on legal immigration.

When word of Buchanan's interest in the race spread in November, Bush's press secretary cracked a joke. "We've already had one Buchanan," Marlin Fitzwater said, referring to James Buchanan, one of history's least well-regarded presidents. But dismissiveness wasn't going to deter this Buchanan.

On December 10, 1991, he declared his candidacy at the statehouse in Concord, New Hampshire. The state's first-in-the-nation primary was barely two months away and the race looked unwinnable for Buchanan. A poll put him fifty-one points behind, 64 to 13 percent. But the contrast was irresistible. Bush, with his faith in international coalitions and calls for a "new world order," embodied a philosophy that made Buchanan bristle. His blue-blooded heritage made him an even more ideal foil for Buchanan, who fancied himself the workingman's tribune.

"My friends," Buchanan said in his announcement speech, "we are the sons and daughters of the men and women who brought America through the Depression and crushed fascism on two sides of the world. We ourselves are the men and women who won the Cold War with Communism. We can win the future and we can hand down to those who come after us a country as great and grand and good as the one that was given to us. But first we must take America back.

"So we are taking this campaign not just to Republicans, and not just to conservatives. Every American is invited to join, the middle class of both parties, and of no party. For the establishment that has dominated Congress for four decades is as ossified and out of touch with America as the establishment that resides in the White House."

He acknowledged Bush's combat heroism in World War II and saluted the decency of his character. "But," Buchanan said, "the differences between us are now too deep.

"He is yesterday and we are tomorrow. He is a globalist and we are nationalists. He believes in some Pax Universalis. We believe in the Old Republic. He would put Americans' wealth and power at the service of some vague New World Order. We will put America first."

To most ears, though, it was Buchanan who was sounding like yesterday. The "America first" theme called back to the days just before World War II, when isolationism had been the dominant foreign policy strain on the right. Back then, leading conservatives had argued against American intervention in Europe, arguing that Hitler's advance didn't directly threaten the United States. A half century later, history didn't look kindly on those arguments, and Buchanan was practically alone among prominent Republicans in his advocacy of what some now called paleoconservatism. Even conservatives who were frustrated with Bush tended to distance themselves from his new challenger.

As he rolled out his candidacy, a guest spot on ABC's *This Week with David Brinkley* accompanied the campaign rollout. The subject turned to immigration. In the mid-sixties, immigration laws had been liberalized, opening the door to the millions of non-Europeans who had previously faced barriers. Buchanan was asked about his contention that the influx of arrivals since then was threatening American culture, which was arousing charges of racism.

"I think God made all people good," Buchanan said, "but if we had to take a million immigrants in, say Zulus, next year, or Englishmen, and put them in Virginia, what group would be easier to assimilate and would cause less problems for the people of Virginia?"

Again, he was standing virtually alone among major conservative voices. Buchanan, wrote George Will, "evidently does not understand what distinguishes American nationality—and should rescue our nationalism from nativism." The Republican Party's most recent national platform included a commitment to "welcome those from other lands who bring to America their ideals and industry" along with a cursory nod to border security. This posture stemmed from the immigration reform signed by Reagan in 1986. It aimed to accommodate those who had come to the country illegally but lived otherwise lawful lives with a promise to curtail future illegal crossings. Nearly three million people, many of them from Mexico, had since qualified for green cards under the law, but the flow of illegal entries hadn't stopped. To Buchanan, a populist fire was waiting to be lit.

"Take a look at what's happened to the people of California," he said. "One in five felons in a federal prison is an illegal alien. The immigrants are coming in such numbers that they're swamping the schools, and you have to raise taxes. I think this is a valid issue."

Accusations of anti-Semitism had long haunted Buchanan. Most recently, he'd suggested on television in the run-up to the Gulf War that the U.S. was being forced to fight for Israel. "There are only two groups that are beating the drums for war in the Middle East," he said, "the Israeli defense ministry and its amen corner in the United States." He'd also made a pet cause out of the case of John Demjanjuk, who'd been accused decades later of overseeing atrocities at a Nazi concentration camp.

As Buchanan launched his campaign now, the *National Review*'s William F. Buckley pronounced it "impossible to defend" his friend Buchanan against claims that anti-Semitism had infected his Gulf War rhetoric. Buchanan protested: "I don't have a bad heart and I have nothing to apologize for." Michael Kinsley, who'd replaced Braden as *Crossfire*'s liberal half a few years earlier, went to bat for his cohost, saying he'd never witnessed any signs of anti-Semitism from Buchanan—and that, as a Jew, he wouldn't have tolerated it if he had.

Within his party, though, Buchanan was an isolated figure. His brand of conservatism, and the controversy he aroused, discouraged mainstream Bush critics from attaching themselves to him. The likelihood that he'd lose—badly—was further incentive to stay away. Buchanan would run without any significant endorsements from elected officials and against vociferous objections from conservative opinion shapers. Asked if he might back Buchanan, Newt Gingrich scoffed, "He's a temptation. But let's be frank. He's not a serious candidate for president." There would be no elite Republican donor to bankroll the Buchanan campaign, and Bush would refuse to debate him or even to acknowledge his candidacy. Buchanan was entirely on his own.

For the next two months, he raised money through direct mail and poured it into cheaply produced television ads that ran on New Hampshire's two main commercial stations. (He would cede the lead-off Iowa caucuses to Bush.) After tearing into Bush, one spot closed with a group of fed-up locals shouting in unison, "Read OUR lips!" Geographically compact and politically engaged, New Hampshire wasn't a bad place to be a candidate living off the land. Buchanan drew sizable crowds and reeled in a big fish when the *Union Leader* newspaper, the state's dominant conservative voice, threw its support behind him. He played the role of underdog with relish.

"A lot of people who were young lieutenants and captains in the conservative movement in the seventies have become courtiers at the Versailles of King George," he said. "They care about their

White House invitations to dinner. . . . But they know better than I do that the president of the United States, Mr. Bush, has walked away from us time and again."

Polls showed him creeping into the twenties, threatening to break 30 percent. The Bush strategy was to appear above politics. He stayed away from the state, and from all campaigning, until the week before the primary, when he formally announced his candidacy for reelection. Finally, the weekend before the vote, he came to town. He still wouldn't utter his opponent's name, but now he admitted: "I know I'm in a tough race, but I've been in tough races before—many of you at my side in those battles."

Bush left the attacks on Buchanan to his sidekick on this campaign swing, Arnold Schwarzenegger. The action star was one of Hollywood's few outspoken Republicans and a Bush family friend. His most recent movie, *Terminator 2: Judgment Day*, had been a blockbuster hit the previous summer, bringing in more money than any other film in 1991. With Bush seated near him, Schwarzenegger invoked his signature line from the flick: "Send a message to Pat Buchanan: 'Hasta la vista, baby!'"

To Buchanan, it was proof he was breaking through. "The Buchanan brigades," he predicted, "are going to meet the hollow army of King George and cut through it like butter."

The first batch of exit polls on February 18 suggested the unthinkable: a dead-even contest. Buchanan might actually win the primary, reporters realized. Ballots were still being cast and the numbers weren't released publicly, but the tone of the coverage was already taking shape. A sitting president was facing a shocking rebuke from his own party. Minds were already thinking back a generation to 1968, when LBJ struggled to hold off Eugene McCarthy in New Hampshire's Democratic primary, a performance so feeble that the president aborted his reelection effort days later.

The first precincts to report when the polls finally closed were from the most conservative pocket of Manchester. Buchanan country. For a while, he and Bush swapped the lead. Finally, Bush began building some distance from the challenger. When the networks called it, Bush's edge was still in single digits—"a bone-rattling surprise," David Brinkley said on ABC.

On CBS, prime-time coverage of the Winter Olympics in Albertville, France, was interrupted with a bulletin: "Good evening. Dan Rather, CBS News, with a New Hampshire primary update headline of tomorrow. President Bush wins, but not by much. Pat Buchanan mounts a very strong and startling challenge to the Bush-Quayle re-election campaign. Buchanan gets well over 40 percent of the Republican primary vote."

Buchanan didn't wait any longer. "I think King George is getting the message!" he crowed from his Manchester headquarters. The media was treating this as a moral victory for him, and he provided them with a victory speech. "We are going to take our party back from those who have walked away from us and forgotten about us," he cried, "and when we have taken our party back, we are going to take our country back.

"When we take America back, we are going to make America great again, because there is nothing wrong with putting America first."

After the official tally, Buchanan ended up with 37 percent, a strong showing still, but not as strong as it had once seemed. Bush's tally was 53 percent. No matter. "Patrick J. Buchanan's strong showing in the New Hampshire primary," the next morning's *Washington Post* reported, "sent a jolt yesterday through the Bush establishment, which acknowledged that the president who once anticipated easy renomination faces a series of battles that could splinter the party, highlight his shortcomings and leave him a weakened candidate in the fall."

In the exit poll, more than half of Buchanan's voters said they'd chosen him to send a message of dissatisfaction to Bush. He was their protest vehicle. Buchanan held a press conference the morning after the primary. Winning the Republican nomination, he said, was now "possible—but still a long shot."

"We've got to find a state where we can go head-to-head with the president and beat him cold," he said.

He locked in on Georgia, which would hold its primary in two weeks. Even with his momentum, the obstacles were enormous. The state was three times the size of New Hampshire, and this time he wouldn't be able to camp out for two months before the vote. His resources were thin, too, and Bush continued to shun his call for a debate.

But there were new signs he was breaking through. When Bush faced reporters after his New Hampshire scare, he promised to "take this guy on in every single state." It was the most direct reference to his opponent Bush had made the entire campaign. Moments later, he made it more personal: "All I did was lay back and get hammered by these Democrats and, to some degree, by Pat. And so it's a new ball game. And we're coming out strong." For the first time, he'd mentioned Buchanan by name.

Still, the Bushies were hoping to avoid direct engagement with Buchanan. There was "no plan to unload on him right now," one of the president's campaign chiefs said. Then, they looked at their polling. Buchanan had credibility now, and was moving up in Georgia. They reached for the strongest weapon in their arsenal, enlisting P. X. Kelley, a decorated former marine commandant, to star in a television ad. "When Pat Buchanan opposed Desert Storm," Kelley told the camera, "it was a disappointment to all military people, a disappointment to all Americans who supported the Gulf War, and I took it personally."

Buchanan was livid, and thrilled. This was the kind of brawl he'd always wanted. "George Bush has got to realize he's not dealing with a Michael Dukakis here," he said. "He's up against a conservative Republican who when he's hit will fight back."

He pressed his usual attack on Bush over taxes, but also introduced new cultural themes. Over the last generation, Georgia's fraught racial politics had pushed its white voters toward the Republican Party, and Buchanan now aimed hard for them. A few months earlier, Bush had signed a civil rights bill. He'd told conservatives it was a victory because it didn't include quotas. but it did allow for affirmative action, which Buchanan now argued was just as likely to make blue-collar whites the victims of "reverse discrimination." He framed it as an issue of class—Bush and the "Exeter-Yale GOP Club" trying to "salve their social consciences at other people's expense."

The White House was on edge as the vote neared. The president spent the weekend before the primary barnstorming Georgia. He appeared to be ahead, maybe even solidly, but Buchanan was striking a nerve. The situation was volatile enough that Bush decided to do something he'd adamantly refused to do since making his tax deal with the Democrats in the fall of 1990: he said he'd made a mistake.

"Listen, if I had it to do over," the president told a Georgia newspaper, "I wouldn't do what I did then, for a lot of reasons, including political reasons."

The Bush team hoped it would reassure wobbly conservatives. It was an extraordinary statement, given that Bush had until now portrayed the tax increase as a leadership moment, when he rose above politics and did what was best for the country. It was also a curious one. By now, it was an article of faith on the right that the tax hike was directly responsible for the sluggish economy. Bush seemed to be trying to acknowledge that sentiment without actually agreeing with it. "He doesn't consider that it was an economic blunder of the first magnitude," Buchanan countered. "He's just talking about the politics of it."

Gingrich, back in his home state and officially supporting the president, searched for a positive spin on Bush's lukewarm apologia. "I think it is very hopeful that President Bush has recognized that the Democratic leadership of Congress is addicted to higher spending and higher taxes," he said. "I think it's a clear sign we will veto any tax increase this year."

Bush did get his win, but once again that wasn't the headline. The final margin was 64 to 36 percent, confirming that the intraparty discontent was not a one-state story. More damning was the exit poll finding that one in three Buchanan voters said they wouldn't vote for the president in November. Overall, two-thirds of Georgia's Republicans agreed the country was on the wrong track.

Bush remained tucked away in the White House as the results came in. If there was good news, it was that it wasn't worse. Any possibility of Buchanan parlaying his New Hampshire surprise into an outright victory that would threaten the president's grip on the nomination was now dashed. But that was little solace for Bush, who released a statement acknowledging to his party's voters that "I hear your concerns and understand your frustration with Washington. I am committed to regaining your support."

For Buchanan, the Georgia result meant the show would go on, clear through the primary season and all the way to the convention in Houston. He wasn't going to win the nomination, but he'd demonstrated a depth of support no one expected, making his the kind of symbolic candidacy that could continue attracting grassroots money and media attention. His party's leaders badly wanted to ignore him, but they couldn't now. "We are going to go all the way!" he exclaimed.

Buchanan would fail to win a single primary or caucus the rest of the way. In fact, he never again matched his New Hampshire and Georgia showings. But he kept at it, collecting votes and delegates, his lingering presence a constant reminder of the disgruntlement Bush faced inside his own party. As the spring wore on and the nomination battle wound down, Bush found himself in a peculiar position. He'd been embarrassed, humiliated even, by his pundit challenger; and yet, his prospects for winning a second term were brightening. For this, he could thank the ever-reliable Democrats, whose hapless dysfunction had produced an almost laughably flawed candidate. For all the president's problems, the consensus was universal: the Bush machine was going to steamroll Slick Willie.

Buchanan claimed credit, arguing that his candidacy had sharpened the president. He'd called Bush "unelectable" before, but now predicted he'd win a second term—as long as he could unify the Republican Party. And to do that, Buchanan said, Bush would need the full support of the Buchanan brigades. It was clear what the television star turned candidate was angling for: the biggest audience of his life. If Bush didn't give him a prime-time speaking slot at the August convention, Buchanan declared, it "would be a terrible mistake."

NINE

The primary season wrapped in early June, with easy wins for Clinton and Bush, each now mathematically assured of his party's nomination. Not that anyone was all that interested in how they'd match up with each other—not with a brand-new candidate suddenly running ahead of both of them.

It had been building all spring, at first a curiosity and now a political phenomenon without modern parallel. Henry Ross Perot was not officially a candidate yet, but already he had assembled a vast political coalition that defied labeling. In the polls, he'd just passed Bush and had long ago left Clinton in the dust. Most independent candidacies of the past had been factional rebellions, drawing concentrated strength from one pocket of a party or one corner of the country—Strom Thurmond in 1948 and George Wallace in 1968, for example, flexing their muscle in the South and nowhere else. Not so with Perot.

He was now in first place in liberal Massachusetts, conservative Texas, middle-of-the-road Ohio, and a host of states in between. "President Perot?" the cover of Newsweek asked. Richard Nixon, now rehabilitated as a sage elder statesman, counseled a group of Republicans to take the Perot candidacy seriously. "I wouldn't bet against him," Nixon said. Newt Gingrich fired off a memo warning his party that "a 45 percent Perot, 30 percent Bush, 25 percent Clinton race is very plausible." Pundits talked of a deadlocked election, with no one winning a majority in the Electoral College. Others wondered if Perot might just run away with the whole thing.

Desperate to siphon some of the energy, Clinton was now calling himself "a radical moderate in a way." But no one was buying it. Not since 1912 had a major-party candidate finished third in a presidential race. But at this moment, just five months before Election Day, this was beginning to look like Bill Clinton's destiny.

"If nothing else," the *Washington Post* wrote, "during his long quest for the Democratic nomination, Clinton followed what he thought were the rules of modern politics. The shaft was to wake up this morning to find that he was an asterisk in the day's events and that even Democratic voters had told exit pollsters they might junk him for a political maverick, Ross Perot."

"Right now, Clinton is where no Democratic presidential nominee in this century has been six weeks out from his party's convention," a *Chicago Tribune* columnist wrote. "And that's in third place. Clinton is third in most of the polls, third in terms of attention from the news media, third in the minds of voters when they contemplate the general election."

It had begun with barely anyone in Washington noticing, two nights after the New Hampshire primary. Perot was a guest on CNN's *Larry King Live*, which, in seven years on the air, had become the cable network's signature show.

On February 20, King's guest for the full hour was Perot. Plenty of people already knew his name. He'd been dubbed America's first populist billionaire, a folksy Texan who'd created a data processing company at the dawn of the computer age, sold it to General Motors for $2.4 billion, then gotten GM to cough up $750 million more to get him off its board. Perot had a deep sense of patriotism, a flair for showmanship, and bottomless disgust with government bureaucracy.

During the Vietnam War, he paid for Christmas gift baskets for Americans being held captive, then flew to Hanoi and stood outside a prison demanding to deliver them personally. The North Vietnamese rebuffed him, but back home he was lionized. When Iranian revolutionaries seized two of his employees in Tehran a decade later, Perot grew fed up with the Carter administration's failure to secure a speedy release, and he organized a team of commandos who brought both men home alive and well. The rescue became the subject of a bestselling book, *On Wings of Eagles*, which then became a network television miniseries. The actor who portrayed Perot, Richard Crenna, looked and sounded nothing like the five-foot-six Texan, but it hardly mattered. The miniseries aired over two nights on NBC in May 1986, with more than a third of televisions in use tuning in.

The legend of Ross Perot was only growing, and as the 1992 campaign took shape, it intersected with a powerful and unpredictable political current.

At first, there was a gadfly named Jack Gargan, a financial planner living in Tampa. He'd dabbled in local politics, but now had more time on his hands and was determined to use it to blow up Washington. In 1990, just after Democrats and Republicans in Congress voted to raise their own pay, Gargan took forty-five thousand dollars from his own retirement fund and bought full-page ads in major newspapers around the country. "I'm mad as hell and I'm not going to take it anymore," they declared.

He started an organization called THRO, for Throw the Hypocritical Rascals Out. His goal was a coast-to-coast rebellion against every single incumbent member of Congress, Democrat and Republican. Gargan asked for donations and got more than he expected, almost a million bucks, in a time when that still meant something. He earned some attention, mostly as an oddity—*Time* magazine ran a small item saluting him as its "hero of the week"—but his dream went unfulfilled in the 1990 election, with more than 98 percent of incumbents keeping their jobs. Still, he did put up a few wins—sixteen of them, more than double the number of incumbents taken out in the previous election. It validated his conviction: the energy was out there to deliver a body blow to the system. He just had to find the right way to channel it.

He had a new idea. Forget all those individual House and Senate races; what about the big one? In the presidential election year of 1992, Americans would yet again be forced to choose between two creatures of the system, a Democrat and a Republican each offering the country the same heads-we-win-tails-you-lose proposition. But what if there was a third choice? A *real* third choice—someone who blended the stature of an insider with the pique of an outsider. He had exactly one man in mind. "I think Perot is it, in spades," Gargan said. "When I mention him, it gets the biggest applause of anything I talk about that day."

In November 1991, Gargan called a convention at a high school gym in Tampa. He invited Perot, and to his surprise, Perot accepted. Two thousand activists showed up to hear him. They were convinced that neither party had their back, that the Democrats and Republicans running Washington were getting fat while the economy suffered and the country lost its edge. "Unless the average citizen gets upset," Perot told them, "that three-act comedy up there is going to continue." They chanted, "Run, Ross! Run!" but he told them he wasn't looking for any official role in politics: "In a minute, I'd go up there to Washington and spend the rest of my life working night and day to fix it, just a private citizen."

And that was that. He went back to Dallas and there was no follow-up—no speaking tour, no round of high-profile interviews, no quasi-campaign apparatus, absolutely no indication that Perot was actually interested in running for president. By February, when he sat down across from Larry King, no one in the political and media world was seriously contemplating the possibility that Perot could be a candidate. Nor did that change during the broadcast.

When King brought up the idea of running, Perot was dismissive.

"I wouldn't be temperamentally fit for it," he replied. But King persisted and Perot finally issued his terms: "If you're that serious—if you, the people, are that serious, you register me in fifty states."

What Perot was describing, though, wasn't the sort of thing that just happened. Getting on the ballot in all fifty states was an extraordinarily complex undertaking. While the requirements were minimal in some states, others had erected all but prohibitive barriers. Texas, for instance, demanded that any independent candidate collect signatures totaling 1 percent of all votes cast in the previous presidential election—more than fifty-four thousand, for the 1992 election. If that wasn't enough, none of the signatures could come from voters who participated in either the Democratic or Republican presidential primaries, rendering more than two million Texans off-limits to petitioners. And the deadline for collecting all of these signatures was the earliest in the nation, May 11. There was a reason that even established and well-organized third parties, like the Libertarians, with tens of thousands of true believers spread out across the country, still had trouble making all fifty state ballots.

So to the political world, Perot's answer to King sounded only like a roundabout way of saying no. Out in the country, though, many of the same folks who'd responded to Jack Gargan's newspaper ads were also watching and heard in Perot's words the challenge they'd been waiting for. While the press focused on the looming Bush-Clinton contest, calls began pouring into Perot's Dallas office, where fifty new phone lines were soon installed to handle the flood. There was no Internet to connect them, but within weeks they had him on the ballot in Tennessee. The threshold was easy there, though, with just 275 signatures required, so few took note. Not until the end of March did pollsters finally start asking about Perot. More than half of the country still hadn't heard of him, but when his name was added to the mix against Bush and Clinton, he grabbed 21 percent. If he replicated that total on Election Day, it would be the best showing by an independent candidate since Teddy Roosevelt eighty years earlier. Now Perot had the whole world's attention.

By the end of April, he was over 30 percent, pushing past Clinton and challenging Bush for first place. Perot was starting to feel it. When the Texas ballot deadline came, he marched to the statehouse in Austin with a team of volunteers. They needed fifty-four thousand signatures to make the ballot. They handed in 225,000. By May, he was stealing the show in the final primaries, embarrassing both Clinton and Bush with every write-in vote and every exit poll triumph.

Now, as Clinton clinched the Democratic nod, came the clearest sign yet of just how real and just how big this had all become. To captain his ship, Perot was enlisting two brand names. Ed Rollins had run Ronald Reagan's reelection campaign in 1984 and, until recently, the Republican National Committee's political operation. But he'd turned on the Bush White House after the tax hike and now he was teaming up with Perot. So was Hamilton Jordan, who'd transformed Jimmy Carter from a peanut farmer turned governor into president of the United States, then served as White House chief of staff. Together, this bipartisan power duo would now seek to channel one of the largest spontaneous groundswells in American history into something the political world still couldn't believe it was even contemplating: an independent presidency.

A few days later, Gallup released its first poll after the conclusion of the primary season. Perot was sitting in first place, with 39 percent; Bush was in second, with 31 percent; and Clinton was far back in third, with just 25. It was the second straight Gallup poll to show Perot leading. Noted the *New York Times:* "No previous independent or third party candidate has ever placed second, let alone first, in nearly six decades of Gallup's nationwide polling for president."

N ow everyone in politics was scrambling to crack the code. What, exactly, was driving all of this? The better question, though, might have been: What *wasn't* driving it? The ingredients for Perot's rise could be found all over the political map.

There were the Jack Gargan types, blindly furious with the system, yearning to throw every last bum out. The economy certainly had something to do with their disgust. If jobs had been plentiful, wages had been increasing, and growth had been surging, they might not be so preoccupied with Washington's failings. But everywhere you looked, it seemed, there were signs of American decline. The real estate market had gone bust. Hundreds of savings and loan institutions were failing. A booming Japan was expanding its footprint in the American economy. Against this backdrop, the news out of Washington could be enraging.

Take the congressional pay raise, a 23 percent hike jammed through the Senate the previous summer in a surprise late-night vote, a surprise jointly sprung by both party leaders. "A SWIFT, STEALTHY COUP," the *Washington Post*'s headline called it. There was the savings and loan crisis, too, in which shoddy regulation, risky junk bond investments, and the collapse of the real estate market in the late 1980s led to the failure of more than fifteen hundred thrifts—or about half the total of all S&Ls in America. This, in turn, bankrupted the federal agency that insured S&Ls, and when Washington stepped in with a $124 billion bailout, there were shrieks from Americans of all political stripes. Soon thereafter came the Keating Five scandal, resulting in formal rebukes of five senators who'd been showered with gifts and cash by Charles Keating, the head of one of the largest S&Ls to go bust. Now, the new uproar was over the bank that members of the House used, which allowed them to write rubber checks with impunity.

For Ross Perot, it was all rocket fuel. What was tantalizing to so many voters wasn't the possibility that he'd move the country to the left or to the right, or in any particular ideological direction at all. It was something simpler, and bigger: that he could move the country at all. They looked at Perot, a son of Texarkana who'd started as a computer salesman and made himself into one of the richest men in the country, and saw in his story the antithesis of Washington. Here was a results man, someone to get things done. He could talk, too, with a country store folksiness that belied his billionaire riches. To members of Congress, Perot offered this advice: "Go to the airport, fly commercial, get in line, lose your baggage, eat a bad meal, face reality."

"An interview with Perot is a wild ride," wrote John Mintz in the *Washington Post*. "Before the first question, he is into a full gallop, denouncing, in his East Texas twang, 'the handlers' who want to run his campaign, the pampered government elite that has lost touch with America, and 'the hitmen from Washington' hired to do him in. He anticipates most questions long before they're asked, tells jokes about himself, laughs loud and long, and whispers intimacies about his life. He jumps from his desk to examine his Americana bric-a-brac, Norman Rockwell paintings and Frederic Remington bronzes of cowboys.

"He said he bases his campaign on his love of country—gained, he said, in a loving family in Texarkana, Tex., during the Depression. His rhetoric comes mostly from his Naval Academy training as an engineer, and his experience in the computer industry. At times, he divides issues into parts and then makes them go away, chop-chop."

The Perot platform tapped the disgruntlement in both parties. He denounced global trade deals and cheap Japanese imports—the same America-first protectionism Pat Buchanan was selling in the Republican primaries. But he aligned himself just as passionately with the signature issue of Paul Tsongas. "Can we agree," Perot asked, "that going four trillion dollars into debt did not create utopia?" There was plenty of overlap, too, between Perot's excoriations of D.C. and Jerry Brown's call for a political revolution against the "bought and paid-for" Washington establishment.

Together, Tsongas and Brown had won nearly eight million votes in the Democratic primaries. Tsongas said he'd support Clinton in the general election, but added: "People come up to me at airports and say they are sorry I got out and they are intrigued by Perot. They are projecting my views onto him."

Simply put, Perot was hard to pigeonhole. He'd been a Republican, yes, but he also supported abortion rights, called for stricter gun control and more education funding, and was open to raising taxes.

Whatever your politics, if you were at least intrigued by Ross Perot, then he had something to sell you.

In time, Republicans would come to think of Perot's candidacy as a kamikaze mission aimed at Bush. In terms of motive, at least, there might have been something to it. Perot and Bush were hardly strangers to each other. Decades before, they'd both made names in Texas, then gone on to national fame and, eventually, a collision. It came in 1987 over claims that some American prisoners of war had been left behind in Southeast Asia after the Vietnam War. There was no evidence, but plenty of suspicion, and Perot was among those convinced it was true. He wanted the Reagan White House to put him in charge of a search effort, and Bush, the ever-loyal vice president, was deputized to tell him no, an answer Perot wasn't used to hearing and didn't appreciate at all. Now, a few years later, here he was savaging Bush's presidential leadership and soaring past him in the polls. As revenge plots go, it wasn't bad.

And yet, as the spring of 1992 progressed, it was Clinton, far more than Bush, who was paying the price for Perot's rise. Perot was cleaning up with independent voters while also grabbing 35 percent from Democrats and 28 percent from Republicans. What did they all have in common? They were anxious about the economy; two-thirds of them said it was getting worse. And they were fed up with the incumbent president; fully 96 percent of them said that Bush's policies were either harming the country or just not helping it. Here was the heart of the Perot coalition: the great, vast middle of the electorate. They'd been with Bush in his landslide win in 1988, but now they were ready to throw him out.

In a normal election, they'd be low-hanging fruit for the opposition-party candidate. But here was the catch: they didn't like Bill Clinton—at all. He'd dodged enough land mines to secure the Democratic nomination, but only with his image in ruins. Now, asked to choose between a failed president and Slick Willie, they were going instead with Option C.

If anything, in fact, Perot was keeping Bush in the game. The president's approval rating sat in the thirties. Unemployment was climbing each month; it would peak at nearly 8 percent over the summer, the highest level since the early 1980s. After a dozen years of GOP presidents, the country was souring on Republican rule. A poll found Americans more sentimental for Jimmy Carter than for Ronald Reagan. For Bush, winning an outright majority in this climate would be challenging. But with Perot scrambling the math, he wouldn't need an outright majority. "Perot," the *Washington Post* reported, "is now hurting Clinton much more than Bush."

It was so bad for Clinton that the top Democrat in the country's biggest state, longtime California state assembly speaker Willie Brown, mused about his party dumping its presumptive nominee in favor of Perot. "I am pursuing a winner," he said, "and right now I don't see Clinton in that role."

That wasn't going to happen, though. Clinton had the delegates, and if he'd shown anything so far it was an almost superhuman willingness to plow forward in the face of devastating political news. There was frustration from the candidate, who now lashed out at the media for focusing on "diversions" like his scandals.

There was also new opportunity. The uprising Perot unleashed through one Larry King appearance had his opponents rethinking old assumptions about how to reach voters. Television had long been the dominant medium, but its filter was stodgy—buttoned-down staples like NBC's *Meet the Press,* CBS's *Face the Nation,* and ABC's *This Week.* Throw in an occasional interview on one of the Big Three broadcast network newscasts or—a more recent innovation—an appearance on the erudite Ted Koppel's *Nightline* and that was about the limit of where you might find a presidential candidate on your television dial. Serious newsmen (and the rare newswoman) asking serious questions of candidates who would answer in serious tones: this was the television campaign the country had come to know.

He was born before color television even existed, but Clinton was made for how personal and confessional the medium was becoming. In the primary campaign, Clinton had excelled in town hall settings.

Perot's breakthrough hinted at how transferrable those skills might be on the proliferating circuit of talk shows. Clinton joined in, making his own appearances with King and sitting for that tense talk with Phil Donahue before the New York primary.

He went further than Perot when he accepted an invitation from MTV, the ten-year-old cable network known mainly for playing music videos, to field questions from a live audience of college-age voters. For ninety minutes, Clinton held forth on a range of topics that never would have come up on the Sunday talk shows, like whether, if he had to do it over again, he would have inhaled. "Sure, if I could," he replied. "I tried before." These were the sorts of queries that the older generation in politics—Bush's generation—regarded as unworthy of a presidential candidate, but Clinton handled them without a hint of self-consciousness. On Arsenio Hall's syndicated late-night show, a ratings hit with younger viewers and African Americans, Clinton showed up with shades and a saxophone, regaling the audience with his rendition of "Heartbreak Hotel."

The old guard of the media world winced at the rise of the talk-show campaign, warning that it was letting candidates off the hook for vital policy questions. "People covering the White House on a day-to-day basis are familiar with every comma in a speech or legislation," said Bernard Kalb, a veteran network correspondent. "Talk-show people are rediscovering the world."

Bush tried to hold out, but his refusal became a sensitive subject. At sixty-eight, he was one of the oldest presidents in history, and now he risked appearing out of step with—or, worse, hostile to— the changing times. Finally, he relented. When spokesman Marlin Fitzwater announced that the president would consider talk-show appearances, he noted that Arsenio Hall's would not be among them, suggesting the show's content was too lowbrow. Hall, who'd been critical of Bush's response to the recent riots in Los Angeles, responded on-air: "Excuse me, George Herbert, irregular-heart-beating, read-my-lying-lipping, slipping-in-the-polls, do-nothing, deficit-raising, making-less-than-Millie-the-White-House-dog-last-year, Quayle-loving, sushi-puking Bush! I don't remember inviting your ass on my show. My ratings are higher than yours!"

For months, much of Clinton's image had been defined through tabloid headlines. Now, the talk shows were humanizing him in a way that emphasized his best attributes, just as they were making Bush look like a fuddy-duddy.

But Clinton was still running in third place, weighted down not just by the scandals but also by the peculiar path he'd taken through the primaries. He'd been forced left by necessity, rallying the old Democratic coalition of liberals, blacks, and labor. This had never been the plan, not for a candidate who'd entered the race as "a different kind of Democrat." Now, Republicans were raring to depict him as the next Mondale or Dukakis, another product of the broken "Democrat machine." Clinton was desperate to prove he was something else, but how? When he looked at his calendar for June 13, he found the inspiration he'd been looking for.

TEN

Jesse Jackson also had June 13 circled. Bill Clinton would be appearing before his Rainbow Coalition, and Jackson had a list of demands—and, he was quite certain, the juice to make Clinton accept them.

It was the same spot Mondale and Dukakis had been in coming out of their primaries. It didn't matter that Jackson hadn't run this time around; by now his status as the chief tribune of black America was assumed. And Clinton, already running behind both Bush and Perot, could hardly afford to alienate one of his party's biggest and most reliable blocs of voters. Jackson was playing this just like he had in the past, accusing the soon-to-be-nominee of ignoring his issues, demanding concessions and promises, threatening to withhold his endorsement. If anything, though, he was doing it with extra relish this time.

And why wouldn't he? There was a history between Jackson and Clinton, one laced with slights and skirmishes. Some of it was political. Clinton was the candidate of the DLC, an outfit Jackson despised. Its very existence, he believed, was a response to his own rise, a way

for a certain type of Democrat to communicate distance from him
to a certain type of voter. He'd accused DLC leaders of "traveling
around the country as a group of all whites" and attacked them as
frontmen for corporate interests. "DLC," Jackson liked to say, stood
for "Democrats for the Leisure Class." When Jackson was invited to
address the 1990 DLC convention, he delivered what members took
as a taunting, sarcastic speech. The next year, he wasn't invited back,
with the group's spokeswoman claiming that Jackson "is representative of an old style of politics that has not done much to help our party
win the White House." Jackson called it a "personal, provocative and
unprovoked attack."

Two months later, Jackson took his revenge. The Rainbow Coalition, which he'd created after his 1984 campaign, was holding its
own conference, a must-attend event for any Democrat with White
House aspirations. There was no request for the presence of the ambitious governor of Arkansas, though. "He's not a contender," Jackson said.

In 1992, Jesse Jackson found himself in a difficult position. This
time around, he wasn't a candidate, choosing instead to accept an offer from CNN to host a talk show. And when Wilder, the only black
candidate in the race, flamed out before the first votes were cast, it left
an all-white field to chase the black vote. Jackson endorsed no one,
then watched Clinton gobble up four out of every five black votes,
powering him to a decisive series of primary routs that locked up
the nomination. No one, Jackson very much included, believed this
signified any sort of bond between Clinton and black voters; it was
merely a reflection of how inadequate the other choices—Tsongas and
Brown, basically—were.

There was even another flare-up between Clinton and Jackson.
It came after Jackson issued a blanket offer to campaign with any
of the candidates. When Tom Harkin took him up on it, a reporter
mistakenly informed Clinton that Jackson was actually endorsing
Harkin and asked for a comment. Clinton, hearing this for the first
time, mouthed a formulaic response, but when he thought the camera

was turned off exploded in fury. "It's an outrage!" he said. "A dirty, double-crossing, backstabbing thing to do. For him to do this—for me to hear this on a television program is an absolute dishonor."

Clinton later phoned Jackson with an apology, which Jackson accepted, but only after noting that he was "disturbed by the tone of the blast at my integrity, my character." This was not exactly a healthy political relationship.

Now, here was Clinton, stuck in third place, wheezing his way into his June 13 appointment with the Rainbow Coalition. Jackson had him just where he wanted him, and he wasn't afraid to say it. "Those campaigns that had been programmed for the suburbs, they've begun tilting to the urban vote, the labor vote," he told the press.

Jackson was framing the conference as a response to the recent Los Angeles riots, the most deadly domestic turmoil the country had seen in a generation. The impetus had been the decision of an all-white jury in suburban Simi Valley to acquit four white LAPD officers of brutality charges after they were videotaped beating a black man after a high-speed car chase. The unrest started in South Central on April 29, spread across the city and Los Angeles County, and finally ended six days later when the National Guard was called in. Buildings were destroyed by the dozens, shops were looted, business owners were harassed, abused, even killed. Arson was rampant. Long-standing ethnic grudges were acted out in violent assaults. Fifty-five people were killed, more than two thousand were injured, and more than ten thousand were arrested.

Jackson's plan was to use the conference as a call to action, and a chance to extract a major commitment from Clinton. He would be unveiling a plan to "rebuild America" and asking Clinton to adopt it as his own campaign's platform. Some of the details of Jackson's program were new, but his overall prescription was the same as ever: a hefty infusion of federal money into education, job training, infrastructure, public works projects, and safety net programs. It was, in other words, a very liberal platform, and now he would use Clinton's weak political standing as leverage to force him to sign on.

It was early in the afternoon of Saturday, June 13, when Jackson and Clinton walked onto the stage together at the Washington Sheraton, more than an hour later than planned. Neither man was ever known for punctuality. A choir was leading chants of one of Jackson's signature calls: "Keep hope alive! Keep hope alive!" They took seats next to each other, Jackson immediately to the left of the podium and Clinton to the left of him, and chatted while lunch was served. Then Jackson rose to give his introductory remarks.

He spoke first of racial reconciliation. The Rainbow Coalition would that night be honoring the white man who had videotaped Rodney King's beating and made it public, along with four black Los Angeles residents who during the riots rescued a white truck driver who'd been pulled from his vehicle and set upon by a group of young black men. Good Samaritans reaching across the racial divide to help their fellow man, the essence of the Rainbow Coalition's mission, Jackson said. With pride, he brought up a panel discussion he'd organized the night before—"a generation of youth sat around a common table trying to take the pain of this generation and turn it into power."

Then he got down to business. After telling Clinton he respected him for surviving the primary process and for winning, Jackson argued for the political wisdom of embracing the Rainbow Coalition's agenda for the fall. "We are keenly aware," Jackson said, "that this year the arithmetic of American politics have changed substantially—that in a two-person race, one hundred people voting, fifty-one is the threshold. In a three-person race, thirty-four, thirty-five is the threshold."

Implicit in Jackson's logic was an assumption that Clinton was a weak candidate, incapable of making inroads beyond the Democratic base but also extremely fortunate to be running in a three-man race, where a unified and energized base might be just enough to win anyway. Hence the urgency of Jackson's platform, with its appeal to "urban America and workers and women and youth."

Now it was Clinton's turn. Jackson was handing him a blueprint and telling him to run with it. Would he play along? The assumption was yes. Mondale and Dukakis had been terrified of alienating Jackson, and Clinton was hardly in better political shape than they'd been.

He started on safe ground, condemning Bush's response to the riots and a recent speech from Vice President Dan Quayle that linked Democrats to a "cultural elite" that looked down on traditional values. "I'm tired of people with trust funds telling people on food stamps how to live," Clinton said.

The room applauded. So did Jackson. Clinton continued. Even amid the riots, he said, there was proof traditional values were alive and well. "Let's not forget, folks. Most people who live in that city did not burn or loot or riot. Most little children were home with their parents. Even the poorest children were sitting in their houses when they could have been looting goods, because their parents told them it was wrong to steal from neighbors."

He acknowledged the platform Jackson was presenting to him and said he shared the goal of rebuilding the country, but went no further. Instead, he would put forward his own thoughts, Clinton told the room, "and you can measure for yourself how much it does or does not square with the program you have outlined."

For the next twenty minutes, he held forth. He spoke of the lack of bank lending in inner cities and promised to connect aspiring business owners with the capital they badly needed to get started. He vowed to "invest in America" and bemoaned that the deficit had ballooned under Bush and Ronald Reagan even as public investment shrank. He talked about gun control and education, even statehood for the District of Columbia, an issue of importance to Jackson. He repeated the centrist-sounding themes of his campaign ("We should reward those who play by the rules and do the reverse for those who don't") and even issued a call to overhaul the welfare system and pair benefits with a work requirement—a major point of separation with Jackson that was greeted with near silence as Clinton delivered it.

These were not controversial sentiments in this room. There was plenty of applause and no taunting or booing. If he'd ended it there, it would have been deemed a fairly bland attempt at balancing deference toward Jackson with a modicum of independence. The kind of speech Mondale or Dukakis would have delivered.

But that's not where he ended it. Before this appearance, the Clinton team had taken note of an article in the *Washington Post* that had caused a small ripple in May but otherwise passed without notice. It concerned one of the young people who'd taken part in the panel discussion the night before—the one Jackson had spoken so fondly of during his own remarks. Her name was Lisa Williamson, but she went by Sister Souljah, the persona she'd adopted when she merged her work as a political activist with a new role as a hip-hop artist. Along with some other rappers, she expressed sympathy during the riots for the black Angelenos who turned to violence, their "revenge," she argued, against a system that oppressed them. The *Post* story had been conceived as a straightforward profile of a new addition to the national conversation, but, David Mills wrote in his article, "during an interview in Washington last week, Souljah's empathy for the rioters reached a chilling extreme."

"I mean, if black people kill black people every day, why not have a week and kill white people?" she asked. "You understand what I'm saying? In other words, white people, this government and that mayor were well aware of the fact that black people were dying every day in Los Angeles under gang violence. So if you're a gang member and you would normally be killing somebody, why not kill a white person? Do you think that somebody thinks that white people are better, or above dying, when they would kill their own kind?"

With no social media to throw her words into every American's face, the blowback to the article was brief and contained, and when Souljah took her seat on the youth panel at Jackson's Rainbow Coalition, there was no commotion or controversy. Few even knew what she'd told the *Post*. But the Clinton campaign did, and now, nearly a half hour into his speech, the candidate was about to make Souljah infamous.

"Finally," Clinton told Jackson's group, "let's stand up for what's always been best about the Rainbow Coalition, which is people coming together across racial lines."

He brought up the youth panel. "You had a rap singer here last night named Sister Souljah. I defend her right to express herself through music, but her comments before and after Los Angeles were filled with the kind of hatred that you do not honor today and tonight."

The room was hushed. What was he doing? Jackson stared straight ahead as Clinton continued.

"Just listen to this, what she said. She told the *Washington Post* about a month ago, and I quote, 'If black people kill black people every day, why not have a week and kill white people? So if you're a gang member and you would normally be killing somebody, why not kill a white person?'

"Last year," Clinton continued, "she said, 'You can't call me or any black person anywhere in the world a racist. We don't have the power to do to white people what white people have done to us. And even if we did, we don't have that low-down dirty nature. If there are any good white people, I haven't met them. Where are they?'"

He looked up at the audience. "Right here in this room," Clinton said, "that's where they are."

"I know she is a young person, but she has a big influence on a lot of people. And when people say that—if you took the words white and black and reversed them, you might think David Duke was giving that speech."

More silence. Still no reaction from Jackson, no reaction from anyone. Clinton wasn't directly attacking his host, but by implication he absolutely was. But even if anyone was miffed, what could they do? The words Clinton was citing weren't exactly easy to defend. The crowd wasn't about to applaud him, but they couldn't really jeer, either.

Perhaps anticipating criticism, Clinton now offered his own act of racial contrition. "Let me tell you. We all make mistakes and sometimes we're not as sensitive as we ought to be, and we have an obligation—all of us to call attention to prejudice wherever we see it.

"A few months ago, I made a mistake. I joined a friend of mine and I played golf at a country club that didn't have any African American members. I was criticized for doing it, and you know what? I was rightly criticized for doing it. I made a mistake and I said I would never do that again.

"And I think all of us have got to be sensitive to that. We can't get anywhere in this country pointing the finger at one another across racial lines. If we do that, we're dead and they will beat us."

In the back of the room, reporters weren't sure what to make of what they were witnessing. It was hard to tell if Jackson was transfixed with interest or stewing with rage. The same went for the audience. One thing was obvious: neither Mondale nor Dukakis had ever given a speech like this. Clinton had more to say, and now challenged the electoral strategy Jackson was promoting.

"Even in Reverend Jackson's new math of this election, it's hard to get to a 34 percent solution or a 40 percent solution if the American people can be divided by race.

"I have seen the hatred and division of the South that Jesse Jackson and I grew up in, and I was just lucky enough to be raised until I was four by a grandfather with a grade-school education who believed we were all created equal—who showed me by the life he lived how to treat people without regard to race, and told me that discrimination and segregation were morally wrong.

"I was lucky. I learned more from my granddaddy with his grade-school education about that and how to live than I did from all the professors I had at Georgetown and Oxford and Yale. In the wisdom of a simple workingman's heart, I learned something that many of our young people today who are role models no longer believe."

Clinton had started with Souljah's words, but now he was making a much more expansive statement about America's journey from slavery and segregation to the present day.

"We don't have a lot of time," he said. "And so I say to you: Let's all one more time reaffirm the kind of life I'm trying to have my family live. My little daughter is a seventh-grade student in a public school in Little Rock, Arkansas, where she is in the minority. But she's getting a good education, in life and in books. She's learning about the real problems of real people, but she's able to do what Martin Luther King said children ought to do—judge people by the content of their character.

"So that is what I close with. We have to think big and be big and do big things in this election. We've got to give this election back to the people of the United States and reinvest in our country again. But we also have to have the courage to make government work and to challenge people to come together."

Clinton finished his speech to a warm reception and Jackson stood to shake his hand, then returned to the podium. He seemed uncertain. There'd been a rebuke from Clinton, he knew that. But it had been packaged in a fulsome testament to healing and reconciliation. In this moment, Jackson chose to err on the side of generosity. He thanked Clinton for coming and noted that there'd be a reception that night featuring Gladys Knight and the Pips. Clinton, he joked, would be welcome to attend and play the saxophone for a song if he wanted. Then, joining hands with Clinton, he delivered the closing prayer. The two men left the stage together.

When Jackson was asked immediately afterward about the Souljah comments, he replied, "I don't know what his intention was." Among the reporters in the room, though, there was no murkiness. The Clinton team had been debating how to handle Jackson for months. They recognized the potential downside of angering him, but had calculated that the payoff from a public display of independence would be worth the gamble. This, they now made clear to the reporters in tow, was what they had just seen.

CLINTON STUNS RAINBOW COALITION was the *Post*'s Sunday headline. "The challenge to Jackson was deliberate," wrote Thomas Edsall. "Clinton's frank remarks seemed designed to demonstrate his willingness to challenge core Democratic constituent groups and to begin to break his image in the public as a 'political' person who would bend to pressure from major forces within the party."

It took time for the reaction to take shape—and for Jackson's irritation to build. It was only after seeing Clinton off and making his initial comments to the press that Jackson learned of how the Clinton team was spinning this to the press. He'd been used, he realized, shown up on his own turf by a crafty politician he'd crossed swords with before. He came out to talk to reporters again, ready to let Clinton have it.

Jackson demanded an apology from Clinton to Sister Souljah for his "unfair attack on her, her character, her reputation." He claimed she'd been misquoted, even though the *Post* had an audiotape of her comments. Mentioning Souljah at all, Jackson said, was "very bad judgment" and "a diversion" aimed at "an audience that was not here."

"The rippling effect through the conference has been traumatizing," he said.

Now it was the top national story, vastly bigger than Souljah's original comments had been. The presumptive Democratic presidential nominee was instigating a feud with one of his party's most powerful leaders. Clinton held his ground.

"This was the best audience for that message," he said as he left the Sheraton. "If not here, where would have been better?"

"I did not attack her personally," Clinton noted. " I pointed out that she has a lot of influence on young people and that I understand how alienated and divided people are. But it is simply wrong to suggest that there are no good white people or that under any circumstances that people of one race should kill others just because they are a different race. It's just not right."

Souljah entered the fray. "Mr. Clinton took my comments completely out of context," she said. "In the quote he referred to I was speaking in the mind-set of a gang member."

She accused Clinton of "using me as a political football, the Democratic version of Willie Horton." That was a reference to a 1988 ad from Republicans that, critics charged, sought to scare white voters away from Dukakis with the menacing image of a black criminal.

Clinton, according to Souljah, "was trying to scare white people to mobilize them into action." She called him "a hypocritical, draft-dodging, pot-smoking womanizer."

The back-and-forth continued for days. Clinton claimed that Souljah and other critics were "essentially taking the position, I guess, that because I'm white I shouldn't have said it. And I just disagree with that."

Jackson invoked Clinton's chief liability. Executing a nakedly political stunt revealed "a character flaw," he said. He called Clinton's speech "a very well-planned sneak attack, without the courage to confront but with the calculation to embarrass," designed "purely to appeal to conservative whites by containing Jackson and isolating Jackson."

Clinton reminded reporters of Jackson's immediate reaction to the speech. There'd been no outrage then, only an invitation "to come back that night and play the saxophone. He went back and had a very cordial meeting with me. So all these discoveries of things that are after the speech are for his own purposes."

Jackson simply hadn't seen it coming. He'd been sure that his endorsement was too valuable, and that Clinton was too feeble a candidate, for Clinton to try anything like this. Now, he declared, it was a real possibility he'd endorse Perot instead. "This is a very strange political year," Jackson said.

Jackson continued to fume, but the backlash from black leaders and black voters never did materialize. Polls showed no erosion in Clinton's black support following the Souljah uproar. Just as he had in '84 and '88, Jackson would spend the weeks leading up to the 1992 Democratic convention in New York withholding his endorsement and generating headlines with his demands. But Clinton would pay him only minimal attention. Jackson had been a featured attraction at the last two conventions, but now he was barely a sideshow.

Veterans of the Mondale and Dukakis campaigns marveled that Clinton had found a way to accomplish what they'd been scared to even try. "I watched two good men, Walter Mondale and Michael Dukakis—men who really care about equality and justice—grapple with Jesse Jackson and struggle with a relationship to make it work,"

Dukakis's old political lieutenant, John Sasso, said. "They were un-successful. This is a moment of historical importance. You have got to treat him like any other politician and tell him when you agree and when you disagree. Bill Clinton has handled this better than we have in the past."

Jackson's endorsement finally came the weekend before the con-vention. It was grudging. Clinton, he said, had "insulted and infuri-ated a lot of people." But he had no choice. The army he'd assumed would march into war—any war—with him had already defected to Clinton. For coming on board, Jackson was given a speaking slot, safely out of prime time and under strict conditions.

"The political reality," concluded the *New York Times*, "is that Jesse Jackson, for now, is not the commanding force he once was, and he is political pro enough to know it."

After months of languishing in Perot's shadow, the Jackson war returned Clinton to the front pages, this time on his own terms—not as some beleaguered candidate trying to fend off another scandal but as the "different kind of Democrat" he'd always wanted to run as. Along with his blitz of the talk-show circuit, the course correction was erasing the Slick Willie caricature and injecting some-thing brand-new into his campaign: momentum.

What's more, the headlines were turning on Perot. Zooming up in the polls, he'd been treated as a supernatural political force, cele-brated for his populist flair, business prowess, and system-shattering potential. Now the scrutiny was coming, and he wasn't handling it well.

Perot was long on colorful denunciations of Washington dysfunc-tion and corruption, but short on specifics, and the press's patience was waning. On ABC's prime-time newsmagazine *20/20*, Barbara Walters pressed him on the issue of gay rights. "What people do in their private life is their business," Perot said. But what about the military's ban on openly gay service members, Walters asked—would he repeal that? "I don't think that that's realistic," he replied. Walters asked if he'd consider an openly gay person for his cabinet and the answer came back no: "I don't want anyone there that will be a point of controversy with the American people."

For the first time, Perot was thrust into a media firestorm. Public opinion on gays in the military was in 1992 evenly divided, so Perot was hardly out of the mainstream on that front. But his cabinet answer rubbed people as a lot closer to discrimination than common sense; a poll showed a majority of voters thought Perot was wrong. Reporters followed up with more questions and Perot struggled to avoid another trip wire. At one point, he said that, as far as he knew, he'd never even met a gay person. Finally, he reversed himself, meeting with gay leaders and pronouncing himself open to gay cabinet members and service members. That abated the controversy and aligned him more closely with public opinion. It also looked just like the kind of calculated political maneuvering that he claimed disgusted him.

In their interview, Walters also brought up reports that he had fired employees of his company found to be cheating on their spouses. "I put a very strong store on strong moral values," he said. Did that mean that anyone who'd committed adultery would be unwelcome in a Perot cabinet? "The American people deserve better than that, and no, I wouldn't have people like that," he said.

That set off a whole different uproar, which intensified with reports that Perot was fond of using private investigators to scare up dirt on business and personal associates. Bush, in particular, had apparently been in his sights since their 1986 encounter, when the Reagan administration had deputized Bush to tell Perot he wouldn't get to lead a hunt for Vietnam POWs. In one instance, the *Washington Post* now reported, Perot paid a Washington law firm to compile two thick binders about a Texas business deal involving an old Bush business partner, touting it as a "mini–Teapot Dome scandal" and passing the results to Bob Woodward, the *Post* reporter famous for breaking the Watergate story open. Another report, from *The New Republic*, claimed Perot commissioned investigations of his own children's social habits, and there were claims he monitored his employees, too.

Perot wasn't used to anything like this, and it was showing. Each new accusation prompted a new round of questions, which he couldn't avoid. Did he have people snooping around on workers to find out if they were cheating on their wives? Did he really believe, as the *Post* reported, that the CIA was lying about POWs to cover up its own drug-running activities? "Everybody is writing every Froot Loop story in the world, without regard to the facts," Perot fumed. "What would you expect? Ninety-nine percent of these stories are just elves across the ceiling."

But the stories weren't going away. Neither the Bush nor Clinton campaigns had known how to handle Perot; now they pounced. Bush said it wasn't American, what Perot was accused of, and Quayle, his vice president, said: "Imagine having the IRS, the FBI, and the CIA under his control. Who would be investigated next?"

"Hitler's propaganda chief would be proud," Perot responded. He accused the Bush team of launching an orchestrated attack against him, which was true enough. Perot seemed surprised to discover that this was how politics worked. For the first time, his negative poll numbers were rising. Cracked one analyst: "There haven't been that many thin-skinned, short guys, running on anti-adultery platforms, elected president in the last 200 years."

On a Monday night at the end of June, ABC devoted 150 minutes of programming to Perot. First came a one-hour prime-time special, *Who Is Ross Perot?* Hosted by Peter Jennings, the top-rated network anchor, it was split into four segments, each putting a supposed Perot selling point under the microscope. The first was about an airport Perot had built in Texas, which he'd been touting as proof of his executive skills. But what it really was, Jennings told viewers, was a testament to Perot's relationships with powerful politicians— that "in Fort Worth and in Washington, Mr. Perot was considered a very special interest."

And what about the story that had made Perot famous in the first place, the hostage rescue mission he'd engineered in Iran? The ABC special acknowledged it, but then told a different hostage story, one few Americans knew of, in which Perot also went around the government, but this time with disastrous results. There was also a look at the "sophisticated" political operation he'd built in Texas, and his tumultuous tenure as General Motors' chief stockholder, which ended with GM's executives deciding he was a bloviating showboater and ridding themselves of him with an obscene buyout.

The Jennings special was followed by a ninety-minute town hall featuring Perot himself. A live studio audience was on hand to question the candidate, along with voters in nine other cities linked in by satellite.

"It is certainly consistent with the age that town meetings are now held on television," Jennings said as he opened the broadcast. Perot was upset after watching the earlier special, and for the first ten minutes, he and Jennings sparred over the reporting. When it was their turn, the audience members weren't any easier on him.

"Mr. Perot has had a great history of investigating people, and I want to know that if he's president of the United States, will he take the cameras to our bedrooms?" asked one. "Will he take the cameras to our church basements, where sanctuary happens? And I want to know that."

Perot said he wouldn't dig into anyone's private life, at which point another audience member interrupted. "Unless they're gay or lesbian, Mr. Perot," he said. "What about gays and lesbians, huh? We pay—we're taxpayers, too, in this country, you know. And you're a bigot for what you said so far about gays and lesbians, and America better watch out, because this week it's gays and lesbians you attack, next week you go after the next group that's suffering. You're a bully, Mr. Perot, nothing but a pint-sized bully."

Almost one in three televisions sets in use was tuned in for the Perot town hall. This wasn't how it had been for Perot in those months immediately after his Larry King interview, but it was his new reality, and he didn't like it one bit. There were reports of internal discord, too. Ed Rollins and Hamilton Jordan, the big-name hired guns whose additions had supposedly shown how serious Perot was, grew frustrated as Perot diminished their authority and ignored their counsel. They wanted Perot to stop flirting with the race and formally enter it—to tell the country not just that he might run for president, but that he was ready to serve. They wanted an organization in place, television ads, a campaign schedule, and they were pleading with Perot to get moving immediately. He wouldn't hear of any of it.

The roles were reversing. Now it was Perot on the defensive, and Bill Clinton who was resurgent. At the start of July, Clinton's favorable rating finally moved ahead of his unfavorable score. Democrats who'd been playing footsy with Perot were coming home. Perot's numbers were moving in the other direction; he was still above water, but suddenly he wasn't any more popular than Clinton.

The Democratic convention was to start on July 13 in Madison Square Garden, and Clinton needed to pick a running mate. Conventional wisdom said he'd look north, for geographic balance, and to his left, to appease the party's liberals. The last winning Democratic ticket, back in 1976, had featured this same recipe, with Jimmy Carter and Walter Mondale teaming up—"Grits and Fritz."

But that was the old playbook. On the Thursday before the convention, Clinton appeared at the Arkansas statehouse with his selection, a senator from a neighboring state who was a year younger than him. "Throughout American history, each generation has passed on leadership to the next," Al Gore of Tennessee said. "That time has come again, the time for a new generation of leadership for the United States of America."

Gore and Clinton had been rivals, two southern moderates with designs on the White House. In 1988, Gore, then thirty-nine years old, had run for president only after Clinton removed himself at the last minute. In the '92 race, it was Clinton who'd only entered after Gore said no. Now they'd be partners, the first all-boomer ticket, challenging an incumbent who'd served in World War II. Explained *USA Today*: "By choosing someone who comes from the same ideological pod—they helped found the moderate Democratic Leadership Council—Clinton again showed he believes Democrats can win only by renouncing some liberal beliefs."

The Gore pick earned a rebuke from Jesse Jackson ("Now we start from a narrow base"), but the very next day Clinton and Gore were given a hero's welcome at the NAACP's convention in Nashville. "I don't want any more us versus them!" Clinton told the crowd, which gave him several standing ovations.

Perot showed up at the NAACP convention, too, but to a more restrained welcome. He spoke of equality and opportunity, but this was not a man accustomed to addressing black audiences, and soon it showed. "Financially, it's going to be a long, hot summer," Perot said. "Now, I don't have to tell you who gets hurt first when this sort of thing happens, do I? You people do. Your people do. I know that. You know that."

There were groans from the crowd. "Your people?" someone shouted. Onstage, one of Perot's top black supporters put his head in his hands. A few minutes later, Perot did it again: "Now good, decent people all over this country, and particularly your folks, have got bars on the windows and bars on the doors and they're sitting up at night with a shotgun across their knees." He was losing the crowd, and with it his chance to shatter the Democrats' grip on the black vote.

"A person of his stature in America shouldn't be saying 'your people' at a NAACP convention," one of the delegates said after. "That's insulting. That's insulting to everybody, and no one can say it's not."

"It never occurred to me they'd be offended," Perot explained.

"In all the years I have worked with the black community, I have never run into a sensitivity problem," he said.

By the time the Democratic convention started, Clinton was tied for first with Bush, with Perot falling behind them both. The party on display in the Garden was the picture of unity, a made-for-television parade of big-name Democrats reciting Clinton's script. Liberals had once dreamed of Mario Cuomo accepting the nomination in his hometown, then taking the battle to Bush. Instead, it was Cuomo who delivered the nominating speech for Bill Clinton, singing the praises of the DLC man he'd excoriated only a month before. The delegates, now smelling a winner, roared with delight. A poll showed Clinton opening up a twelve-point lead over Bush, 42 to 30 percent, with Perot fading to just 20 percent.

Clinton's acceptance speech was set for the next night. But hours before he was to deliver it came breaking news from Dallas. "I believe it would be disruptive for us to continue," Perot announced.

Technically, he wasn't dropping out, since he'd never officially entered. He said he'd concluded there was no plausible path to the presidency for him: "If we cannot win in November, the election will be decided in the House of Representatives. Since the House of Representatives is made up of Democrats and Republicans, our group would be unlikely to win." And anyway, Perot added, "I don't have any drive to be president of the United States."

The decision wasn't a complete shock, given how dramatically the race had turned, but the timing sure seemed like a boost to Clinton— and a parting swipe by Perot at his old nemesis the president. He said he wasn't endorsing anyone, then added that "the Democratic Party has revitalized itself. They've done a brilliant job, in my opinion, in coming back."

Bush, on a fishing trip in remote Wyoming, hustled back to civilization to make a statement before cameras. It took hours for even a sound bite of his reaction to reach any viewers. Clinton, meanwhile, strode onstage in New York shortly after 10 P.M., beaming as thousands of frenzied Democrats waved pennants with his name on them and tens of millions of Americans watched in their living rooms.

"In the name of the hardworking Americans who make up our forgotten middle class, I proudly accept your nomination for President of the United States," Clinton said.

The first poll after the convention put Clinton ahead of Bush by twenty-seven points. Even the lead Dukakis had famously blown four years earlier was nothing compared to this. Clinton was now more popular than Bush, his party was unified, and unemployment was still climbing. The Republican lock on the White House was getting awfully loose. It was Bush, now, who was scrambling to shore up his base, and it led the president to make a decision that would reverberate for years to come.

ELEVEN

The president came back to his adopted hometown, Houston, for his party's convention. He was raring to punch back at Clinton and promising "the most stirring political comeback since Harry Truman gave 'em hell in 1948." This was the talk of every underdog, but the hole Bush was facing was especially deep and stubborn. Clinton's convention had wrapped a month earlier, but his bounce endured. As the Houston festivities came to order, the Democratic challenger's lead was still around twenty points.

He was addressing a divided party. The polls showed soft Republican support for Bush. Morale was low. Pat Buchanan, despite his quick flameout after the New Hampshire and Georgia primaries, had come to symbolize all the intraparty disgruntlement with the president, and the Bush team concluded they had no choice: he would get the marquee speaking slot he'd been demanding for months.

Not long after the networks picked up their live coverage at 10 P.M. on Monday, August 17, Buchanan was called to the Astrodome's stage. His delegates waved BUCHANAN BRIGADE signs and he smiled with delight.

"Listen, my friends," Buchanan began. "We may have taken the long way home, but we finally got here to Houston. The first thing I want to do tonight is to congratulate President George Bush and to remove any doubt about where we stand. The primaries are over, the heart is strong again, and the Buchanan brigades are enlisted all the way to a great Republican comeback victory in November!"

The delegates whooped and hollered. On television, First Lady Barbara Bush was shown rising to her feet and applauding. She turned to the woman next to her and said a few words. It was Buchanan's wife, Shelley. They smiled and nodded their heads as the ovation continued around them. Chants of "Four more years!" could be heard. It was the unifying scene the Bush campaign had been banking on.

Buchanan turned to the Democrats. The convention that nominated Clinton had been a "masquerade ball," he said, where "twenty thousand liberals and radicals came dressed up as moderates and centrists in the greatest single exhibition of cross-dressing in American political history." More cheers. The camera cut away to Michael Kinsley, his *Crossfire* sparring partner, watching from the CNN set in one of the skyboxes. He was grinning. The Buchanan style was hardly new to him.

But how many people watching at home were experiencing it for the first time? To the viewers of pundit shows, Buchanan was a brand name, but the audience for this speech was far larger than he'd ever commanded. It included not just political junkies but casual consumers of news. This was their introduction to who Pat Buchanan was, and the oratory they would absorb for the next half hour would be as masterful in its delivery as it was polarizing in its content.

The theme he chose involved culture. This was part necessity; Buchanan remained a critic of the Gulf War, so he couldn't celebrate Bush's signature first-term achievement, nor was he about to praise the president's commitment to free trade or his decision to raise taxes. Culture offered common ground. That decadent liberalism was undermining traditional values was a long-standing Buchanan conviction. But months earlier, the Bush administration had trodden on the same turf, when Vice President Quayle delivered a speech on the importance of fathers and two-parent families that included a shot at the CBS sitcom *Murphy Brown*. On the show, the title character, a network newswoman, found herself impregnated through a one-night stand—"mocking the importance of fathers," Quayle bemoaned, "by bearing a child alone and calling it just another lifestyle choice." It set off a frenzied national debate, pitting feminists and liberals who felt insulted against religious conservatives and traditionalists who rallied to Quayle's defense. At first ambivalent, the Bush team had ultimately decided to embrace the controversy and to emphasize "family values" at the convention.

If that phrase sounded fuzzy, Buchanan now moved to define it as the fault line separating the two parties. Clinton's Democrats, he said, believed in "unrestricted abortion on demand"—so much that they'd refused a speaking slot at their convention to one of their party's few prominent anti-abortion voices, Pennsylvania governor Robert Casey. "Yet," he continued, "a militant leader of the homosexual rights movement could rise at that same convention and say, 'Bill Clinton and Al Gore represent the most pro-lesbian and pro-gay ticket in history.' And so they do!"

Buchanan singled out Hillary Clinton, the Democratic nominee's "lawyer spouse," for her "radical feminism" and warned of the agenda "Clinton and Clinton" would impose: "A litmus test for the Supreme Court. Homosexual rights. Discrimination against religious schools. Women in combat units."

"That's change, all right," Buchanan said. "But that's not the kind of change America needs, that's not the kind of change America wants, and that's not the kind of change we can abide in a nation that still calls itself God's country."

It segued to the passage that would define this speech for years to come.

"There is a religious war going on in our country for the soul of America," Buchanan said. "It is a cultural war, as critical to the kind of nation we will one day be as was the Cold War itself. And in that struggle for the soul of America, Clinton and Clinton are on the other side, and George Bush is on our side!"

The Republican Party had been aligned with religious conservatism since Reagan's ascendance, but this was something different. The Gipper had given it a soft, gentle touch, and Bush tried to do the same. Buchanan was drawing a line and telling Americans to pick a side. On CBS, Dan Rather called it a "raw meat" offering to the base. R. W. Apple in the *New York Times* wrote that Buchanan "epitomized the scowling face of conservatism." Democrats condemned him. But the discomfort included Republicans, too. "Pat's message," Senator Richard Lugar of Indiana said, "is not a very appealing one for most Americans, and it's not a winning message."

Three nights later, it was the president's turn. He was running from behind and straight into a mighty economic headwind. On ABC's newscast a few hours before the speech, reporter Brit Hume described the challenge Bush faced: "He must combine a defense of the way he's done things for the past four years with a promise, in effect, not to do them that way anymore."

Maybe it would be impossible, but Bush had to try. He needed to refashion his biggest vulnerability, the tax issue, back into the weapon it was supposed to be against a Democrat. His campaign researchers had done all the digging they could into Clinton's Arkansas record, which dated back to the late seventies. They counted up every instance they could find of any kind of tax or fee or surcharge going up, anything that could, however technically, qualify as an instance of the state government seeking more revenue. The research was handed in and the decision was made. Bush would dust off the old Dirty Harry act from 1988 and try to recapture the magic.

The moment came halfway through the speech, after he'd reminded Americans of the military triumph he'd overseen and of the collapse of Soviet Communism that had played out on his watch. He started by painting Clinton as another tax-and-spend liberal, a career politician whose "passion to expand government knows no bounds." Clinton, Bush noted, was promising to hike taxes only for the rich—

"but he defines rich as anyone who has a job." These were rhetorical darts that Republicans had been aiming at Democrats since the Reagan revolution, and if they weren't that innovative anymore, history suggested they were at least effective. But Bush couldn't keep lobbing them without confronting his history. He was accusing Clinton of duplicity even though he was at least as vulnerable to the same charge. He and his team had labored over how to do it, and it was time to put forward the product of their work.

"Now let me say this," Bush said as the delegates quieted. "When it comes to taxes, I've learned the hard way. There's an old saying, 'Good judgment comes from experience, and experience comes from bad judgment.'

"Two years ago, I made a bad call on the Democrats' tax increase. I underestimated Congress's addiction to taxes. With my back against the wall, I agreed to a hard bargain: one tax increase one time in return for the toughest spending limits ever.

"Well, it was a mistake to go along with the Democratic tax increase, and I admit it. But here's the question for the American people. Who do you trust in this election? The candidate who's raised taxes one time and regrets it? Or the other candidate who raised taxes and fees 128 times and enjoyed it every time?"

It brought the Republican crowd to its feet in the Astrodome, although what else were they going to do? Bush was going for the bravado that had played so well four years earlier, but in the service of explaining away a broken promise. Tactically, he was proving the point Gingrich had long argued. Compromise erodes the power of contrast. Clinton didn't even need to hear the speech to have a response ready: "We've heard it all before. We've seen it all before. We've tried it all before. It's read-my-lips all over again, except this time we can read the record."

When the convention was over, Bush did get a polling bounce, but not nearly as much as he wanted. Clinton still led the race, and one reason, it seemed, was Buchanan's speech. "Among the Republican target vote of moderate swing voters—pro-choice women, suburban moderates—Buchanan was death," one Democratic pollster said. "All he did was preach to the converted and scare the rest of America. He made Democrats realize why they're Democrats." The Bush campaign sensed the same thing, junking plans for a campaign swing featuring the president and his former foe. Confronted a few weeks later by a gay voter who objected to what he'd heard at the convention, Quayle sought distance from Buchanan: "I don't think you heard any of that rhetoric coming from me. You didn't hear it coming from the president."

Bush came to the home stretch of his reelection campaign with a message that lacked coherence. As a candidate in 1988, he'd described elaborate scenarios where Democrats would try to force him into a tax hike, and how he'd refuse them. Then, as president, he'd signed off on one anyway, insisting it was just the kind of difficult but necessary choice a leader is supposed to make. But now here he was saying to forget that—that he never should have raised taxes after all, that it was a mistake. He was flip-flopping on the flip-flop.

His planned character assassination of Clinton wasn't sticking either. But Clinton was asking voters to consider a different definition of character, one rooted not in private conduct but in whether you could trust his public commitments, and Bush was playing straight into it. He'd made a public commitment not to raise taxes and then broken it, and now was telling voters to trust him over his opponent.

Through September, Clinton maintained a double-digit lead, but there was still tumult to come. From the moment of his abrupt withdrawal, there'd been chatter that Perot might yet return, fed by his insistence that his name remain on the ballot. His image was now in tatters, with voters judging him thin-skinned and erratic and even many of his supporters believing they'd been duped. Polls that had once shown an overwhelming appetite for his presence in the race now indicated broad hostility to the idea of his reemergence. The populist groundswell of the spring was gone, and yet the chatter only intensified, propelled this time by Perot himself, who was refusing to rule out a reentry. He called the volunteers who'd led his ballot drive to Dallas for a meeting; it would be their call, Perot said, whether he jumped back into the race—never mind that he hadn't consulted them before exiting it back in July.

Perot invited the Bush and Clinton campaigns to send ambassadors to Dallas. Ostensibly, it would be their chance to persuade the volunteers that they were addressing Perot's issues and that there was no need for him in the race. They sensed a setup but complied anyway, more fearful of provoking Perot's wrath in the event he did stay out. On September 28, with every top national media outlet on hand to chronicle it all, a bipartisan who's who of politicians showed up to make their presentations—"a remarkable political pageant," the *Los Angeles Times* called it. When the pitches were done, Perot headed over to a makeshift television set. Larry King was in Dallas for the occasion, his one-hour show expanded to ninety minutes in anticipation of another momentous announcement. The room was packed with Perot's volunteers. King surveyed them. Were they impressed by the Democrats and the case they'd made? Silence. The Republicans? Crickets. What about Perot—did they want him running again? On cue, they erupted. Grinning at the spectacle he'd created, Perot played it coy. He wasn't ready to commit to running, he told King, but he did have news: He had set up a new 1-800 number to solicit opinions from everyday Americans on what he should do. If you wanted Perot to run, you could dial in and register your encouragement. If you didn't want him to run—well, apparently there was no number for you to call. No matter. By the next day, Perot's team was reporting that 1.5 million people had called in.

It was all he needed to claim another grassroots uprising was afoot, and who was he to say no to something like that? In Dallas on October 1, Perot declared that he was a candidate for president. "I thought that both political parties would address the problems that face the nation," he said. "We gave them a chance. They didn't do it."

The election was barely a month away and Perot had almost no campaign infrastructure. He'd driven away the professionals back in the spring and much of his volunteer army had long ago deserted (or rather, concluded that he'd deserted them). A snap poll showed him barely impacting the race. Clinton still led Bush handily, 52 to 35 percent, with Perot at just 7 percent. "There's no great clarion call from the electorate for Ross Perot to reenter this race," said Bill Schneider, CNN's chief political analyst. "My feeling is that the voices he's hearing are coming from inside his head." What Perot did have was money, an unlimited stockpile from which he was ready to spend. He promised an unconventional campaign that would rely heavily on media but avoid the thirty-second ads Americans had grown accustomed to. And, he said, he expected to be in the debates.

This was the immediate question posed by Perot's reemergence. Since 1960, when John Kennedy and Richard Nixon introduced the televised debate to American presidential politics, there had never been a three-person general election showdown. When George Wallace ran as an independent in 1968, there hadn't been any debates at all; and when Ronald Reagan agreed to one that included independent John Anderson in 1980, Jimmy Carter stayed away in protest. (There was ultimately one Reagan-Carter debate, a week before the election.) Now the decision was up to the Bush and Clinton campaigns, whose negotiations had been stalled since the summer. There was no consensus on how Perot's presence would affect the race; the small support he had seemed to be coming from both candidates about evenly. Just as with his stunt in Dallas, though, it was his volatility that unnerved both parties. What would happen if he was locked out and then used his fortune to blame one of them? An agreement was reached. There would be three debates, and Ross Perot would be in all of them.

The first one was in St. Louis on October 10, a Sunday night. Bush was still far behind. It had now been three months since he'd led in any poll. Here was his chance, maybe his final one, to alter the dynamics of a race he now seemed likely to lose. "For George Bush," Tom Brokaw said as NBC's broadcast opened, "perhaps the most crucial ninety minutes of his presidency."

Once upon a time, character was going to be Bill Clinton's undoing. Back in the primaries, Bob Kerrey had predicted that the Republicans would open him up "like a soft peanut" in the fall, and Bush and his team were now trying to do just that. There was the issue of the Vietnam draft, but there was also more. Clinton, it turned out, had also participated in protests against the war while a student at Oxford—on foreign soil, in other words. He had also traveled to the Soviet Union during that same time, a trip on which Republicans were now casting dark aspersions. But there was no evidence yet that any of it was turning voters against Clinton, who was fighting back by accusing Bush and the GOP of McCarthyism, a callback to Senator Joe McCarthy and his discredited red-baiting campaign of the 1950s. Shortly after the debate began, moderator Jim Lehrer asked Bush if he was questioning his opponent's patriotism.

"I was nineteen or twenty flying off an aircraft carrier, and that shaped me to be commander in chief of the armed forces," Bush said. "And I'm sorry, but demonstrating—it's not a question of patriotism. It's a question of character and judgment. They get on me—Bill's gotten on me about 'read my lips.' When I make a mistake, I'll admit it. But he has made—not admitted a mistake, and I just find it impossible to understand how an American can demonstrate against his own country in a foreign land—organizing demonstrations against it when young men are held prisoner in Hanoi or kids out of the ghetto were drafted. Some say, 'Well, you're a little old-fashioned.' Maybe I am. But I just don't think that's right."

If his delivery was disjointed, Bush's point was clear. This was supposed to be one of his few clear advantages, the contrast between his own sense of duty when his country called and Clinton's. But Clinton knew this moment was coming, and had prepared for it. This couldn't be about litigating service records; he'd lose on that score every time. Instead, he ceded the point: Bush was a World War II hero and Clinton saluted him for it.

"But," he said, "when Joe McCarthy went around this country attacking people's patriotism, he was wrong. He was wrong. And a senator from Connecticut stood up to him named Prescott Bush." If there'd been time, the audience might have gasped. With the president just feet away from him, Clinton was taking this in a personal direction no one had foreseen.

"Your father was right to stand up to Joe McCarthy," he said. "You were wrong to attack my patriotism. I was opposed to the war but I loved my country, and we need a president who will bring this country together, not divide it. We've had enough division. I want to lead a unified country."

Historically, what Clinton was saying was accurate. Prescott Bush, the president's father, was a liberal Republican senator from Connecticut who had publicly rebuked McCarthy forty years earlier. But how many casual viewers knew this story? It was the air of principle and defiance that gave power to Clinton's rebuttal, which would be played on television over and over afterward.

The rest of the signature moments in St. Louis didn't come from Clinton or Bush. It was the third candidate on stage, Perot, who repeatedly broke through with a quick wit and arsenal of aphorisms. He set the tone almost immediately, when the subject of experience came up. Bush and Clinton had both been in government for decades, while Perot had never even sat on a school board. "Well, they've got a point," Perot said. "I don't have any experience in running up a four-trillion-dollar debt." The live audience was split into thirds, a batch of tickets for each campaign, but the entire place burst into laughter. He continued.

"I don't have any experience in gridlock government where nobody takes responsibility for anything and everybody blames everybody else. I don't have any experience in creating the worst public school system in the industrialized world, the most violent, crime-ridden society in the industrialized world. But I do have a lot of experience in getting things done. So, if we're at a point in history where we want to stop talking about it and do it, I've got a lot of experience in figuring out how to solve problems, making the solutions work, and then moving on to the next one. I've got a lot of experience in not taking ten years to solve a ten-minute problem."

This was the Ross Perot who'd captured the imaginations of millions back in the spring, Mr. Fix-It with a Texas twang. He'd caught on once as the antidote to everything Americans hated about politics, and he was determined to do it again. He was selling himself as a political product, but also selling the idea of hard choices. The deficit was his main issue, and like Tsongas during the primaries, he framed it as a both an existential threat and an abdication of generational responsibility. Unlike Tsongas, he didn't sound like a wonk, and merged it with his other dominant theme: that America's governing class was awash in staggering incompetence. "We are sitting on a ticking time bomb, folks," Perot said, "because we have totally mismanaged our country, and we had better get it back under control."

Bush and Clinton criticized him only gently, and the format allowed the candidates to speak in generalities without much risk of follow-up from the moderator. Besides the strict time limits, it was a setting ideally suited for a Perot sales pitch. His demeanor was confident, even peppy; his vocabulary was accessible; and his timing was impeccable. Discussing his call for a gas tax hike to pay down the deficit, he nodded to one of his exaggerated physical features: "If there's a fairer way, I'm all ears." They played that line a lot after the debate, too.

Strictly speaking, Perot was the night's big winner. His support doubled, at least, from single digits to the mid-teens, and he repaired much of the damage he'd inflicted on himself over the summer. In one poll, nearly 70 percent said they walked away with a more favorable view of him. Still, no one was talking about Perot winning. Even with his improved standing, he was now considered a spoiler, a stigma he would rail against all the way through Election Day. But this was the price he paid for vanishing for those ten weeks; even his own voters no longer believed he was electable. And that made Clinton the real winner. He came into the night comfortably ahead of Bush and left with the same advantage. Time was running out for the incumbent, and just when he needed them most, all his old tricks seemed to be failing him.

The second debate brought a brand-new format—a town hall, for the first time ever in a general election—but more of the same. In the defining portion, a member of the audience asked the candidates how the national debt had personally affected their lives. It was exactly the kind of question that separated this format from the traditional model, where journalists would try to trip candidates up with pointed policy choices. Instead, here was a demand to personalize federal budget policy. This was a talk-show question, and by his own admission Bush was not a talk-show candidate. "Well, I think the national debt affects everybody," he began.

"You personally," the audience member interrupted.

"Obviously, it has a lot to do with interest rates . . ."

The moderator, Carole Simpson of ABC News, stepped in: "She's saying you personally. Has it affected you personally?"

Bush was clearly flummoxed. "I'm sure it has," he offered. "I love my grandchildren. I want to think that they're going to be able to afford an education. I think that that's an important part of being a parent." He was struggling to figure out what on earth this voter was trying to get at. "If the question—maybe I got it wrong. Are you suggesting that if somebody has means that the national debt doesn't affect them?"

He was confused because he was treating the question literally. Clinton, the master of this new format, knew better. When it was his turn, he gently threw the question right back at the voter. "Tell me again how it's affected you." When she didn't provide an instant reply, he suggested one. "You know a lot of people who've lost their jobs and lost their homes, right?" She said she did, and just like that Clinton had widened the scope of the inquiry to include the entire state of the economy. Now he was telling her about life back in Arkansas, where "when people lose their jobs, there's a good chance I'll know them by their names. When a factory closes, I know the people who ran it. When the businesses go bankrupt, I know them."

"And I've been out here," he said, "for thirteen months meeting in meetings just like this ever since October, with people like you all over America, people that have lost their jobs, lost their livelihood, lost their health insurance." He was on to his usual campaign spiel now, but in his gentle southern drawl, his eyes locked in on this voter, it could all seem spontaneous and heartfelt. There was a reason his campaign had insisted on this format, and why Bush's had resisted it.

By the night of the third and final debate, the president's prospects were dire. The election was just two weeks away and Clinton's lead wasn't budging. Bush would have to land a meaningful blow, but how? Clinton had proven an unusually deft and nimble communicator, all but impossible for anyone—opponents, reporters, voters—to corner. All Bush had left was one desperate shot at reclaiming the issue he'd rode to the presidency four years earlier.

The setting was East Lansing, Michigan, and this time it was just Jim Lehrer and a panel of reporters quizzing the candidates. No more curveballs from undecided voters. Bush used the very first question to go after Clinton and his promise of a middle-class tax cut. "Mr. and Mrs. America, when you hear him say we're going to tax only the rich, watch your wallet," Bush said, "because his figures don't add up and he's going to sock it right to the middle-class taxpayer and lower." Clinton was vulnerable here; Tsongas had shown as much during the primaries when he belittled his "pander bear" opponent for promising more than he could afford.

But Bush had his own history, and almost instantly it overtook the conversation. Clinton brought up the 1990 deal Bush had signed. "He is the person who raised taxes on the middle class after saying he wouldn't." And it was true. Taxes on gasoline, cigarettes, and alcohol had all gone up, affecting millions of middle-income Americans. Bush was offered a chance to respond, and the basic incoherence of his story once again became apparent. It had, he insisted, really been "a Democratic tax increase."

"I didn't want to do it and I went along with it," Bush said. "And I said I made a mistake. If I make a mistake, I admit it. That's quite different than some. But I think that's the American way." One of the journalists on the panel, Gene Gibbons of Reuters, followed up: If it was a mistake to raise taxes, what should you have done instead?

"What should I have done? I should have held out for a better deal that would have protected the taxpayer and not ended up doing what we had to do, or what I thought at the time would help," Bush said.

"So I made a mistake," he continued. "And I—you know, the difference, I think, is that I knew at the time I was going to take a lot of political flak. I knew we'd have somebody out there yelling, 'Read my lips!' And I did it because I thought it was right. And I made a mistake. That's quite different than taking a position where you know it's best for you. That wasn't best for me and I knew it in the very beginning. I thought it would be better for the country than it was. So there we are."

Clinton's rebuttal was much easier to process. "The mistake that was made," he said, "was making the 'Read my lips' promise in the first place just to get elected, knowing what the size of the deficit was." It was all he needed to say.

The gap narrowed some toward the end, but Bush never caught Clinton in any polls, not a single one. The close of October brought, at long last, a burst of positive economic news, with the nation's GDP growing at nearly twice the expected rate for the third quarter. "This is very encouraging for America," Bush declared. "If you think I'm happy, you're right." It was the light at the end of the tunnel he'd been predicting—and predicting and predicting—for more than a year. But it was too tentative, too abstract, and just too late to alter the country's psychology. After twelve years of Republican rule, Americans seemed eager for a fresh start, urged on by a Democratic candidate who was telling them that anything was possible "if we have the courage to change."

Bush kept on predicting an upset. What else could he do? But when he made one last campaign swing in Pennsylvania, a state he'd carried easily the last time out, the stage was emptier than expected. Arlen Specter, the two-term Republican senator locked in his own

reelection race, had decided it was better to stay away from his party's
president and fend for himself. There were no such embarrassments
for Clinton, whose party could taste it now. With a victory at the top
of the ticket, they'd control all of Washington. Finally, the Permanent Democratic Congress would have a president to sign its bills
into law.

Perot remained the wild card. He'd followed through on his vow
of an unconventional campaign, declining all television interviews
and instead buying up thirty-minute chunks of broadcast network
airtime. Instead of their favorite sitcoms, viewers were now tuning
in to find Ross Perot seated behind a desk, ticking through a series
of handheld charts documenting the country's economic decline.
Stranger still, a lot of them didn't change the channel; the audience
could exceed ten million for a Perot infomercial. In one month, he'd
revived his image, but there were limits. His temperament remained
a question, even to his fans, who still wondered what drove him from
the race back in July. He only added to the apprehension when, late in
the campaign, he claimed the real reason he'd withdrawn was his discovery of a scheme by Republican operatives to sabotage his daughter's wedding. He provided no evidence, couldn't name any of the
operatives, struggled to describe how the plot might have worked,
and attributed the information only to a "top Republican." The White
House called him "loony."

Try as he might, Perot just couldn't reach the same heights he'd
attained in the spring. He told Americans that if they all just voted
their hearts, he'd win in a landslide, but he couldn't break past 20 percent and into contention. The real choice, most voters had concluded,
was between Clinton and Bush.

And that choice wasn't a hard one for them to make. One of
the earliest poll closings was in Kentucky, where Bush had crushed
Dukakis in 1988. When the returns there put Clinton ahead, it was
clear that the coalition that had carried Nixon, Reagan, and even Bush
was crumbling. Clinton swept the Northeast, the Rust Belt, and the
Pacific Coast, reversing what in many states had been double-digit

Bush victories four years earlier. And he made incursions in what had become the most Republican regions of the country, carrying the Old Confederacy states of Georgia, Tennessee, Louisiana, and Arkansas and even a handful of rural western states where Bush had obliterated Dukakis. Here was the national victory that before this campaign—and even during it—Democrats had come to believe one of theirs couldn't achieve anymore. Clinton carried thirty-two states, along with the District of Columbia, good for 370 electoral votes. The previous three Democratic nominees combined had won just seventeen states (and D.C.) and 188 electoral votes. Clinton's tally was the highest for any Democrat since LBJ's landslide in 1964.

His coalition was broad. The base was there for him—massive margins with nonwhite voters, solid support from women and voters on the low end of the income scale. The bigger story was the voters he wasn't supposed to win, at least when the race started. Republicans had been running up the score with men, winning them by an average of nineteen points over the last three elections. Now, Clinton carried them by three points. He won independents, carried every age group, and for all the scrutiny of his own military record, he even finished first with veterans. He won almost half of the counties in America, doubling what Dukakis had managed, and tripling what Mondale and Carter had done. And he did all of this with turnout rising significantly from four years earlier.

America was turning out a sixty-eight-year-old incumbent who'd served in World War II for a man who would now become the first commander-in-chief born after that war. It was, Clinton told his supporters in Little Rock, "a clarion call for our country to face the challenges of the end of the Cold War and the beginning of the next century."

In Dallas, Perot threw what felt like a victory party. With 19 percent, he'd won a larger share of the vote than any independent since Teddy Roosevelt. It was a far cry from where he'd stood earlier in the year, but also far better than anyone thought he'd do when he reentered the race. His support spanned the spectrum, from Buchanan's

America First crowd to the commuter-class deficit hawks who'd flocked to Tsongas. Exit polls found Perot drawing evenly from Clinton and Bush. At his celebration, Perot sang "America, the Beautiful" with the crowd, danced with his wife, and talked of turning his campaign into a permanent political force. His supporters wondered what might have been if only he hadn't lost his cool over the summer. Others marveled at the populist energy he'd conjured—and shuddered at how it might shape future uprisings. "The next time the man on the white horse comes, he may not be so benign," warned Jim Squires, who'd been Perot's spokesperson early in the year. "He could be a real racial hater or a divider of people."

Bush's 38 percent was the worst for an incumbent president in eighty years. He'd been prepared for defeat, and his concession was graceful. "We have fought a good fight and kept the faith, and I believe I have upheld the honor of the presidency of the United States," he said. His loss meant that power within the Republican Party would now shift away from the White House and back to Capitol Hill, where it would be left to congressional Republicans to deal with the new Democratic administration. Nominally, this meant Bob Dole and Bob Michel, the Senate and House party leaders, would be in charge. In reality, something quite different was brewing.

TWELVE

Bill Clinton was forty-six years old when he was sworn in, the youngest president since John F. Kennedy. It was a comparison he enjoyed. Over the summer, his campaign had unearthed an old newsreel from 1963. The silent footage showed delegates to the American Legion's Boys Nation convention visiting the White House, with the teenage Bill Clinton shaking hands with his idol, the president. The Clinton team turned it into one of the emotional high points of a well-received biographical documentary, *The Man from Hope*, that aired at the convention and in fall television spots.

Kennedy, the first president born in the twentieth century, had used his inaugural to declare a changing of the guard, serving notice that "the torch has been passed to a new generation of Americans." Kennedy and the six presidents who followed him were all part of the World War II generation, but with Bush's defeat, that chain was now broken, possibly for good. January 20, 1993, would mark a new milestone: the installation of America's first baby boomer president. Addressing his countrymen for the first time as their leader, Clinton reached for the forward-looking spirit of JFK: "You, my fellow

Americans, have forced the spring. Now we must do the work the season demands."

After a campaign dominated by themes of national decline, the possibilities of a new presidency were awakening a hesitant sense of optimism among the public. Sixty percent of Americans now said they viewed Clinton favorably and 80 percent said they liked how he'd handled the transition. The country was pulling for its new president. As his motorcade made its way down Pennsylvania Avenue after the inauguration, there were shouts for Clinton to get out and walk. It was a populist gesture that Bush and Reagan had refused; Clinton ordered his driver to stop and obliged the people.

But did Clinton's election signal something more? The possibility teased the Democratic imagination. Key to Clinton's message had been a populist indictment of the Reagan-Bush years, when "the rich got richer while the forgotten middle class—the people who work hard and play by the rules—took it on the chin." For all his centrist packaging, Clinton had called for a more active role for government and for the wealthy to pay more in taxes. Reagan's famous inaugural decree that "government is not the solution to our problem. Government is the problem" had set the tone for American politics for the past dozen years, tormenting the party of FDR. Now Clinton had won back many of the Reagan Democrats, and his party dared to wonder: Had he just assembled the governing coalition they'd been dreaming of since 1980?

The new president would have plenty of reinforcement on Capitol Hill. The Permanent Democratic Congress had, yet again, survived easily and would now live to see its fortieth year—and, surely, many more beyond that. For the last generation, Democrats had been playing almost nonstop defense in Washington, their congressional power checked by one Republican president after another. Now they would control everything. "An aggressive Democrat in the White House working with a Democratic Congress could break the political gridlock that has frozen the federal government for much of the past four years," the *Chicago Tribune* wrote.

A few weeks after the election, Clinton and his new partner, Al Gore, made the rounds on Capitol Hill. They paid a courtesy call on the Republican leadership and said the right things about working together and bridging the partisan divide for the good of the country. But everyone knew the numbers. The GOP would have well under two hundred members in the House and barely more than forty in the Senate. Stick together and Clinton and the Democrats would be able to do pretty much whatever they wanted, without begging for a single Republican vote.

Richard Gephardt, the second-ranking Democrat in the House, walked out of his meeting with the incoming president and declared, "Both new and returning members understand what an important responsibility we have been given." Dan Rostenkowski, the crusty Chicago Democrat who ruled the Ways and Means Committee, had been slow to warm to Clinton as a candidate. But now he was sensing the possibilities. "Bill Clinton is now the eight-hundred-pound gorilla," Rosty told the press. "He is American politics and he's going to have to call the tune. I'll tell him to be bold."

"If you listen carefully," Gore proclaimed, "you can hear today the first sounds of gridlock loosening."

There was a lot more you could hear, at least if you wanted to. Clinton's share of the popular vote was just 43 percent. Not even close to half the country had voted for him. You could chalk that up to the unique three-way nature of the 1992 race, but then again, hadn't Clinton's deficiencies been one of the sources of Perot's strength? Even on Election Day, voters still expressed real doubts about Clinton's integrity and his character. To read the overall result as a loud cry for change, and a repudiation of Bush, was obvious enough. But was it actually an endorsement of Clinton and Clintonism? "The simple reality is that a 43 percent plurality constitutes a fragile base for governing," David Broder wrote in the *Washington Post*. "It falls far short of the public support Clinton will need to achieve the ambitious policy changes he outlined in the campaign."

And what, exactly, was Clintonism anyway? Viewed from one

angle, he'd made a clear break with liberal orthodoxy—support for the death penalty, a work requirement for welfare recipients, and a tax cut for the middle class. He'd separated himself from Jesse Jackson, too, although not in a way that forced him to renounce any of Jackson's policy agenda. Then again, it was also true that he'd won the Democratic nomination by running to the left, not the center, and by rallying the traditional party base, not expanding it. Now, it would be his party's entrenched congressional chieftains pushing him—and expecting him—to govern more as the politician he'd been in the primaries, not the general election. The cautionary tale of the last Democratic president would loom large. When Jimmy Carter, another self-styled southern moderate, tried to bend the gavel wielders of the Permanent Democratic Congress to his will, they resisted, and his party dissolved into dysfunction. From two leaders of the DLC, the same group the president-elect had used as his national springboard, came a warning: "Clinton did not have the opportunity to encounter and overcome the most forceful advocates of the traditional Democratic Party. The absence of a clear-cut, climactic battle for the nomination leaves Bill Clinton with a coalition that may be less committed to his [agenda] than he is himself."

The Republicans would serve as the loyal opposition, but when they looked at Clinton's victory, they weren't nearly as cowed. There was his unimposing plurality, along with hints that beneath the burst of goodwill that accompanied Clinton's triumph remained nagging misgivings about his honesty, his judgment, his character. The day after the election, Bob Dole declared himself "a watchdog for the 57 percent" who hadn't voted for Clinton. "If Bill Clinton has a mandate," Dole said, "then so do I." It became a favorite Republican refrain, the ever-present reminder that Clinton hadn't even come close to winning a majority, implying that his presidency was accidental. And it meshed with another fashionable reading of the election on the right: that Bush lost by going wobbly on conservatism and getting cozy with Democrats.

This part was key, because it positioned Gingrich as a vindicated man inside his party. As Bush packed his bags, Republicans found themselves thinking back to that budget war of 1990. We need to team up with the Democrats to raise taxes, their president had told them, so that we can reduce the deficit and charge up the economy in time for '92. Michel and Dole and the rest of the top Republicans on Capitol Hill had gone along, but not Newt. And now look what had happened. "Bush's conservatism didn't cost him the election," the head of the conservative Heritage Foundation concluded, "it was his lack of conservatism. The voodoo Bush economics of tax-spend-and-regulate crippled the economy; the bad economy cost Bush the election." It was what Gingrich had been saying since his first days in the House: contrast wins elections, not compromise.

Officially, Bob Michel remained the leader of the Republicans in the House, but his days felt more numbered than ever. It was Newt's party now.

Gingrich said nothing at all on Clinton's big day in January. But he'd already spotted an opening that, so far, the rest of Washington had missed. It was just like back in his freshman year in the House, when he shamed his colleagues into holding an expulsion vote on Charlie Diggs. The clubby culture of Capitol Hill, he'd recognized then, contained all the ingredients for a populist backlash. Now, fourteen years later, the new president and the norms of Washington were inviting him to do it all over again, on a much bigger scale.

At issue was Clinton's pick for attorney general. Zoë Baird was supposed to represent all that was fresh and forward thinking about the new administration. She was forty years old, another baby boomer, and an accomplished legal professional, plucked by Clinton from her perch as general counsel for the insurance giant Aetna. Her biography—dual Berkeley degrees, a stint in the Carter White House, an ascent to the top of the corporate legal ladder, marriage to an esteemed law professor—seemed the fulfillment of the modern feminist ideal. Clinton had excited cultural liberals with a vow that his cabinet would "look like America," and now, after an uninterrupted parade of white males, he was proposing to place a woman atop the Justice Department for the first time ever. "I believe that the first female attorney general will be the best attorney general," Clinton said.

Baird wasn't a household name, or a controversial one. Women's groups cheered and the rest of Washington shrugged. Confirmation was sure to be a cakewalk, an assessment that didn't budge when the January 14 issue of the *New York Times* hit the streets. Their scoop: Baird and her husband, Yale professor Paul Gewirtz, had been employing two illegal immigrants from Peru, a husband and a wife, as household help for the last few years. The woman was the nanny for their son; the man acted as their chauffeur. It wasn't something she was trying to hide. Baird had told Clinton's team about the arrangement and the information had been included in her submissions to the Senate Judiciary Committee, the venue for her confirmation hearings.

Technically, this put Baird and her husband in violation of the law, which mandated that all employers verify the eligibility of their employees to live and work in the United States. But as a practical matter, they'd had nothing to worry about. The penalty for knowingly employing an illegal immigrant for domestic work was only a few thousand dollars, and the provision had never actually been enforced in the state of Connecticut, where Baird and Gewirtz lived. They also hadn't paid any Social Security taxes for the couple, something employers were also required to do. It would be pointless to try, their lawyer had told them, since the Peruvian couple wasn't legally permitted to work in the United States. Still, when her nomination was announced, Baird quietly cut a check to the government for the rough amount she and her husband would have been on the hook for had the couple been legal employees.

If it all sounded complicated, even absurd, it's because it was. Immigration laws had been overhauled in the Reagan years, but they remained a clunky and incoherent mess. Enforcement was arbitrary, with the government striking an implicit deal with businesses and well-to-do families: dependable workers were hard to come by and immigration laws were almost impossible to change, it was agreed, so why don't we all just look the other way and go about our business?

Consequently, arrangements like the one between Baird and her husband and their Peruvian employees were understandable, maybe even familiar, to the power couples of Washington. Here were two driven and accomplished professionals striving to balance the grueling time demands of their careers with their desire to raise a child in a stable home in a quality neighborhood. This was the sort of thing that people like them just did.

Which is why the bipartisan reaction to the *New York Times* article was one big yawn. Clinton's press secretary assured the public that he was familiar with Baird's situation and didn't see a problem. The chairman of the Judiciary Committee, Delaware senator Joe Biden, talked it over with the panel's senior Democrats and announced, "This is not a deal-breaker. This is not anything that should keep her from being attorney general." Republicans gave Baird cover, too. Utah's Orrin Hatch, the top Republican on the committee, called it "no big deal."

"No one is above the law," he said, "but people make honest mistakes, and that should not deprive her from serving her country."

"This is not something sinister," agreed Wyoming's Alan Simpson.

Gingrich couldn't believe what he was seeing. Sure, to the citizens of official Washington, everything that Baird and her husband had been doing was perfectly understandable. But what about the rest of America—the men and women who had never confronted a situation like this in their own lives, who didn't have the kinds of jobs and income where they'd ever face a dilemma over whether and how to employ household help of dubious legal status? What about all the men and women—the tens of millions of them, probably the majority of all men and women out there—who didn't know anyone like Zoë Baird and Paul Gewirtz? The House would have no actual role in Baird's confirmation, but Gingrich did have a platform, and now he was determined to use it to show his party exactly what it was supposed to do here.

"She has crossed the line," he said. "You can't have a person who ought to be prosecuted serving in the cabinet."

That was it, a simple statement that cut to the heart of what the Baird story looked like to millions of Americans—but to almost no one, apparently, in Washington. The incoming president's choice to be the top law enforcement officer in the country had knowingly broken the law, and Clinton didn't seem to mind at all. It was a classic Gingrich issue—elites protecting elites, making excuses for behavior that many ordinary people figured they could never get away with.

Until Gingrich spoke up, no Republican was calling the Baird revelations grounds for her rejection, but now the ground shifted in a hurry. The controversy was grist for the exploding genre of conservative talk radio. Hosts railed against Baird, and their listeners flooded the Capitol Hill switchboard with calls opposing her. The story was permeating popular culture, too—newspaper front pages, network morning shows, Jay Leno's monologue. Gingrich's gut was right. Forget all the complexities of immigration law; this story touched a very different nerve.

"She thought she could do something illegal and get away with it," a woman working at a pet store in Maryland told the *New York Times*. "I don't think it's fair. I raised my kids while I was working. I worked days. My husband worked nights at the post office. Our in-laws filled in when they had to. This makes me mad."

Baird came before the Judiciary Committee the day before Clinton's inauguration. She offered an apology, of sorts. She'd been worried about finding quality child care for her son, she explained, and relied on a lawyer who'd downplayed the significance of hiring illegal aliens for household work. "I gave too little emphasis to what was described to me as a technical violation of law," Baird now said. She and her husband also paid a twenty-nine-hundred-dollar fine, levied by the Immigration and Naturalization Services immediately after the *Times* story landed.

It had only been a few days since that initial *New York Times* story, but the same senators who had been so quick to dismiss the revelations now suddenly turned on her. When Baird noted that she'd only hired the Peruvian couple after struggling for two months to find suitable help, Biden interjected. "It is amazing," he said, "the ability of the human condition to rationalize, to justify what you know is not right." Alan Simpson, one of Baird's many Republican defenders just days before, now told her about feedback he was getting from his constituents—"much of it from women who, I think, feel that they, as single parents or working mothers, did not have this advantage. There's something stirring there that is very real." Ohio Democrat Howard Metzenbaum, the committee's most liberal member, reported that his office had received fifty calls against Baird and none for her. Pennsylvania's Arlen Specter said he'd gotten "very few" calls supporting Baird—and eight hundred against her.

The hearings took a break for Clinton's inauguration, then picked up the next day. By now, eleven senators—six Republicans and five Democrats—were formally asking for the Baird nomination to be withdrawn.

At the White House, where the Clinton team was now settling in, the president's spokesman was asked if—knowing everything he now did—Clinton would still choose Baird if he had it to do over again. "I can't answer that hypothetical," George Stephanopoulos replied. Baird took the hint and late that night sent a letter to Clinton asking him to withdraw her nomination. "I am surprised at the extent of the public reaction," she wrote, "but face the reality that this situation affects my ability to achieve the goals we both have for the Department of Justice." At 1:30 A.M. the White House released Clinton's response. He was accepting her request.

The Clinton presidency wasn't forty-eight hours old and Newt Gingrich had already drawn blood. What really rattled official Washington, though, was how blindsided they were by it. They hadn't even flinched at the first reports about Baird and her nanny and driver, but the rest of the country saw something else entirely. "It may be that we live a life so unlike average citizens that we're really not very attuned to what they're thinking about," suggested liberal columnist Al Hunt. "Once the call-in shows started to weigh in, senators found out that people in Kansas and Iowa and Missouri don't hire illegal aliens and don't think it's like a speeding violation."

It was the Gingrich method. He wanted his fellow Republicans to see what he saw. There was a conservative majority in the country. It was patriotic, it was suspicious of government, it revered traditional values. And what these Americans really couldn't stand, he was certain, was the arrogance and condescension of the elites who ruled them. Show these Americans that this new president, for all his emoting and feeling their pain, was just another creature of that elite world, and show them that the opposition party was dedicated to dismantling it, and Republicans would finally find their way out of the wilderness.

He was helped by the pent-up ambitions of the new president's party, which had been waiting a dozen years for the opportunity it now had. One demand that came to the fore quickly was over the issue of "homosexual rights," the phrasing still favored by many media outlets. Never before had a candidate cultivated support from the gay community as publicly as Clinton had, and one of his centerpiece promises was to use the executive powers of the presidency to lift the military's long-standing prohibition on openly gay service members.

The ban, in some form, had been in effect since shortly after World War I, when branches of the armed forces first drafted policies requiring the discharge of anyone committing sodomy. The formal modern policy went into effect in 1949, when the newly created Department of Defense decreed, "Homosexual personnel, irrespective of sex, should not be permitted to serve in any branch of the armed forces in any capacity, and prompt separation of known homosexuals from the armed forces is mandatory."

It was hardly controversial, reflecting a culture in which homosexuality, when it was acknowledged at all, was treated as a morally deviant disorder. That would change with time, but the process was slow and still incomplete. Most discharged service members left quietly, but by the 1970s more were speaking out publicly and fighting back through the legal system. With the courts scrutinizing the ban for the first time, the Pentagon in 1981 prepared a revision, with specific justifications for the ban. Having gay military personnel, it argued, would undermine troop morale and cohesion, hinder recruitment, and create security risks, with gay soldiers susceptible to blackmail. "Homosexuality," the revised policy stated, "is incompatible with military service."

One decade and tens of thousands of additional discharges later, this remained official policy as Clinton set out to run for president. The slow evolution of public attitudes continued apace, with more and more people telling family members and friends—and, in the case of a growing number of celebrities, the whole world—that they were gay. The gay community itself, more visible and organized than ever, was becoming a player in politics, its clout now centralized in the Democratic Party. This was partly a legacy of the Democrats' sharp turn to the left on cultural issues in 1972, but it also spoke to the corresponding rise of evangelical Christian conservatism within the Republican Party and the reluctance of the Reagan administration to address the AIDS crisis in its early days. As a bloc, the gay vote was hard to define, with only around 2 percent of voters identifying as such in exit polling. But as a financial force, Clinton and other Democrats were beginning to realize what an untapped resource the gay community represented.

Clinton leaned on an old friend, David Mixner, whom he'd first met through the antiwar movement during Vietnam, to act as his liaison to the gay community during the campaign. Mixner advised him that ending the military's ban should be at the top of his list of promises, and the candidate complied. He said it privately and publicly, without equivocation or wiggle room. It was a basic matter of equality, Clinton argued, and if elected he would get rid of the ban "with the stroke of a pen."

Gays reciprocated. In May 1992, Clinton traveled to Los Angeles for what became the largest campaign fund-raising event ever organized by the gay community. It brought in a hundred thousand dollars. By the end of the campaign, gay groups had raised over two million dollars for Clinton, back when that was still a lot of money. On Election Day, Clinton took nearly three-quarters of the gay vote. It was a turning point. In the years ahead, the gay community would grow into a far more significant source of campaign cash for the Democratic Party, but Clinton's outreach was what started it. Just four years earlier, the Dukakis campaign had shunned offers of financial help from gay leaders, fearing a broader backlash.

Now the bill was due. Clinton had made a clear promise and gay leaders were intent on holding him to it. Clinton's victory, proclaimed the National Gay and Lesbian Task Force, represented "a rite of passage for the gay community: from the political margin to the center, from social pariah to political partner. We are now part of the governing coalition of this country, and that's unprecedented."

It seemed straightforward enough. The ban was Pentagon policy, not an actual law, meaning the president had the authority to reverse it with an executive order. This was how Harry Truman had integrated the armed forces back in 1948. Little more than a week after the election, at the conclusion of a Veterans Day event in Arkansas, Clinton took questions from the press and was asked if he would follow through on his promise. He said he would. "My position is that we need everybody in America that has got a contribution to make, that's willing to obey the law and work hard and play by the rules."

Clinton was calculating that the politics were now on his side. This wasn't just a matter of appealing to a relatively small constituency; more and more voters were ready to reward politicians who championed equality for gays and to punish those who resisted. Two polls during the campaign showed majority support for ending the military's gay ban, and Gallup in the summer of 1992 found that nearly eight in ten Americans favored equal employment opportunities for gays. It also seemed that Republicans had suffered a backlash by emphasizing "family values" at their national convention, where featured speakers—most famously Pat Buchanan—used prime-time hours to denounce the political agenda of gays, feminists, and others. So there was reason for Clinton to believe he'd be on the right side of public opinion if he went ahead.

Where was that tipping point, though, where people would stop seeing this as a basic matter of fairness and start worrying it was something else? You didn't have to look far for clues. In that same Gallup poll that showed overwhelming opposition to job discrimination, 53 percent also said that homosexuality was not an acceptable lifestyle; 45 percent called it a threat to American families and their values. What about trusting gays around children—should they be permitted to be schoolteachers? An even split. On Election Day in Colorado, a state Clinton carried, voters ignored boycott threats and passed a referendum invalidating all local gay rights ordinances. In Oregon, another Clinton state, a ballot question asked voters to declare homosexuality "abnormal, wrong, unnatural and perverse." It didn't pass, but the "yes" side did get 45 percent. There was broad support for the idea of equality, but this was plainly a matter that stirred complicated, and often contradictory, impulses.

The immediate hurdle Clinton faced was the military itself, a revered institution, especially with the voters he'd just coaxed back into the Democratic fold. Already, there was skepticism, and plenty of hostility, among military brass when it came to Clinton. Their politics tended to be conservative, and most of them had little regard for the modern Democratic Party. To many career officers, especially his generational peers, Clinton's avoidance of service during Vietnam remained an affront. That a man who had paid no apparent price for his scheming would now be their commander in chief awakened some of their most deeply held resentments. They were not eager to make the new president's political life easy.

Plus, the gay ban was popular with military leaders and, to the extent that it could be polled, rank-and-file soldiers, too. It was true that there already were—and always had been—gay men and lesbians serving, many in silence, some with the quiet support of their units and commanding officers. But uniformed voices expressing that view were few and far between, and when Clinton reiterated his intentions, he confronted a parade of highly decorated objections. Admiral Thomas Moorer, a former Joints Chiefs chairman, appeared on ABC's *This Week* and said that the spectacle of "men kissing each other and hugging" would foment chaos and disunity among soldiers. "I defy you to find a military man who has had the responsibility of molding a unit, taking men into combat, who won't agree with me," he argued. A marine general told the *New York Times:* "We were standing in this shower tent, naked, waiting in line for thirty-five minutes for a five-minute shower. Would I be comfortable knowing gays were there standing in line with us? No. It just introduces a tension you don't need."

Clinton was challenging the military's culture and tradition, shaky ground for any politician, let alone one with his personal history. He was also placing himself on a collision course with one of the most admired men in America.

General Colin L. Powell was, to many of his fellow countrymen, an authentic American hero. As the nation's highest-ranking military officer, the chairman of the Joint Chiefs of Staff, he was one of the faces of the Gulf War triumph. As a man, he embodied the kind of story people liked to believe could only play out in their country. Born in Harlem to Jamaican immigrant parents at a time when the military was a strictly segregated institution, Powell joined up with the ROTC while in college, then slowly over three decades climbed his way all the way to the top, becoming the first black Joint Chiefs chairman when Bush appointed him in 1989. He was cerebral but authoritative, and in the wake of the war now rated as one the most admired men in America. He had all the makings of an appealing political candidate, and there'd been chatter in the campaign that Bush might dump his vice president, Dan Quayle, in favor of Powell. Now the talk was in full gear that Powell might run for president in 1996. His political views, and even his party preference, were a complete mystery—but then again, that had not stopped Dwight Eisenhower.

For the moment, though, Powell still had his day job. His term would run into the new president's, so Clinton would have to deal with him. Powell was already on record with his support for the ban, and with Clinton now reiterating his vow to overturn it, Powell dug in for a fight. "The military leaders in the armed forces of the United States—the Joint Chiefs of Staff and the senior commanders—continue to believe strongly that the presence of homosexuals within the armed forces would be prejudicial to good order and discipline," he said. "And we continue to hold that view."

Bill Clinton vs. Colin Powell. The boomer president who'd protested Vietnam and maybe dodged the draft against the decorated general who'd just won a war. Gays in the military had been a minor issue in the campaign, but now everyone was watching this fight—and taking sides.

Liberals and gay groups had Clinton's back, and found some surprise allies here and there, including the now eighty-four-year-old Barry Goldwater, who said he had no problem with gay soldiers as long as "they shoot straight." Christian conservatives, meanwhile, were joining forces with veterans groups. The Southern Baptist Convention, the American Legion, and the Veterans of Foreign Wars pooled their resources to launch a campaign to keep the ban in place. The debate entered popular culture. *Saturday Night Live* featured a sketch set in the Civil War about a fictitious all-gay unit, "the Gloria Brigade," a collection of effeminate men who minced around their encampment talking about fashion and design and trading catty insults. Urging Clinton on, a *New York Times* editorial drew a comparison to the discrimination blacks once faced: "Surely the military chiefs can achieve the same progress toward acceptance of homosexuals if they put their minds to it." It was a comparison that opponents of the ban frequently made, and one that infuriated the ban's supporters—particularly Powell.

"I'm well familiar with the black experience in the military," he said. "I need no lectures."

The warnings were emphatic. So were the pleas for tolerance from gays and their allies. It was a passionate argument that played out in public for the entirety of Clinton's presidential transition, and it was starting to give Americans pause. A poll by the *New York Times* and CBS News just before the inauguration put support for "permitting homosexuals to serve in the military" at 42 percent. During the campaign, when the issue wasn't being so vehemently litigated, polling had put that number well over 50 percent. Clinton's argument was losing steam. More people now said they preferred to leave the ban in place than scrapping it. The top Republican in the Senate, Bob Dole, made a prediction: if Clinton went ahead and lifted the ban on his own, "it would blow the lid off the Capitol."

Actually, it was more of a threat—and one Clinton was now realizing he had to take very seriously. The gay ban wasn't technically etched into law right now, which is why as president he would have the power to get rid of it on his own. But if it wanted to, Congress could still play a major role in the debate, in a way that could undermine Clinton not just on this one issue but on his entire agenda. What Dole was threatening was to take the current Pentagon language and turn it into legislation. Presumably, most Republicans would be for it, which by itself wouldn't count for much, since they were so outnumbered in the Senate and House. But there were plenty of moderate and conservative Democrats on Capitol Hill, too, who might be persuaded by the combined weight of Powell, the military establishment, and public opinion. Lose enough of them and things could get very ugly, very fast for Clinton. In the extreme scenario, both houses of Congress could pass a bill banning gays from the military and then override his veto. More realistically, a gay ban could be slapped on as an amendment to a much bigger piece of legislation—something vitally important to the new president, like, say, a health-care plan. Then Clinton would face the impossible choice of accepting the ban or blowing up one of his own top agenda items.

Clinton tried to make concessions to the changing landscape without abandoning his core promise. Instead of doing away with the ban in one fell swoop, the plan was now to end it in two phases, spread out over six months. Then a memo leaked. It was to Clinton from Les Aspin, his new defense secretary and before that a two-decade veteran of the House. Aspin was a man who knew exactly how Capitol Hill worked, and he reported to the president that—at best—only thirty of the fifty-seven Democrats in the Senate were with them on the gay issue. Ambushed with the memo on *Face the Nation* on January 24, Aspin said: "The point you've got to understand is that as a practical matter we are not going to be able to force this down the throat of the Congress. If the Congress does not like it, it isn't going to happen."

Aspin's memo did offer Clinton a way to turn it around "The key to a successful vote," he wrote, "is the active leadership of Senator Nunn."

As the chairman of the Senate Armed Services Committee, Sam Nunn had nurtured relationships with generals and was treated by the D.C. press as one of the Democratic Party's most credible voices on national security. If he was on board with the Clinton plan, he'd bring plenty of powerful reinforcements with him. But Nunn felt slighted. This was supposed to be his turf, but it felt like Clinton was trying to go around him. He announced that he'd hold hearings on the issue, and not until March. The process was moving too fast, he said, and he was taking the initiative to slow it down. "I think something is fundamentally flawed when the men and women in the military have an issue that is vital to them, that affects them, and they never have been heard from," Nunn said.

Nunn's comments came on a Monday, the start of the first full week of Clinton's presidency; the plan was for Clinton to issue his first executive order—the one suspending discharges for six months—by the end of the week. All systems were still go, the White House said.

On that same day, January 25, Clinton held his first meeting as president with the Joint Chiefs of Staff. The White House described the two-hour session as "respectful, frank, cordial, honest," which was an anodyne way of saying that the generals gave the president an earful. The intensity of their feelings varied, but not a single one was in favor of lifting the ban. They said nothing in public, but Powell was well connected in the Washington media world, and his damning sentiments soon made their way into the *New York Times:* "What worries General Powell, friends and associates say, is his belief that Mr. Clinton and, to a lesser extent, Mr. Aspin do not fully understand or respect his cherished institution."

There were some howls that Powell was crossing a sacred line, from advising Clinton to actively opposing him. "He's trying to submarine the president politically!" protested Representative Barney Frank, one of the House's two openly gay members. But even Democrats who privately agreed recognized how futile a public battle with a man as revered as Powell would be. Soon there were reports that Powell, whose term was up at the end of September, was getting ready to call it quits early. A president who'd sat on the sidelines during the Vietnam War driving out the architect of an American military triumph, all because of gays in the military? Powell denied the stories, but the contrast was political gold for Clinton's opponents. Said Arizona senator John McCain, a navy veteran who'd spent more than five years in a North Vietnamese prison camp: "I think his lack of military service emphasizes his need to consult on matters he knows nothing about."

Clinton held a meeting with congressional Republican leaders, too. He meant to outline his big-picture agenda, but he never got the chance. Instead, the conversation was hijacked by the second-ranking House Republican, who gave the new president what one paper described as "an abrupt lesson in political reality." Gingrich told Clinton his two-step plan would change nothing. Republicans remained ready to fight him, and with public opinion on their side, they'd have plenty of help from Democrats, too. To push forward in the face of this, Gingrich warned, would undermine his presidency and poison his relationship with Congress going forward. Nunn's unhappiness was also now a weapon for Gingrich, and he deployed it when he talked to the press later. "I think if a senior member of your own party publicly serves notice, it's pretty foolish the first week or second week of your term to decide to pick a fight with the senior member of your own party."

On January 27, the White House was still signaling that an executive order was just days away, so Nunn went to the floor of the Senate. He argued that the issue was far more complicated than anyone appreciated. Sure, there were the questions of whether it was fair to exclude gays from service and if military culture could easily adjust to open expressions of homosexuality, but there were more basic, practical matters, too—important ones that no one had yet considered. Like: If the ban was lifted, would the partners of gay service members be eligible for spousal benefits? If they were, would that mean unmarried partners of heterosexual service members would be, too? How about joint exercises with American allies that didn't allow gays in their armed forces—how would that be handled? Would separate showers and bathrooms have to be built? An executive order would sidestep all of these delicate questions and stir chaos, he claimed. It was why there had to be hearings.

"When the interests of some individuals bear upon the cohesion and effectiveness of an institution on which our national security depends, we must move very cautiously," Nunn said. "This caution is prudence, not prejudice."

Clinton was cornered. Constituent calls were flooding congressional offices—nearly half a million that week, "five times the normal volume and far more than recorded over the failed nomination last week of Zoë Baird as attorney general," the *Boston Globe* reported. "This has touched off a firestorm unlike anything I've ever seen," said the head of the Veterans of Foreign Wars. Maybe the public wasn't as ready for this as they'd thought, gay rights leaders admitted. Dole was now vowing to attach an amendment codifying the gay ban to the next piece of legislation Democrats brought up in the Senate. "I'm not a gay basher," he said, "but I think this is an issue that since he brought it up, we have to deal with it."

If Clinton went ahead with his order, a humiliating congressional rebuke now loomed, and all for an issue he'd barely talked about in the campaign. It was the economy he'd promised to focus on "like a laser beam," not a cultural battle. Nunn was called to the White House after his speech. The president was ready to fold.

January 29 was a Friday, the end of Clinton's first full week in office—the day he was supposed to put his signature on an executive order ending the gay ban. Instead, it was the day he and Nunn held a press conference to announce their deal. Clinton would issue a temporary order placing openly gay service members on standby reserve status, meaning they'd be separated from active duty and receive no pay or benefits. Practically speaking, it was no different than a discharge, but technically it was a change in policy. Recruits would no longer be asked preemptively about their sexual orientation, the biggest concession Nunn was willing to allow. The policy would remain in effect until a permanent solution could be reached. It was, the *New York Times* concluded, "a victory for Senator Sam Nunn of Georgia, the chairman of the Senate Armed Services Committee, who wanted to make clear that avowed homosexuals were still not welcome in the military."

Months later, after the issue had receded from the front pages, the final deal was struck and the "Don't Ask, Don't Tell" policy was codified. It was a product of a political climate that put Clinton at Nunn's mercy, and it would endure for nearly two decades.

Clinton had won the election by convincing millions of voters that he'd reformed the Democratic Party, pulling it away from the pressure groups and their liberal checklists and refocusing it on the bread-and-butter concerns of the middle class. Now all of that painstaking work was in danger. Kevin Phillips, one of the architects decades earlier of

Richard Nixon's southern strategy but now a respected nonpartisan analyst of politics, warned that Clinton was risking the same "values problem" that had sunk so many other Democrats over the past generation.

"Somehow he's let the issue be transformed from one of defending civil liberties into an example of his dubious national priorities—of rewarding an ambitious voting bloc, of legitimizing homosexuality as a culture and of overruling the reluctant Joint Chiefs of Staff (who remember that he ducked the military draft)," Phillips wrote.

More alarming was the narrative rapidly taking hold about this new administration. "A number of its actions—be they political correctness in Cabinet selection, animal rights parties at the inaugural, Hillary's role as first mate (not First Lady) or casual dismissal of promises to the middle class—hint at a 20th reunion of 1972's left-leaning George McGovern presidential campaign," Phillips argued. "The suggestion that the Clinton administration is leading a general campaign for gay culture fits easily into this scene."

It was a bum rap, the Clinton team swore. The last thing they ever wanted, they insisted, was to launch the new administration with a draining culture war eruption. Privately, they seethed at Nunn and Powell, convinced both had chosen a public standoff with the new president when conciliation and collaboration were possible.

Republicans, meanwhile, basked in the latest affirmation of their assessment of this president. He'd only gotten 43 percent of the vote. Then there was Zoë Baird. Now this. To Gingrich, it had to be heartening, witnessing his fellow Republicans, even some of the ones who'd once laughed at him, coming around to his theory of the game, even reveling in it. But so far it had all been a warm-up act, a few small victories to boost morale, build resolve, and prime the troops for the big one—because the new Democratic president was about to kick off the battle Newt Gingrich had been waiting for his entire career.

THIRTEEN

When candidate Clinton called himself "a different kind of Democrat," he pointed always to the issue of taxes. The tax-and-spend liberal Democrat had proven a politically toxic caricature, and Clinton was intent on erasing it. What he came up with was the essence of Clintonism, a simultaneous bow to the right and left. The Reagan-era GOP had demonstrated the political appeal of tax cutting, and Clinton sought to co-opt it, but with a populist twist. His tax cut would be for the middle class only, to be paid for by a tax hike on the wealthy, whose rates had been slashed by Reagan and whose share of the economic pie had been expanding ever since.

"Millions of Americans are running harder and harder just to stay in place," Clinton said on the campaign trail. "While taxes fall and incomes rise for those at the top of the totem pole, middle-class families pay more and earn less."

It was a promise Clinton couldn't keep. The same runaway deficits that had compelled Bush to abandon his "Read my lips" pledge were now staring President Clinton straight in the face. As he took office, his team projected that the red ink would reach nearly $350 billion in five years—well over one hundred billion dollars more than the estimate candidate Clinton had relied on to make his numbers add up during the campaign. The lost revenue from a middle-class tax cut—his campaign plan called for sixty billion dollars' worth of relief for families making under eighty thousand dollars a year—would only make a bad fiscal situation worse.

This wasn't without warning, either. Several times during the campaign, deficit projections had been revised upward thanks to the sluggish economy and increasing health-care costs. It was clear the problem was getting worse, not better. As a candidate, Clinton would only acknowledge this in subtle ways, deemphasizing the tax cut in speeches toward the end of the campaign and trying to carve out wiggle room for the future. It was his natural style, inserting lawyerly fine print into his rhetoric in the hope that it might exculpate him from future charges of dishonesty. Pressed in a debate to guarantee middle-class Americans that they would actually see a tax cut under him, Clinton pointed to Bush's own example. "I think even he has learned that you can't say, 'Read my lips' because you can't know what emergencies might come up. But I can tell you this. I'm not going to raise taxes on middle-class Americans to pay for the programs I've recommended."

Now that he was president, he couldn't stall any longer. The impetus went beyond the simple fact of accelerating deficits, though those numbers alone could feel staggering. In the first two centuries of its existence, America had accrued just under one trillion dollars in debt; since 1980, that figure had more than quadrupled. Interest payments on that debt were eating up an ever-larger share of federal spending. There were some new facts on the ground, though. People weren't quite feeling it yet, but the lousy economy that had played such a decisive role in Clinton's election was already revving to life. The first signs came just before the election, too late to do Bush any good, but just in time to rebalance the incoming president's priorities. A strengthening economy, Clinton was persuaded, would mean much

THE RED AND THE BLUE

less need for direct government intervention—stimulus—and much more space for private investment. And that, in turn, would require low interest rates, which were set by the Federal Reserve, whose chairman, Alan Greenspan, was quite clear that he considered reducing the deficit to be of paramount importance.

He would break the bad news to voters in two parts, starting on the night of February 15 with a nationally televised address from the Oval Office. This was not to be a speech about details, just a diagnosis of a dire problem, and an admission that the treatment would be far more painful than anything he'd described as a candidate.

"I had hoped to invest in your future by creating jobs, expanding education, reforming health care, and reducing the debt without asking more of you," Clinton said that night. "But I can't—because the deficit has increased so much beyond my earlier estimates and beyond even the worst official government estimates of last year. We just have to face the fact that to make the changes our country needs, more Americans must contribute today so that all Americans can do better tomorrow.

"But I can assure you of this: you're not going alone, you're not going first, and you're no longer going to pay more and get less."

He borrowed from Perot's stagecraft, too, holding up a series of charts behind his Oval Office desk just as Perot had done in his thirty-minute infomercials. "Look at this!" Clinton exclaimed. "The big tax cuts for the wealthy, the growth in government spending, and soaring health-care costs all caused the federal deficit to explode. Our debt is now four times as big as it was in 1980." He was issuing "a call to arms," Clinton told Americans—an unmistakable echo of Tsongas, who'd called his own economic plan "A Call to Economic Arms."

Besides going back on his word, the risk for Clinton was looking like what he'd strained to persuade Americans he wasn't: just another tax-and-spend liberal. That's why the next phase of the rollout was so crucial. Clinton would address a joint session of Congress and present his comprehensive economic plan. Here was his chance to ground his shattered campaign promise in context, to convince the public that as painful as it would be now, the long-term payoff for them and for their country would be more than worth it.

Two nights after his Oval Office show, Clinton walked into the House chamber for the first time as president. From the Republican side came a courteous greeting; from the Democrats, a hero's welcome. It had been more than a decade since one of their own had addressed the members of the Permanent Democratic Congress. Here for them was a moment to savor, even if they were apprehensive about the message they were about to hear. Clinton would get to the question of the night—which taxes he wanted to raise, and by how much—but only after anticipating and answering the other big concerns percolating in the room and around the country.

There were liberals who'd believed all through the campaign that the combination of a feeble economy and all-Democratic rule would translate into expansive new public investments. Now they worried that their president's new embrace of deficit reduction might leave them out in the cold. They cheered the loudest when Clinton called for an immediate federal stimulus program "to create jobs and guarantee a strong recovery."

Other Democrats feared that Clinton might be narrowing his focus too much. For years, they'd piled one ambitious policy goal on top of another as one Republican president after another ignored them. Now, finally, everything was supposed to be in place—unless Clinton had suddenly decided he was more interested in reducing the deficit than doing big things. This was Clinton's audience as he wove in plans for childhood immunizations, Head Start funding, worker retraining, and gun control, not to mention what for Democrats going back generations had been the ultimate elusive pursuit.

"All of our efforts to strengthen the economy will fail—let me say this again, I feel so strongly about this," he said. "All of our efforts to strengthen the economy will fail unless we also take—this year, not next year, not five years from now, but this year—bold steps to reform our health-care system."

This wasn't the grand unveiling of a universal health-care plan, just a reminder from the president that one would be coming after he heard back from the task force he had just appointed, one that was being chaired, in a historic arrangement, by the First Lady.

There were offerings for the DLC crowd, too—talk of national service, community banking, and small-business loans, plus a vow that "later this year we will offer a plan to end welfare as we know it." The opposition party was ready to accuse him of taxing too much and cutting too little. To head them off, Clinton committed himself to eliminate one hundred thousand jobs from the federal bureaucracy during his term and to cut one-quarter of his own White House's staff. With the Cold War over, the Pentagon would get a haircut, too, he said, but so would a host of domestic programs and agencies— a total of $246 billion in savings, he claimed.

Only then did he finally turn to the dirty work. He was calling for an income tax hike from 31 to 36 percent targeted to the wealthy. An additional surtax on income over $250,000 would raise the top effective rate to 39.6 percent.

"I want to emphasize the facts about this plan," Clinton said. "Ninety-nine point eight percent of America's families will have no increase in their income tax rates."

This was true enough, and broadly consistent with his campaign platform, which called for the wealthy to pony up more. The next part wasn't, though. Clinton was now also asking to raise taxes on Social Security benefits, taking him into treacherous political territory. The hike would be aimed at higher-income retirees and would apply to only about one in five beneficiaries. But it was not something Clinton had campaigned on, and putting it on the table now risked riling up a vast and politically powerful constituency.

"My fellow Americans, the test of this plan cannot be what is in it for me," he said toward the end. "It has got to be what is in it for us."

Would the country buy it? The perils were obvious. "The package embodied a word—sacrifice—that was never spoken during the Clinton campaign," the next morning's *Washington Post* noted. There was reason for optimism in the White House, though. Yes, Clinton was going back on his word, and yes, Democrats had paid dearly before for their tax-and-spend image. But Tsongas and Perot had just demonstrated that tax hikes could also be framed as tough but virtuous medicine. The true aim of Clinton's speech was to convince Americans that the money he needed from them wasn't to feed some bloated bureaucracy; it was to pay down the obscene bills that both parties had recklessly run up. He didn't want big government, he was trying to tell them—just a solvent government and a sound economy.

"The person giving that speech is not the same person I campaigned against," Tsongas told the *Post*. "And I think the country is better off for it."

Polls taken hours and days after the speech all showed similar, overwhelming backing. Was it even possible to call for $274 billion in higher taxes—more than any president had ever asked for—and still be "a different kind of Democrat"? Maybe, just maybe, Bill Clinton was that good a salesman.

Newt Gingrich had been telling Republicans for years that they've been giving away the game. Let the Democrats be the party of higher taxes, he would say, and let our party harness the backlash and win big at the polls. Now here was another president sounding the alarm over deficits, prescribing a mix of spending cuts and tax hikes, and calling for bipartisan cooperation. This time, though, Republicans were ready to listen to Newt—all of them.

It had everything to do with 1990. Caving on taxes and compromising with Democrats, Bush had told Republicans back then, might be painful in the moment, but it would produce a big payoff—lower deficits that would yield higher growth, which would revive the economy and lift the president to a second term in 1992, carrying lots of Republicans along with him. But look what had happened. After pronouncing the final deal "good medicine," Bush watched as the economy went from bad to worse and the deficit climbed even higher. The view among Republicans was now universal: instead of saving his presidency, Bush's tax hike had wrecked it. The effect on the old guard was profound. They'd long recognized that Gingrich's antitax fervor was catching on with their base; now they were starting to see it as smart strategy, too.

Bob Michel epitomized that old guard. As the House Republican leader, he'd been instrumental in the '82 tax hike and a key player in the negotiations that produced Bush's initial deal with Democrats. Even when the Gingrich rebellion killed that pact, Michel stayed with Bush, casting one of just forty-seven yes votes on the Republican side for the plan that was ultimately enacted. The other 126 Republicans joined Gingrich in defecting. Now, days away from his seventieth birthday, Michel lived with the knowledge that Gingrich had won the battle for Republican hearts and minds and that his own days as leader were numbered. His response to Clinton would represent his acquiescence to this new reality; he would be reading from the Gingrich script now.

"There are those who say some taxes are a necessary evil," Michel told the nation. "The difference is that Democrats stress the word 'necessary' and we Republicans stress the word 'evil.'" He accused Clinton of trying to "orchestrate the biggest propaganda campaign in recent political history" and mocked the president's rhetorical employment of various euphemisms for taxation.

"'Investment' now means big government spending your tax dollars," Michel said. "'Change' now means reviving old, discredited big-government tax-and-spend schemes. 'Patriotism' now means agreeing with the Clinton program. The powerful, evocative word 'sacrifice' has been reduced to the level of a bumper-sticker slogan. And my favorite, 'contribution,' is now the new word for 'taxes.' On April 15, just try telling the IRS that you don't feel like 'contributing' this year."

Michel's remarks, the *Washington Post* wrote, represented "an unusually sharp response for him."

Gingrich held a press conference the next morning. "It was a good speech," he said, "and a very destructive program." Other veterans of the '90 rebellion were there too. "My opposition to the plan is that it won't work," said Representative Dick Armey. "The good news is that it will cost Clinton his presidency." On the Senate side, Bob Dole, the Republican leader, pointed out that for all Clinton's talk of cuts, the president had also inserted more than $150 billion of new spending into his program. Dole, like Michel, had been on the side of tax increases in '82 and '90. At one point, Gingrich even branded him "the tax collector for the welfare state." But now Dole announced that if Clinton wanted his program to pass, "I think it will be up to Democrats to get it through Congress."

From the op-ed page of the *New York Times* came another Republican voice, this one warning that the new president "has begun to sound like an 'Old Democrat.' That's the kind who does not understand one simple fact: the problem is not that the people are taxed too little, the problem is that the government spends too much." The headline on the column: "There They Go Again." The author: Ronald Reagan.

Clinton had tried in his speech to head off the critics. "If this package is picked apart," he warned, "there'll be something that will anger each of us, won't please anybody. But if it is taken as a whole, it will help all of us." As a practical legislative matter, though, it couldn't all be considered at once. Action would have to proceed on two separate tracks. There was the budget itself, with all the tax hikes and spending cuts, which would now be subjected to a months-long gauntlet of committee hearings and floor votes. Then there was the much shorter track: the immediate economic stimulus program Clinton had called for in his speech. Here was something that could be done in a few weeks.

It was cobbled together by early March, cash for extended unemployment benefits, summer jobs programs, research projects, block grants for cities, infrastructure spending. The price tag when it hit the House floor was $16.3 billion. A few dozen moderate Democrats objected, along with every Republican, but it passed anyway, and without much stress, and then it was on to the Senate—where all hell broke loose.

On paper, the votes were there. Even if every Republican voted no, there were still fifty-seven Democrats in the Senate. But these Democrats weren't nearly as liberal a bunch as their House counterparts, and they were far less eager to open the purse strings. Oklahoma's David Boren demanded the stimulus be lopped in half, with any additional stimulus money after that contingent on equivalent spending cuts. He was joined by Louisiana's John Breaux and Nevada's Richard Bryan. Eventually, they relented, assuaged by a written promise from Clinton that he would enforce deficit reduction targets, but their protest stalled action for a week.

Republicans, in the meantime, were having a field day portraying the stimulus as a pork-riddled payoff to Clinton's base. They ridiculed specific line items—fifteen million dollars for the production of "fish atlases," for instance—and touted an estimate that each job created would cost ninety thousand dollars. The media's skepticism was building, and soon there were more cracks on the Democratic side. "It's not good for us to be voting for things we're not prepared to fund," Nebraska's Bob Kerrey said as he announced his opposition.

The momentum the White House was depending on was draining away, but it still had a majority, however reluctantly, ready to vote for it. The stimulus came to the Senate floor, but when Democrats prepared to call a vote they were confronted with a rarely used tactic. Led by Dole, Republicans were refusing to end the floor debate indefinitely—a filibuster. Now Democrats wouldn't just need fifty votes to pass the stimulus; they'd first need sixty to shut off the filibuster. "There's almost unanimous feeling we shouldn't pass the stimulus package and we have the votes to prevent that from happening," Dole said.

Suddenly, the new president was staring at serious and unexpected defeat. He offered to knock off four billion dollars from the package, cutting its total size by 25 percent, but didn't even get a nibble. In desperation, he turned to the bully pulpit. It had been a rough start, but he was still a new president with plenty of goodwill across the country, especially in the places he had won. Some of those Clinton states were also represented by moderate Republican senators, the kind who didn't have deep philosophical objections to the idea of a stimulus. So, it was decided, Clinton would head to their backyards and turn up the heat. The first target: Arlen Specter of Pennsylvania.

Clinton had carried the Keystone State by nine points, the first Democrat in sixteen years to do so. Specter, meanwhile, had just barely won reelection. A onetime Democrat, he was known to break regularly with the GOP's leadership—especially when he was far out from his next election. On the first weekend in April, Clinton flew into Pittsburgh for a campaign-style rally. "You've heard that old saying, 'It takes two to tango?'" he asked. "It also takes two to un-tangle the gridlock in Washington. And I came here today asking you to ask Senator Specter to help me." The speech was impassioned—the big dog back in his element. The crowd numbered in the thousands. Every media outlet in the state was there. And Specter's reaction to it all? "I think this package is bad for America."

There were other targets, too—Jim Jeffords in Vermont, John Chafee in Rhode Island, Al D'Amato in New York, big Clinton states all of them. But Clinton was banging his head against the wall. Democrats tried again on the Senate floor and the result was the same. Every one of the forty-three Republicans held firm. "I don't see any way out of the impasse," Dole said. "We have a right under the rules to try to prevent this bill from passing."

With that, the stimulus was dead, slain by a Republican Senate leader who hadn't played politics like this before. Dole had refused to compromise, insisted on party discipline, and used every available tool to fight the Democrats. "It's the best spring of Bob Dole's politi-cal career," Gingrich proclaimed. And didn't that say it all? Even Bob Dole was playing Newt's game now—and winning.

I n 1990, they'd fought themselves to a bloody pulp over taxes. But now Republicans looked around and found nothing but unity. After sinking the stimulus, they turned next to a much bigger target: Clinton's first budget—the one with the surprise tax hike. Already, the GOP had settled on two themes of attack. The tax increase—the largest in history, they never forgot to mention—would sock it to the middle class, kill jobs, and wreck the fragile recovery. And the spending cuts Clinton was offering were a woefully insufficient joke. If the budget passed, voters would find only Democratic fingerprints on it. "It's not our package, folks," Representative John Kasich, the House GOP's point man on the budget, boasted.

The Republican assault was resonating. Shortly after Clinton's speech to Congress, a poll showed voters supporting his plan by a two-to-one margin. Now, three months later, the public was evenly split. From the moment he unveiled it, Clinton had been emphasizing that most of the tax hikes were weighted to the upper end of the income scale. Still, two-thirds of voters said he was raising them too much. A majority also said that he wasn't cutting spending enough and that they would be hurt personally if the plan was enacted. Could the budget—the centerpiece item of the new president's agenda— suffer the same fate as the stimulus? For Clinton, the possibility of a devastating political failure was growing by the day.

Republicans got another boost when Perot, his popularity revitalized after his strong finish in the fall campaign, bought thirty minutes of airtime on NBC to discuss the Clinton plan. He'd complimented it in broad terms immediately after Clinton's speech to Congress and had advocated higher taxes in his own campaign, but now Perot denounced the budget in no uncertain terms. "We, the people, have to step in and just say no to new taxes," he said.

Clinton fumed at the attacks. "When you hear people say 'No, no, no,' ask them where they were the last twelve years," he told a rally. His problem wasn't just the opposition party, though. Democrats were getting skittish—too many of them. Some were deficit hawks who wanted more spending cuts. One group was demanding curbs on entitlement programs like Medicare and Social Security. An energy tax, conceived by the Clinton team as both a giant revenue-raiser and tool for encouraging conservation, pushed Democrats from oil and gas states into open rebellion. Others were simply reading the polls—including the ones showing Clinton's job approval rating edging into the low forties—and wondering what the upside to going along with something like this was.

The budget landed on the House floor just before Memorial Day, practically dead on arrival. Democrats were dozens of votes short as the floor debate opened on May 27. Republicans smelled blood, stepping forward one by one to shred the budget. Three years earlier, Michel had been here trying to sell his side on a tax hike. Now the Republican leader told the House that Clinton "wants to thrust upon middle-income Americans an enormous tax increase. Let's face it: The Clinton White House is out of touch. It's out of sync. It's out of ideas. It's out of excuses. And it's out of control."

From the White House, Clinton worked the phones. For too many Democrats, the energy tax was just too much to swallow—especially with rumors that the provision would end up stripped from the bill to get it through the Senate. "It's a very dangerous thing for a Texas Democrat to vote for a very unpopular tax and then have Republican senators going back to the state saying, 'Charlie voted for the tax and I saved you,'" Representative Charlie Wilson explained. Clinton told Wilson and other wavering Democrats he was committed to the energy tax—their painful votes for it, he assured them, wouldn't be in vain. That brought Democrats to the brink of the magic number, then they leaned on a blunt message to push them past it. "If we don't vote for the president, we cut him off at the knees early in his term," New York representative Charles Schumer told his colleagues. "We can't do that."

When the vote was called, thirty-eight Democrats defected, enough to embarrass the president but not to derail his signature program. By a 219–213 tally, it lived to fight another day. Now it was the Senate's turn. Once again, every Republican was ready to vote no. The good news for Democrats was that, this time, Dole couldn't use the filibuster, since a Senate rule barred it from being employed on deficit legislation. The bad news was that they were still short. The single biggest culprit, again, was the energy tax. This time, the holdouts weren't ready to settle for a promise from Clinton that he'd walk the plank with them. They wanted it gone, or they'd vote no. The White House blinked, and just like that the BTU tax was out, replaced by a hike in the gasoline tax. That, along with a new series of spending cuts, was just enough to get to fifty votes, putting Vice President Gore in position to break the tie, which he did in the wee hours of Friday, June 25.

The weeks dragged on. A conference committee merged the House and Senate versions, and finally, as August arrived, the final package was ready to meet its fate. It had been nearly six months since Clinton first outlined his plan, and all the noise—the attacks from Republicans, the skepticism from some Democrats, the horse trading on Capitol Hill—had taken a profound toll. It had cleared the House the first time around, but that was back in May. The politics were clearer now: this was not a popular plan. Dozens of Democrats who'd voted for it the first time were now ready to vote no. Just as he'd done back in February, when he broke his tax cut promise, Clinton scheduled a prime-time address from the Oval Office. He would try to close the deal himself.

Once again, it was staged like a Ross Perot infomercial, with Clinton holding up charts and graphs to illustrate his points. "America faces a choice," he said. "We can continue on the path of higher deficits and lower growth or we can make a fundamental change to improve our nation's economy by adopting my economic plan." The vast majority of new taxes, he pointed out, would be paid by the wealthy. Only the top 1.2 percent of income earners would be subject to a higher income tax, and only one in eight Social Security recipients would pay more. The only hike that would be felt across the board would be a 4.3-cent-per-gallon jump in the gasoline tax—less of a hit to consumers than the broader BTU tax Clinton originally proposed, but still a hit. "For working families making less than $180,000 a year, there will be no income tax increase," Clinton said. "I repeat, for working families making less than $180,000 a year, there will be no income tax increase."

Clinton spoke for twenty minutes, then it was time for the Republican response. This time, the party tapped Dole, who started his remarks with an assurance that "like nearly all Americans, I want President Clinton to succeed. I want him to lead America in the right direction—more jobs, more opportunities, and reduced federal deficit." Then he went to work slashing the Clinton plan apart.

Dole, too, covered familiar ground. The size of the tax hike was historic, he said, and no matter what Clinton said about the wealthy footing the bulk of the bill, it remained true that middle-class Americans were being asked to pay more.

"The world will not end if this bill is defeated," Dole told viewers.

The dozens of Democrats in the House now threatening to bolt weren't united by any specific concern about the bill. Some said it taxed too much, others said it simply raised the wrong taxes, others still said it cut too little. What all of them did agree on was that it was politically toxic and that voting no would be a lot easier to explain to the voters in 1994 than voting yes. A snap poll underscored the Democrats' dilemma. Sixty-eight percent of voters now said that the middle class—not the wealthy—would bear the biggest tax brunt. A majority also said the plan wouldn't do anything to reduce the deficit. Most said it wouldn't improve the economy, either, and that Congress should simply defeat it.

Two nights later, it was time for the House to vote. The reality was by now obvious. Selling the holdouts on the merits of the plan wasn't going to work. Putting the prospect of a derailed Democratic presidency on their shoulders was the White House's best—and only—hope. "This is the Super Bowl of pressure," said one of the Democrats on the receiving end, Pennsylvania representative Marjorie Margolies-Mezvinsky.

Margolies-Mezvinsky was in an agonizing bind. A freshman, she'd won a seat from the traditionally Republican—and staunchly antitax—Philadelphia suburbs the previous November. It had been a squeaker; she'd barely cracked 50 percent, and already Republicans had identified her as a top target for the 1994 elections. Recognizing these realities, she'd voted against the plan on its first pass through the House in May. Democrats could afford her defection then, but not now. Too many of the yes votes from the first round had now declared themselves nos. Margolies-Mezvinsky had at least left a few inches of wiggle room, making it known that the possibility of handing Clinton a massive political defeat weighed on her. As the House vote began late that night, she still wasn't sure what she'd do.

At around 10:30 P.M. the vote board in the House showed there were 216 yes votes and 214 nos. Four members still hadn't cast their votes. All of them were Democrats, and all of them had been hoping against hope it wouldn't come to this. Margolies-Mezvinsky was one of them. The others were Pat Williams from Montana, Minnesota's David Minge, and Ray Thornton from Arkansas. Two hundred eighteen votes were needed for passage. Either two of the four would vote yes or their party's president would suffer one of the most crippling rebukes in memory. There was a pause, then Margolies-Mezvinsky and Williams walked to the front of the House and cast their votes for the plan. The two others voted no and it was over. By the thinnest of margins, 218 to 216, the Clinton plan was still breathing.

On the Republican side, it felt like anything but a loss. Here was a tax deal that looked every bit as unpopular as the one that had torn their party apart in 1990. But this time, they had nothing to do with it. Not a single Republican had even been tempted to vote for it. Now and forever, this would belong entirely to the Democratic president and the Democratic Congress. As Margolies-Mezvinsky cast her fateful vote, a chant came forth from the Republicans, who were already thinking ahead to what this might mean for the midterm elections: "Bye-bye, Marjorie!"

Gingrich rejoiced. "We've been given back the antitax position." The plan had passed by a single vote. That meant that every Democrat who had voted for it could be branded the deciding vote. The Clinton plan, Gingrich now predicted, "will lead to a recession next year. This is the Democrat machine's recession. And each one of them will be held personally accountable."

It still wasn't over, either. There remained the matter of the Senate, where an old thorn in Clinton's side, Sam Nunn, had just declared himself against the budget. That brought the number of Senate Democrats opposing it to six, the maximum number of defections Clinton could survive. Now there was more bad news: Nebraska's Kerrey was saying he was undecided, too. Kerrey had been a yes vote the first time around. If he switched to a no now, the plan would fail in the Senate.

Thus did Bob Kerrey become the star of a last-second drama that stretched through the day and into the night. Many, especially inside the White House, sensed cynical motives, with Kerrey still nursing a grudge from his primary campaign against Clinton—and perhaps eyeing another run in 1996. On the phone, Clinton warned Kerrey that sinking the budget could derail his presidency, which enraged Kerrey, but not quite enough to follow through.

"My head, I confess, aches with all the thinking," he told his colleagues. "But my heart aches with the conclusion that I will vote yes for a bill which challenges America too little, because I do not trust what my colleagues on the other side of the aisle will do if I say no." He then addressed Clinton directly and added: "I could not, and should not, cast a vote that brings down your presidency."

The roll call was fifty to fifty. As he did back in June, Gore then cast the deciding vote, and with that the fourth and final hurdle was officially cleared. Clinton had won. His budget would be enacted. But at what cost?

"The message that came through loud and clear in Washington was, 'No-Clout Clinton,'" Bill Schneider, CNN's lead political analyst, wrote in his syndicated column. "That will seriously weaken the president's position in Congress. It's bad enough when a president's job approval ratings go down. It's worse when he's seen as poison at the polls." It wasn't just Republicans who'd decided Clinton was in over his head anymore. The media was now reaching the same conclusion. In the *Washington Post,* David Broder wrote that "bit by bit and piece by piece, Republicans see themselves rebuilding their political strength on the foundation of eroded public hopes in President Clinton." The cover of *Time* featured a tiny picture of Clinton underneath an enormous headline: THE INCREDIBLE SHRINKING PRESIDENT. No president since World War II, the article inside explained, had broken the 50 percent disapproval mark in polling as early as Clinton now had. "If he fails to adjust quickly, he will confirm the widespread belief that the biggest problem with the Clinton presidency is Clinton himself. Unless he can, as he likes to say, make change his friend, he is in for a decidedly unfriendly 3½ years."

FOURTEEN

At the start of October, Mr. Nice Guy announced his retirement. Bob Michel insisted it wasn't the threat of a challenge from Newt Gingrich that was compelling him, but at the very least he was surrendering to reality. "My style of leadership, my sense of values, my whole thinking process . . . is giving way to a new generation, and I accept that," Michel said. "It's probably for the best. I was really more comfortable operating under the methodology that we did when I first came here." From the press, generous appraisals of the House Republican leader issued forth. Michel, according to the *Los Angeles Times*, was "an outstanding practitioner of a conciliatory style of politics." David Broder called him "a legislator first and a Republican partisan second." Democrats joined in as well, including President Clinton, who hailed Michel as a man who "would never put his party's interest above the national interest."

No sooner had Michel made his announcement than Gingrich trotted out a list of supporters totaling two-thirds of the Republican House membership. There would be no succession dispute. That war had long ago been fought and won, in a rout, by the gentleman from

Georgia. Michel would serve out his term, through the 1994 election, but his reign would be primarily symbolic now. When Gingrich came to the House in the late seventies, he could have counted his allies on one hand. Now he held a victory rally, with scores of Republican members chanting his name. "Our generation must replace the welfare state with an opportunity society," he said.

Over the years, Michel had done what he could to thwart Newt's rise, reminding members of what they risked losing with intense partisanship and even publicly opposing Gingrich in the '89 whip's race. But his pleadings were no match for the nerve Gingrich was tickling. For far too long, he'd tell Republicans, they'd been a feeble minority, trampled over and condescended to by an arrogant "Democrat machine" that never seemed to pay a price for its misdeeds. Michel was asking these Republicans to suck it up and play nice anyway. Gingrich was telling them to fight back with every weapon they could find. This was the future they wanted. At Gingrich's victory rally, John Boehner, a second-termer from Ohio, said: "We standing here today see Newt as the next generation of Republican leadership in the House."

Besides, Gingrichism was working. It was the inescapable conclusion of the Clinton presidency so far. On Zoë Baird, gays in the military, the stimulus, and the tax hike, Republicans had banded together in opposition, creating the clear definition and contrast that his brand of politics depended on. And the results were something close to astonishing. The Clinton honeymoon lasted perhaps a week, his approval rating falling farther and faster than that of any of his modern predecessors. The defeats on Capitol Hill were piling up, embarrassingly, and even the wins seemed to be having the effect of a loss, as with the budget. The Republican strategy was rendering Clinton indistinguishable from the Democratic Congress. Welfare reform hadn't come up at all. Fewer and fewer voters now believed the president really was "a different kind of Democrat."

The fall of 1993 brought further signs of a Republican revival at the ballot box. To their wins in the Georgia and Texas special elections and the Los Angeles mayor's race they could now add a new trifecta: governor's races in Virginia and New Jersey and the mayoralty of the nation's largest city, New York. All three reinforced doubts about Clinton's political health—espccially New Jersey. It was a state Clinton had won back from the GOP in 1992, a marker of the inroads he'd made into America's suburbs. But now the state's Democratic governor, Jim Florio, fell to Republican Christine Todd Whitman after a state income tax hike incited a popular backlash. For Democrats, the parallels between Trenton and Washington were ominous. Was this a precursor of a national uprising against them for their own tax increase? Taking stock of all of the November results, the *Washington Post* called the Republican conquests "part of a trend over the past year decisively favoring the GOP." They were part of another trend, too, of Republicans recognizing that minimal cooperation with Clinton was yielding maximum political success.

The rewards weren't just coming from policy debates, Republicans were finding. Clinton had come to power facing a unique challenge. Never before had so many Americans had so many doubts about the character and integrity of a president they'd just elected. For the opposition party, there was opportunity in this, and it meshed with a particular component of the Gingrich playbook. From the Charlie Diggs expulsion vote in his first weeks in Congress to the Jim Wright drama, he had demonstrated not just that scandal could be a mighty political weapon, but that it could be manufactured and intensified by a determined political party. As Clinton now flailed politically, Republicans were coming around to a view that he was both unusually weak and unusually exposed. They had reason to think this. The campaign-season scandals suggested an undisciplined leader who relied on his own clever wordsmithery to extricate himself from the fruits of his own personal recklessness. Surely, though, it was out there somewhere, the scandal that even Slick Willie wouldn't be able to talk his way out of.

Actually, Republicans were starting to think they'd found it. It was rooted in an old real estate transaction from the Clintons' Arkansas years, one that had briefly flared up during the campaign. It was on March 8, 1992, that Jeff Gerth of the *New York Times* reported that both Bill and Hillary "were business partners with the owner of a failing savings and loan association that was subject to state regulation early in his tenure as governor of Arkansas." The partnership was called Whitewater Development, a money-losing venture launched in 1978 by the Clintons and their friends Jim and Susan McDougal. Their idea was to buy up about two hundred acres of Ozarks land, divvy it up in lots for vacation homes, then turn a profit by selling them.

The story made for a dense and dull read, a hard-to-keep-track-of web of characters and associations and dealings that couldn't be reduced to a concise allegation of wrongdoing. Jim McDougal, who'd worked in Clinton's administration in Little Rock, had left government to start a savings and loan institution, Madison Guaranty. Madison, in turn, played host to a 1985 fund-raiser that helped then-governor Clinton retire his campaign debt. Also around that time, McDougal's S&L retained the Little Rock law firm where Hillary Clinton was a partner. Eventually, McDougal ran afoul of federal regulators over lending practices and Madison was shut down in 1989—one of the many failed S&Ls that American taxpayers bailed out that year. When all their Whitewater lots were eventually sold off, the Clintons' losses totaled around forty thousand dollars. Had Whitewater funds been used to pay Clinton's campaign bills? Did he as governor try to protect Madison from federal investigators? Why did it appear the McDougals had been on the hook for most of the losses in what was supposed to be an equal partnership? The story hinted at impropriety without establishing any.

It landed two days before Super Tuesday, when Clinton built an insurmountable lead in the race for the Democratic nomination. He brushed it off. Whitewater, he said, was "nothing but a big money loser for me." And for the moment, that was that. Nothing stuck with voters or the media, which wrote it off as an illustration of the murky intersection of politics and real estate in a backwater state. Quietly, some regulators began poking around, but the topic of Whitewater played no meaningful role in the '92 race.

What it did, though, was plant the seed of an idea: that in the Clintons' Arkansas past there might be things that weren't quite on the level—maybe a lot of things. After all, he'd been the governor for twelve years. She'd been a partner in the state's most powerful law firm. They'd gone into business with a politically connected banker. And it had all transpired far off the national radar, in the outpost town of Little Rock, where the clubby elite had plenty of incentives to look the other way. Suspicion was widely shared by the media in D.C., and would be fed by the Clintons and their own behavior when they moved to the White House.

It started in May 1993, when—seemingly out of nowhere—the entire staff of the White House Travel Office was fired. The firings smelled fishy right away. The travel office was an obscure piece of the executive branch bureaucracy. Its main purpose was to handle hotel and flight arrangements for members of the media covering the president on his trips. Its staff was small—a director and six deputies—and it wasn't partisan. Billy Dale, the director suddenly facing termination, had been with the travel office for thirty-one years, through Democratic and Republican presidents. With all of its other major priorities, why on earth was Clinton's White House training its fire on such a tiny operation?

Dee Dee Myers, the White House press secretary, said a review had turned up "gross mismanagement" and "shoddy accounting practices" and that the situation was so grave, the FBI had been brought in to investigate. Then the real story started trickling out. Dale was being replaced by a woman named Catherine Cornelius. She was twenty-five years old and had worked on the Clinton campaign. A native Arkansan, she was also distantly related to Clinton himself.

Cornelius, it turned out, had written a memo to Clinton's team months earlier proposing that she run the travel office. She also suggested that air charters be farmed out to an Arkansas company that had contributed to Clinton's campaign. And she accused the existing

White House Travel Office—Dale and his team—of being "overly pro-press." On top of that, it then emerged that one of the Clintons' closest friends, Hollywood producer Harry Thomason, had aired concerns about the travel office directly to the president. Thomason, in addition to his television work, also had an ownership stake in an air charter company; it turned out that he had asked the White House to speak with his business partner.

Now it stank to high heaven. As the press picked apart the White House's official rationale, Dale and his lieutenants spoke up to say they were being railroaded. The administration's defense shifted to an emphasis on Clinton's right as president to fire employees at will. But the backlash was severe, and Clinton was forced to authorize an internal investigation, which concluded in early July that the manner of the firings had been "unnecessary and insensitive." The report also said that the White House "did not furnish sufficient evidence" to warrant the dismissal of most of the travel office employees.

Many of those employees were then rehired, although not Dale, who ultimately was accused of embezzlement by the feds, then acquitted in a 1995 trial. Cornelius, the Clinton relative who'd replaced him, was reassigned. And even those sympathetic to the Clinton White House expressed their bafflement at the whole episode. "Put aside the cronyism charge being made by some in the press," liberal columnist Anthony Lewis wrote. "Did no one on the Clinton staff understand how unfair it was to fire long-time employees as suspected wrongdoers without giving them a chance to defend themselves?"

Something else had been slipped into the White House report, though, that would keep the travel office controversy alive well into the future and give rise to countless others. It introduced a new character into the mix: the First Lady. Hillary Clinton, it turned out, had played a role in the firings. The report was vague and took pains to cast her as a minor figure, but it nonetheless revealed that she'd pushed behind the scenes for scrutiny of the Travel Office and been kept apprised of the subsequent investigation—and, at least as notably, that she'd been told that the firings would take place before the president was. This was news. No one had been talking about the First Lady having any involvement. Now they were, and they were wondering: Why would Hillary have been interested in something like this?

Travelgate, as the episode came to be known, had nothing to do with Whitewater—and also everything to do with it. Already, the press corps, and the public for that matter, was conditioned to treat this president's words with skepticism. "I didn't inhale" was still a running gag in popular culture. Now, his White House had been caught dissembling over a matter that reeked of cronyism. If they'd had a general sense before that something might be amiss in the Clintons' Arkansas past, reporters felt a new urgency to pursue it going forward.

Another force was being unleashed, too. In the precincts of the right, it was already an article of faith that Clinton's character was beneath that of the office he held. There remained, too, a conviction that his election was accidental, a plurality enabled by a third-party spoiler. Now, Travelgate told Clinton's enemies that he wasn't just mendacious; he was sloppy. There was no telling what even a little digging would unearth. And would it really take much unearthing to put an end to a presidency that most of the country had never wanted in the first place?

In Washington, Republicans demanded congressional investigation of the firings. There was "possibly real sleaze" to be exposed, Dole said. The *Wall Street Journal*'s editorial page, which enjoyed wide influence with D.C. Republicans, burrowed in on the Arkansas "cronies" who'd followed Hillary Clinton from the Rose Law Firm into the administration. The *Journal* started with Webster Hubbell, a former Rose partner, who had assumed a powerful, if ill-defined, role in the Department of Justice. The position didn't require Senate confirmation, the *Journal* complained, and allowed him to do the White House's bidding without oversight. He was eventually nominated and confirmed to be associate attorney general.

The *Journal* then shifted its attention to another Rose alum, Vincent Foster, who was now the deputy White House counsel. Reserved and uncomfortable with the spotlight, Foster was unknown to most of Washington. On June 17, the *Journal* moved to change that. Its editorial, "Who Is Vincent Foster," accused him of thumbing his nose at a simple request to provide a photograph, something the paper was entitled to under the Freedom of Information Act. "How an administration deals with critics is a basic test of its character and mores, and how scrupulously it follows the law is even more directly significant," the editorial read. "Does the law mean one thing for critics and another for friends? Will we in the end have to go to court to get a reply, or will even that work?"

It turned out that the White House, upon learning publication was imminent, had hurriedly faxed a photo to the paper, but it arrived after deadline. No matter. Over the next month, the *Journal* would make Foster a focal point in a series of broadsides against the administration and its "carelessness about following the law."

On July 20, Foster ate lunch at his office, then drove to Fort Marcy Park, just across the Potomac River in Virginia, and shot himself in the head. Media reports called it an "apparent" suicide, something that three subsequent investigations would all confirm, and noted that Foster's White House colleagues said he'd been disturbed by the negative media attention. But dark suggestions soon emerged from some quarters. The *Journal* called for a special counsel to investigate Foster's death. "We had our disagreements with Mr. Foster during his short term in Washington," the paper wrote, "but we do not think that in death he deserves to disappear into a cloud of mystery that we are somehow ordained never to understand."

The contents of Foster's handwritten suicide note, found shredded in his briefcase, were made public to the media in August. "I was not meant for the job or the spotlight of public life in Washington," he had written. "Here ruining people is considered sport." Conservative columnist William Safire pointed out that only the text of Foster's note was being released, and not the actual physical copy, "presumably out of deference to the feelings of the suicide's family.

"Another reason," he continued, "was to prevent the public from focusing on the missing 28th piece, a triangular piece where the signature would have been. The gaping hole in the page symbolized unanswered questions in the case."

Doubts weren't limited to the right. It wasn't that reporters thought Foster was the victim of foul play. But they wondered: that scrutiny he'd been facing in the final month of his life—was it possible he'd been afraid of what could be revealed if it continued? When the president, struggling to make sense of his lifelong friend's death, responded by saying that "no one can ever know why this happened," it had the effect, Mary McGrory wrote, "of an electric prod on the press. They leapt to the conclusion that Clinton was covering up. They assumed that Clinton felt that the suicide might be a judgment on him, on the prospects of the administration, and was trying to ward off investigation."

This is where Whitewater, that old failed land deal, returned to the fore, in a far bigger way than before. Three months after Foster's death, the federal agency that had been created to dispose of failed S&Ls—the Resolution Trust Corporation—asked the Justice Department to open a criminal inquiry into Madison Guaranty. Specifically, the RTC wanted to know whether Madison funds had been steered to political campaigns in Arkansas. Clinton's attorney general, Janet Reno, deputized a special team to look into it. Republicans demanded hearings, too.

Then, the bombshell Washington had been waiting for: There were Whitewater files in Vince Foster's office. Or at least there had been. Foster had handled the Clintons' personal legal work back in Arkansas, which meant he was intimately familiar with their Whitewater dealings. Just before Christmas, the *Washington Times,* which billed itself as a conservative alternative to the more established *Washington Post,* reported that Bernard Nussbaum, Clinton's White House counsel, had searched Foster's office two days after the suicide, found the files, and removed them. (Engulfed in controversy, Nussbaum would exit the White House within months.)

It was true. The next day, the White House admitted that Nussbaum had indeed taken the files on July 22. There was some context. Nussbaum had been accompanied by law enforcement personnel and had turned the files over to the Clintons' personal attorney. "All the files were handled appropriately," a White House spokesman said. "[The Clintons] are not the subject of any investigation."

But it was also true that no one in the White House had ever mentioned this before. And that while members of law enforcement had been with Nussbaum during the office sweep, none of them had looked at any of the documents. Nor had the files been included in the official inventory of items recovered. It was possible the Clintons were guilty only of undue secrecy and bad optics, but this also wasn't happening in a vacuum, and the press was in no mood to give this White House the benefit of the doubt.

"Mr. Clinton and his associates have been dodging full disclosure about Whitewater and his and his wife's relationship with Mr. McDougal since early in the 1992 campaign," the *New York Times* editorialized. "It's entirely possible that the Clintons have done no wrong. But the evasive tactics by the White House have fueled suspicions that there is something in those files worth hiding."

Bill Clinton vowed to turn over any Whitewater records in his possession to the Justice Department, and the First Lady stepped forward to defend him. "I think my husband has proven that he's a man who really cares about this country deeply and respects the presidency and believes strongly that he's doing the right thing," Hillary Clinton said. "And when it's all said and done, that's how most fair-minded Americans will judge my husband. And all the rest of this stuff will end up in the garbage can where it deserves to be."

But the wheels were in motion now. After Bill Clinton made that promise, his personal lawyer followed up with a request. Instead of the president handing over everything voluntarily, he wanted a subpoena from the Justice Department—his own Justice Department. Why? The rationale was that there were stiff legal penalties for anyone leaking subpoenaed documents to the press. They didn't want anything sensitive to get out that might be misinterpreted. Republicans exploded in protest. Clinton and the Justice Department were colluding to keep the public in the dark, Gingrich charged, and "there's something a little bit sick about that kind of approach, frankly."

Writing in *Newsday*, Susan Page concluded: "Far from quelling the controversy . . . the subpoena has inflamed it, leaving the impression among some that the president and his wife, Hillary Rodham Clinton, may have something to hide in their tangled Whitewater Development Corp. partnership with James McDougal, a former friend and political ally in Little Rock."

The White House was in a jam. The president had no reservoir of trust to fall back on with the press. He and his wife could issue all the pleadings they wanted about Whitewater being an inconsequential business deal that had cost them money. Reporters were seeing smoke, and the Clintons themselves were generating at least some of it. And Republicans were sensing something they'd been waiting a generation for: another Watergate, this one starring a Democratic president.

Republicans had been calling for the appointment of a special counsel. Now, Democrats broke ranks to join them. "Just turn it over to the prosecutor and let him find out," New York senator Daniel Patrick Moynihan suggested. Cornered, the president gave in on January 11, 1994. A special counsel was fine by him, he now said. "All the federal investigators in the world that have looked into this—not a single soul has alleged that I've done anything wrong. But I think we ought to answer questions. I think we ought to be accountable."

It was Reno's job to make the pick, and she went with Robert Fiske, a white-collar lawyer in New York who'd been a federal prosecutor years earlier. A moderate Republican with a reputation for professionalism, Fiske's appointment was well received on the both sides of the aisle. He was given a wide berth. He'd look into Whitewater, but also Travelgate and Foster's death, and anything else that might come up. "I have been told time and time again by the top officials of the Justice Department, including the attorney general, that there are no limits on what I can do," Fiske said.

He went to work, but the White House wasn't nearly done generating smoke. A critical moment came in February 1994. The new head of the RTC, Roger Altman, was appearing before the Senate Banking Committee. Altman was an old Clinton pal taking over an agency that, under its previous leader, had urged the Justice Department to investigate Madison Guaranty. It was supposed to be a routine hearing, unrelated to any Whitewater issues, but the committee's top Republican had gotten a hot tip, and when Alfonse D'Amato got his crack at Altman, he shared it with the world. Was it true, D'Amato wanted to know, that Altman had discussed the RTC's own probe of Madison with the White House? Under oath, Altman had no recourse but to fess up. He replied that he'd spoken to aides about what might happen when the statute of limitations for the investigation expired.

And with that, the specter of obstruction of justice—the charge that brought down Nixon—was on the table. Now Republicans were sure they were on to something, and with D'Amato they'd found an eager and aggressive frontman. Not long before, D'Amato had been running for his life back home in New York, damaged from his own prolonged ethics inquiry. That he somehow engineered a narrow reelection victory in 1992 was a source of amazement to Republicans and rage to Democrats. The assumption when he returned to Washington was that he'd seek cooperation with the new president, mindful that Clinton had carried New York in a landslide. Now, fresh off his Altman coup, D'Amato basked in adulation from his fellow Republicans as his own national profile swelled. There was much, much more to Whitewater, he decided, and he volunteered to take point in exposing it: "Howard Baker, I think, asked the right question during the days of Watergate, and it applies here with a little bit of a twist. What did the president know, and when did Hillary tell him?"

Confronted with Altman's admission, no one was more indignant than the nation's paper of record. "President Clinton and his helpers," the *New York Times* editorialized, "keep saying they have nothing to hide on Whitewater. So some evil genie must be making them act as if they do." The *Times* concluded, "Clinton aides behave as if their president had deep deposits of public trust. In fact, that account was pretty slim when Mr. Clinton got to Washington, and it is just about tapped out now."

This was man-bites-dog stuff. The *Times* was famously liberal; it hadn't endorsed a Republican for president in nearly forty years. Now, here it was unleashing its fury on a Democratic White House, with the kind of language it had once directed at Richard Nixon. More to the point, the *Times* was the nation's premier news organization. Its coverage—and its tone—shaped how virtually every other outlet presented news. This was an editorial that relatively few Americans would ever actually read, but virtually all of them would eventually feel its impact.

Democrats still ran Congress, and they'd been protecting the White House as best they could. But it was an election year now, Clinton was struggling in the polls, and the press was all over this story. The Democrats gave in. There'd be two sets of hearings in the summer, one run by the House Banking Committee, the other by the Senate Banking Committee. They kept the scope narrow, but the top Republican on the House committee, Iowa's Jim Leach, still predicted revelations of "blockbuster proportions."

Believe it or not, there was good news for Clinton in all of this. The Fiske investigation was moving along briskly, and so far it was backing him up. The special counsel issued a preliminary report at the end of June. He had finished two investigations, he said. On the matter of whether there'd been improper contact between the White House and the RTC—the obstruction of justice question—there would be no charges. And on the question of Foster's death, Fiske was adamant: There'd been no foul play. It was a suicide likely brought on by stress and panic attacks. There was no evidence, Fiske added, that Foster's anxiety was in any way related to Whitewater; it was the uproar over the travel office affair that got to him.

There was still the issue of the files that were removed from Foster's office—but that probe, Fiske told the press, was now "in its final stages and should be completed shortly." After that, he'd be done with his investigation of the Clinton administration itself and would focus on Arkansas, where his inquiry would probably finish up sometime in 1995, and where the Clintons already seemed to be on more solid legal footing. "I would characterize it as very good news," said Clinton's new White House counsel, Lloyd Cutler, the septuagenarian Washington superlawyer called in after Nussbaum's resignation.

That was precisely the problem. To the far right, Fiske's report was nothing but a sham designed to protect the First Couple. By now, if you were listening to some of the most prominent conservative media voices, shady land transactions in Arkansas were the least of the Clintons' problems, and the Foster case was central to this frenzy.

Pat Robertson, the televangelist founder of the Christian Broadcasting Network, had run for president as a Republican in 1988, beating out the sitting vice president, George H. W. Bush, in the Iowa caucuses before running out of gas. He'd gone on to create the Christian Coalition and still hosted *The 700 Club* nightly on CBN. "Was there a murder of a White House counsel?" Robertson asked viewers. "It looks more and more like that."

On radio, where his weekly audience approached twenty million, Rush Limbaugh touted a claim—with no substantiation—that "Vince Foster was murdered in an apartment owned by Hillary Clinton" and that his corpse had then been transported to the Virginia park and made to look like a suicide. He later told listeners that "the only difference between Watergate and Whitewater is that Whitewater has a dead body."

And if you wanted to go down that road, well, then why stop there? Jerry Falwell, the Moral Majority founder whose *Old Time Gospel Hour* television show commanded a large national viewership, hawked a video that purported to document numerous crimes committed by the Clintons back in Arkansas. The most explosive—and also unsubstantiated—charge featured the son of a private investigator who'd been shot dead: "I think Bill Clinton had my father killed to save his political career."

All of this was stirring grassroots energy, and that energy was making its way to Washington. Of Foster's death, Gingrich said, "There's a lot that is weird." When Democrats announced that the House Whitewater hearings wouldn't tackle the topic of Foster, Indiana congressman Dan Burton took to the floor to protest. "Why aren't these questions being asked?" Eventually, Burton would stage his own version of Foster's death, firing a gun into a watermelon, trying to prove murder. A congressional aide complained to the *Dallas Morning News,* "We can't get any work done here because the Limbaugh contingent is calling us saying the Republicans on the committee ought to walk out of the hearing because this is a charade."

When Fiske's report dropped, there were two strains of reaction from Republicans. The conspiracy theorists saw a cover-up of the truth about Vince Foster. The establishment types sidestepped all of that and argued Fiske had only skimmed the surface when it came to the RTC matter. Between the two factions, there was consensus: there had to be a whole lot more to Whitewater than Fiske seemed interested in discovering.

Here the timing was critical. Just as Fiske was issuing his report, Congress revived the old independent counsel statute. It was a law that had its roots in the Watergate era. When it came to some politically sensitive investigations, the thinking went, a mere special counsel just wasn't independent enough. After all, special counsels—like Fiske—were directly appointed by the attorney general, who in turn was appointed by the president, and officially, special counsels worked under the Justice Department. An independent counsel law built in more separation. For major investigations that reached into the executive branch, a panel of three federal judges would be empowered to choose a prosecutor, and that prosecutor would enjoy free rein—no deadline, limitless budget, unfireable by the president.

Theoretically, Democrats had the power to block the law's revival. Practically, they knew what that would look like. It passed and was signed by Clinton, and in early July Reno formally requested an independent counsel for Whitewater. She asked the three-judge panel to appoint Fiske; he was already on the job and making progress. Instead, on August 5, the judges gave Fiske the boot. They meant in no way to question Fiske's integrity, their statement said, but since Reno had appointed him in the first place, there were "perceptions of conflict." In place of Fiske, they were choosing a lawyer and former judge named Kenneth W. Starr.

Stomachs knotted in the White House. Starr's name recognition with the general public was nonexistent, but in Washington he was well known as a politically ambitious Republican. He'd started out as the chief of staff to Ronald Reagan's attorney general, then won a federal bench appointment from Reagan. Under Bush, he'd been the solicitor general, arguing the White House's side in cases before the Supreme Court. At one point, he'd considered running for the Senate from Virginia. Conservative activists saw him as a future Supreme Court justice.

Did the new independent counsel have a political ax to grind against the administration he was about to investigate? The White House already thought so and saw alarming connections between Starr and any number of its fiercest enemies. For instance, the presiding judge on the panel that chose Starr was David Sentelle, who sat on the federal appeals court for the D.C. circuit. But before that, he'd been a Republican Party leader in North Carolina, closely allied with Jesse Helms, the state's senior senator and a strident Clinton critic. (A few months later, Helms would muse that the president "better have a bodyguard" if he tried to visit North Carolina.) Sentelle's appointment to the federal court of appeals in 1987 had come at Helms's urging.

It also turned out that, just weeks before the Starr appointment, Sentelle had dined with Lauch Faircloth, North Carolina's other senator. Both men insisted it had just been a meal between old friends, but it added a suspicious context to Starr's appointment. At the time of the meal, Faircloth had been publicly advocating Fiske's dismissal, something that grew into a party-wide effort afterward. Dole argued that Fiske wasn't enough of a bulldog. Leach argued that he wasn't going deep enough. D'Amato said he was too tied to the administration. Burton claimed he was letting the Clintons slide. The chorus of Fiske-bashing from Republicans coincided with the panel's final deliberations. Now the Clintons were convinced they were being set up. Sentelle's two colleagues on the panel were both in their seventies and officially semiretired, while Sentelle himself was fifty-one and in his prime. From the White House's standpoint, a partisan judge with a partisan agenda had just muscled through the appointment of a partisan independent counsel.

Many Democrats saw it the same way, but Starr was getting some important cover, too. Members of D.C.'s insular—and bipartisan—legal community stepped forward to vouch for his integrity. Starr himself said, "I intend assiduously to be fair and objective and even-handed and work as hard as I can." And he received the blessing of the *New York Times*, which called Starr's selection "safe and nonpartisan" and predicted he would "move easily" into the role.

The Clintons were in a bind. Starr seemed an obvious threat, but what could they say? The whole point of the independent counsel law was to shield the prosecutor from any White House influence. If they tried to derail him now, they'd be making their critics' case for them. Delicately, they made one attempt. It was done through Robert Bennett, one of the president's personal lawyers. He told the press that he wasn't trying to force Starr out, and that no one should. But, he then said, "I do believe he should voluntarily decide that under all of the circumstances surrounding the appointment he should not serve."

It was a suggestion Starr was only too happy to ignore. And when he did, there was nothing anyone in Clinton world could do about it. Kenneth Starr was now on the job, and he would stay on the job until he decided he was done.

FIFTEEN

Congress adjourned in the wee hours of Saturday, October 8, 1994, with both parties in agreement on exactly one point: there was no sense sticking around even a second longer. The midterms were a month away, members were itching to head home and defend their turf, and anyway, it wasn't like anything was actually going to happen on Capitol Hill anytime soon, a consensus driven home by one final failure for the majority party. On this occasion, the particular issue involved a lobbying reform bill that Clinton badly wanted to sign, but the pattern was a familiar one. Democrats had the votes to push it through the House, but not without Gingrich and his allies in the grass roots and on the talk-radio airwaves kicking up a mighty fuss and forcing Senate Republicans to decide what was more valuable. They could compromise and watch the White House declare victory, or they could stand together and use their forty-four votes to kill the bill with another filibuster.

By now, it wasn't even a choice. Bob Dole, the Senate Republican leader, had supported the bill on its first pass through the chamber, back in the formative weeks of Clinton's presidency, when even a tough nut like Dole assumed his party would have to find at least some common ground with the White House. That was then. Now, Dole announced he hadn't been paying close enough attention to the bill the first time around, that "the more you look at it, the more questions that are raised." He wasn't prepared to move forward on it, and neither was the rest of his party. The filibuster held, the bill died, and Democrats fumed at what they called an abuse of a legislative tool that was supposed to be reserved for rare circumstances. "I think the strategy is quite clear," George Mitchell, the top Senate Democrat, said. "They don't want anything to pass." Gingrich retorted: "It's not obstructionist. It's interpreting the will of the American people."

It had gone down this way more than two dozen times now since Clinton's inauguration, and what really mystified Democrats was just how popular the Republican strategy seemed to be. If there was a political downside to choosing combat over cooperation, Republicans had yet to encounter it. All of the evidence—Clinton's dismal poll numbers, the swirl of scandal around him, the media's critical coverage—told them they'd made the right call. This was a fluke president who'd be gone at the end of his term, and maybe even sooner.

It was getting to Democrats, too. In those exhilarating weeks between Clinton's victory and inauguration, they'd talked of unleashing a legislative hurricane, entrusted at long last with full control of Washington. Clinton had made cultivating the old bulls of Capitol Hill a priority, and they'd eagerly bought in. "I put my hands on his shoulders," boasted Dan Rostenkowski, a product of Chicago's old Daley machine who ruled the House Ways and Means Committee, "and I said, 'Bill, I'm going to be your quarterback and you're my six-hundred-pound gorilla of a fullback moving these bills.'" There'd been agreement all around. A rare opportunity was at hand—"a great treasure," said Richard Gephardt, the number-two Democrat in the House, "and we want to make it work." Now, as Republicans imagined a midterm bounty, Democrats were left wondering just how badly Clinton's stalled presidency would hurt them at the polls.

The diagnosis from one congressman: "He's just not been able to shake people's doubts about his credibility."

Nothing captured the devolution of those grand Democratic hopes, and the emergence of this new Republican swagger, better than health care. For Democrats, it was the white whale, universal health insurance, a lofty goal that Harry Truman had first proposed. Above all, this was the promise of the Clinton presidency and the Democratic moment it ushered in, to make good at last on an idea that had proven painfully elusive for nearly a half century. As Clinton came to office, there was new urgency, too. Costs were soaring and so were the ranks of the uninsured—a total of thirty-seven million, and counting. The issue now had populist salience, too. In what proved to be a precursor to the presidential race, a little-known Democrat, Harris Wofford, had scored a shocking victory in a special Senate election in Pennsylvania in 1991 by making universal coverage his signature message. "The Constitution says that if you are charged with a crime, you have a right to a lawyer," he argued. "But it's even more fundamental that if you're sick, you should have the right to a doctor." When Wofford won, his campaign brain trust—a pair of consultants named James Carville and Paul Begala—joined up with Clinton, who promised Americans he'd deliver a blueprint for universal coverage in his first hundred days. With his inauguration, the pieces were finally in place for Democrats to realize their boldest policy dream.

Back then, even among Republicans, there was a grudging consensus that the 103rd Congress would pass and Clinton would sign some kind of overhaul of the health-care system. They had their doubts about the nature of the new president's mandate, but they were not immune to those same feelings that were animating their Democratic counterparts—that this had been no ordinary election, that the public was expecting real results from Washington, and that they risked looking like obstacles to change if they simply attacked and opposed Clinton. Here was an issue that Clinton had campaigned on; an issue on which, polls showed, voters overwhelmingly shared the broad goal of universal coverage; and an issue on which, those same polls made clear, voters trusted Clinton and his party far more than they trusted the GOP. There were Republicans, like Gingrich, who were ready to fight Clinton tooth and nail on health care from the beginning. There were many others, though, who believed their hands were tied. When the Gallup organization anonymously surveyed members of Congress just before Clinton took office, two-thirds of them called reforming health care a critical matter. In other words, it wasn't just Democrats who were ready to move.

But Clinton himself, it turned out, wasn't ready to move. He quickly created a White House task force to draw up a plan and picked his wife to lead it. That unusual arrangement—by designating the group as a task force, Clinton sidestepped nepotism rules that prohibited him from formally appointing the First Lady to any position—got everyone talking, but only momentarily. It was the surprise conflagration over gays in the military that dominated those early weeks, blowing the new presidency off course. Then came the broken tax pledge, the stimulus debacle, and the draining slog to get the budget through Congress. Health care was collateral damage in all of this. The pledge from Clinton had been to present his plan to Congress by the end of April, but only when he signed the budget in early August could he even begin to contemplate the major decisions he would need to make. Finally, in late September, he scheduled an address to a joint session of Congress. With dramatic fanfare, Clinton held up what looked like a credit card—a health security card, he said, to be issued to every single American as a guarantee of a lifetime of coverage.

"At long last, after decades of false starts, we must make this our most urgent priority: giving every American health security, health care that can never be taken away, health care that is always there," he said.

It was well received, but something was missing: the details. The actual plan still wasn't ready, and wouldn't be for another month, leaving Clinton to speak thematically about the principles he wanted the final package to address. What they heard, though, voters liked.

An ABC News poll showed support running at better than two to one for the plan Clinton outlined; in another survey, voters by four to one said doing nothing on health care would be more dangerous than adopting Clinton's plan. That last finding was not lost on Republicans, who balanced their criticisms of Clinton with assurances that they would work constructively and compromise. "America is ready for health-care reform and so are we," South Carolina governor Carroll Campbell said in the official Republican response to Clinton's speech. In the Senate, Dole threw his support behind one of the most liberal Republicans, Rhode Island's John Chafee, who was creating a plan that relied on an individual insurance mandate—requiring all Americans to purchase private insurance, with subsidies from the government for those of limited means. Nearly two dozen other Republicans would eventually sign on. It was, David Broder wrote in the *Washington Post*, "a legislative strategy . . . that Republicans hope will give them the power to shape the final design" of the plan Clinton would eventually sign. Dole himself predicted there'd be a compromise by the middle of 1994. Republicans, said one of the party's top pollsters, Bill McInturff, "have to be perceived as supporting dramatic change in today's health-care system."

The plan that was promised by the end of a hundred days ended up taking 281 to deliver, with the First Couple finally dropping it off on Capitol Hill on October 27. Its contents amounted to an extremely elaborate balancing act between maintaining a private health-care system and radically expanding coverage. At its heart, it envisioned nudging Americans into health maintenance organizations, or HMOs, the networks of doctors and providers that had been proliferating since their introduction in the 1970s. It would all be organized under a system of state-based "regional health alliances," which would enforce cost ceilings and ensure that all plans offered new, federally mandated benefits, and would rely on an employer mandate—requiring businesses to provide coverage for their workers. Joined by his party's congressional leaders, the president emphasized his overriding mission. "When it is over," he said, "we must have achieved comprehensive health-care security for all Americans, or the endeavor will not have been worth the effort." Clinton's goal was easy to understand, but his means of achieving it were anything but. The document he handed over to Congress ran 1,342 pages—more than 240,000 words. "People reading the full text of his proposal today were left with a bewildering sense of the immense complexity of the proposal," the *New York Times* noted.

The year 1994, Clinton declared in his State of the Union address that January, would be the year of universal health insurance. "I want to make this very clear," he said as he took hold of a pen. "I am open—as I have said repeatedly—to the best ideas of concerned members of both parties. I have no special brief for any specific approach even in our own bill except this: if you send me legislation that does not guarantee every American private health insurance that can never be taken away, you will force me to take this pen, veto the legislation, and we'll come right back here and start all over again."

Time remained Clinton's enemy, though. He had his own bulky health-care plan, but that didn't exactly make it the Democratic Party's health-care plan. From the left soon came legislation calling for a single-payer system, introduced by Washington's Jim McDermott in the House and Minnesota's Paul Wellstone in the Senate. It had zero chance of passage but served as a reminder of the dissension within Clinton's party. More troubling for the White House was a significantly scaled-back version of the Clinton plan drawn up by Tennessee congressman Jim Cooper. It took out the employer mandate and scaled back cost control measures loathed by the insurance industry. Furious liberals accused Cooper of disloyalty ("He's being used by the Republicans," California representative Henry Waxman charged), but it began building real support from moderate and conservative Democrats and even some Republicans. Even Democrats who supported Clinton's plan were communicating reservations. The congressional process was, after all, a negotiation, so there was little incentive for anyone to offer a wholehearted endorsement out of the gate.

Meanwhile, from the other side of the aisle, there was the Chafee plan, a plan from House conservatives, and an endless supply of criticism of the Clinton plan—"the most destructively big-government approach ever proposed," according to Gingrich.

It was such a massive undertaking—the health-care industry accounted for one-seventh of the American economy—that five separate congressional committees claimed jurisdiction. Month by month through the spring and into the summer, they slogged away, struggling to reach consensus. The Clinton plan cleared two of them, but then came another blow: Rostenkowski, the wily congressional veteran who was supposed to be the president's "quarterback," was slapped with a seventeen-count federal indictment on May 31. He was charged with misusing hundreds of thousands of dollars from his congressional office and campaign committee and was facing possible imprisonment. Rostenkowski proclaimed his innocence and vowed to beat the rap, but he had no choice but to hand over his gavel. Instead of Rosty, a peerless negotiator and arm twister, the House Ways and Means Committee—one of the most pivotal venues for any reform effort—would now be helmed by a comparative lightweight, Sam Gibbons.

With the clock ticking (and ticking and ticking), the forces fighting Clinton's plan were only gaining strength. The health insurance lobby was airing a television ad campaign featuring a fictitious middle-class couple, Harry and Louise, fearfully discussing the potential implications of Clinton's plan at their kitchen table ("Having choices we don't like is no choice at all"). The employer mandate had the business community up in arms. It didn't help Clinton's cause that the Chamber of Commerce and National Association of Manufacturers were opposing him; far more damaging, though, was the engagement of the small-business lobby, which held particular sway with the moderate Democrats the White House most needed. Multiple committees, rival proposals, and well-funded outside groups all confronting an issue of labyrinthine complexity: it was a perfect recipe for confusion, and the polling confirmed it. Voters still said they shared Clinton's bottom-line priority of universal coverage. But a majority now said the issue was more difficult for them to understand than when the debate began, and two-thirds now believed doing nothing would be better than passing Clinton's plan—a complete reversal from when he made his first speech to Congress. It would be best, the vast majority said, if Congress just put the whole thing aside and started fresh next year. You could feel the Republican psychology moving Gingrich's way: maybe, just maybe, they didn't have to say yes to anything after all.

When August arrived, Democrats launched a final, frantic push. If they couldn't finish the job before that month's recess, they realized, they'd never get it done in the fall, and then they'd be left to explain to the voters in November why they had failed to deliver on their signature promise. They were still all over the place, though. In the more liberal House, Democratic leaders were ready to bring to the floor a bill that met Clinton's demand for universal coverage— but they still didn't have the votes to pass it. It was time to turn the screws. There would be no August recess, Speaker Tom Foley announced, until the health-care debate was settled. Meanwhile, on the Senate side, the Democratic leader, George Mitchell, determined that it would be impossible to pass a bill as expansive as the House's, so he trimmed his sails and offered a compromise: a bill that gutted the employer mandate and aimed for 95 percent coverage—"not universal coverage," he admitted, but it would "set us on the road to universal coverage." The challenge was imposing by any standard. Neither the House plan or Mitchell's new plan had the votes to pass either chamber, but they would both have to—and soon—and then somehow be reconciled and passed again. Clinton tried to rally the public. "Don't let the fearmongers, don't let the dividers, don't let the people who disseminate false information frighten the United States Congress into walking away from the opportunity of a lifetime," he pleaded.

However, in the end, growing fears from voters about losing their doctor, paying higher prices, and being subjected to new layers of government bureaucracy won the day. The terms of debate had changed completely. People were less worried now about how bad the system was and more fearful of how much worse it could get. There'd be no compromise from the loyal opposition, and on September 26, Mitchell surrendered on behalf of his party. "The combination of the insurance industry on the outside and a majority of Republicans on the inside proved to be too much to overcome," he said. Clinton pointed his finger at the same culprits and promised that "this journey is far, far from over." But it was over for 1994, and potentially well beyond that. "For all the right reasons, health-care reform did not happen this year," Dole said. "The American people said, 'Slow down.'"

There was other business on the 103rd Congress's agenda, but the pattern was locked in place. Even where there'd previously been Republican support—like on the lobbying bill, or on a global trade pact known as the General Agreement on Trade and Tariffs—there was now only delay and opposition, until finally Democrats threw up their hands and agreed to just go home. "Thus is the year on Capitol Hill drawing to a close," the *Los Angeles Times* wrote in early October, "with neither a bang nor a whimper, but with a partisan rage so deep that many Democrats and Republicans were barely on speaking terms." From the *Washington Post:* "Twenty-one months after coming to town, promising that a unified Democratic government would break the gridlock that had marked the Democratic Congress's relations with Republican presidents, Clinton and his Democratic supporters found themselves caught in the same old legislative paralysis."

Clinton blamed it on the other party and their strategy of "stop it, slow it, kill it, or just talk it to death." He would now have one month to make his case to the voters. Every seat in the House and one-third of the Senate would be on the ballot. There was no question Democrats were going to take a hit, but their numbers were so large that they could afford losses—just not too many. What progress could Clinton and his party point to? They'd whiffed on their biggest promise, universal health coverage, and used a party-line vote to pass a massively unpopular tax hike. NAFTA was less poisonous with voters, but more Republicans had supported that than Democrats.

There was this, though: the core promise of Clinton's 1992 campaign had been to revive the economy, and on this front the news was encouraging. Unemployment, close to 8 percent when Clinton unseated Bush, was now under 6 percent, its lowest level in four years. More than four million new jobs had been added. Inflation was low, too, and the deficit was heading south—from $255 billion in 1993 to $202 billion now, with another projected drop to $162 billion for the next year. When Clinton raised taxes in the summer of 1993, Gingrich had predicted that "this will lead to a recession next year." Now, more than a year later, there was no recession, not even close. "The record is a good one," Clinton said, "and there is ample evidence that if people knew the record, they would respond to it."

The polls said otherwise. But for a few blips, Clinton's standing had never recovered from those traumatic early months of his presidency. As Congress adjourned, Gallup put his approval rating at 42 percent. For all the rosy news about jobs, a majority disapproved of his handling of the economy, and even though the recovery was now two years old, nearly 60 percent said the country was still in a recession. A majority believed the deficit had gone up under Clinton, not down. Was this a failure of Clinton to explain himself and his record? Or was it something deeper?

Gingrich, the Republican leader in waiting, was calling the shots when it came to midterm strategy, and here he had an idea. One of the keys to breaking the Democratic House, he'd always argued, was in finding a way to nationalize congressional elections. In the past generation, Nixon and Reagan had both won forty-nine states in presidential elections, and Bush had taken forty in 1988. That meant that each of them had carried scores of congressional districts that were controlled by Democrats. If Republicans could just convince the voters in those districts to think of their local congressional race as an extension of presidential politics, they'd be poised for a serious breakthrough.

To pull this off in 1994, Gingrich envisioned a national platform, a common set of proposals that every Republican candidate everywhere in America could run on. The proposals would have to be simple, easy to communicate, and have instinctive, populist appeal. He commissioned polling. If an idea didn't have 60 percent support, it wouldn't be included. The planks he came up with included calls for term limits, tax cuts and credits, more defense spending, tort reform, and a balanced budget amendment. He merged all of it into ten pieces of proposed legislation with one big promise: give Republicans control of Congress and each bill will receive an up or down vote in the first hundred days. It would be called the Contract with America, and at the end of September Gingrich unveiled it to the country.

It was, if nothing else, a truly impressive display of party unity. All but five of the 157 Republican House incumbents seeking reelection signed it. Another 185 Republican candidates, challenging incumbent Democrats or seeking open seats, added their names as well. "Our government operates on the party system," Gingrich said. "We are a team. And we're offering you a contract on what our team will do." The reviews were merciless. "The House Republicans' 'Contract with America' is filled with alluring campaign promises—and missing realistic ways to pay for many of them," sniffed the *Washington Post*. "It is a vision that looks backward, Reaganism in a rearview mirror," the *New York Times* editorialized. "What this self-styled 'Contract With America' says to voters is that these Republicans do not speak candidly." Clinton took to calling it the "contract on America."

Referring to the American public, Clinton asked: "Do they really want this contract, which is a trillion dollars of unfunded promises, a contract which will certainly lead to higher deficits, cuts in Medicare, and throwing us back to the years of the eighties, when we lost jobs and weakened our country? Or do we want to face up to the challenges which were not met in this Congress and use the next Congress to keep the economic growth going, to pass health-care reform, to pass welfare reform, to pass political reform, and to deal with these environmental issues?"

He hit the road on Halloween and would stay there for the next eight days, one stump speech after another, until it was time for Americans to vote. There were plenty of Democrats ready to welcome the president's assistance, especially his financial assistance, but from major swaths of the country, especially in the South and mountain West, another message was being conveyed to the White House: it would probably be best if he just stayed away. In 1992, Clinton had made inroads in every region; now, in many of those same places, he was an albatross. In race after race, Republicans were using a new technique in their television advertising, slowly morphing an image of the Democratic candidate into one of Clinton—one more way to nationalize the midterm.

Back in his natural environment, Clinton gained energy, basking in the cheers of the party faithful and honing his final pitch to the masses. He repeated his attacks on the GOP contract, framed the election as a choice between the future and the past, and implored his crowds to "turn on the lights!" He had insisted when Congress adjourned that he had a good record to share with voters, and now he had more to tell them. He'd just returned from the Middle East, where a historic peace treaty between Israel and Jordan had been signed, and the unemployment rate was falling again, down to 5.8 percent for the month of October. "Well, folks, the sun has begun to shine in this election," he said. "Every day, more Americans are beginning to know that the real issue here is who will fight for ordinary Americans."

He could find rays of hope, none brighter than the one in New York, where Mario Cuomo was staging a furious comeback. That it had come to this in the first place was stunning enough. Just three years earlier, it had been Cuomo who paralyzed the entire political universe—and infuriated Clinton—as he agonized over whether to run for president, and for a decade now he'd been the leading voice of American liberalism. But since leaving that plane idling just before Christmas 1991, Cuomo's fortunes at home had darkened dramatically. Chronic budget woes, dysfunction in Albany, and popular fatigue with a governorship well into its third term had all exacted a toll as Cuomo's approval rating fell into the thirties. It had been a surprise when Cuomo announced he'd go ahead and seek a fourth term anyway, and now, on top of all the other baggage, he was contending with the national tide Clinton's presidency had stirred up.

His Republican challenger was George E. Pataki, a state senator and mayor from Peekskill, a small city in the Hudson River Valley about fifty miles from Manhattan. On paper, it was a thorough mismatch: the famed orator who would have been president against an obscure local official. But the blank slate was winning, and by a wide margin—or at least he had been. Then, in mid-October, things began to shift. Cuomo, never better than when he was in front of a crowd, sharpened his attacks and found his opponent's weak spot. Pataki's unlikely political ascent had come by the political muscle of his patron, Senator Alfonse D'Amato. In Washington, D'Amato's stock was rising as his Whitewater crusade inflicted one political wound after another on the Clintons. Back home in New York, he was spreading his wings. Two years before, he'd come close to losing his seat. Now, if he could use Pataki to muscle Cuomo out of the governorship, he'd stand as the dominant political force in one of the nation's most important states. The Cuomo team took to ridiculing Pataki as D'Amato's puppet, even running an ad featuring Paul Simon's "You Can Call Me Al." They were making the Republican nominee seem small, and also stoking a bitter rivalry between D'Amato and the state's other leading Republican, Rudolph Giuliani, who had been elected mayor of New York City the year before. The two men detested each other. At its core, it was a power struggle, and Giuliani had just made a surprise move to gain the upper hand: claiming that

Pataki "personifies the status quo of New York politics," he threw his support behind Cuomo. Now the two men were campaigning together and Cuomo was zooming up in the polls, first drawing even and then pulling ahead. Days before the election, a survey put him up by eleven points.

That was Thursday, November 3, the same day that Clinton arrived for a campaign swing with Cuomo. They'd once looked on each other with suspicion and even hostility, but the friction—and the rivalry—was in the past. Win or lose, Cuomo, at sixty-two, was running his final campaign. They'd patched up the personal relationship, too. Cuomo had been moved when Clinton reached out and asked him to deliver the nominating speech at the 1992 convention, and now Cuomo's son, Andrew, was a rising star in Clinton's Housing and Urban Development department.

The rally was in Albany, where the local television stations carried it live. "They're watching you, New York," Cuomo declared, "to see where you believe the country should go." Clinton hailed the governor as "the real comeback kid" and addressed the mighty current both men were trying to swim against. "Yes, there is frustration in the electorate," he said. "But you know, the first thing you try to teach your children as a parent, once they really begin to be aware of the world, is not to make an important decision when you're mad. If you're mad, count ten before you talk. How many times were we raised with that? What the Republicans want for you to do is go in and vote before you count to two." The crowd loved it. Two weeks earlier, Mario Cuomo had looked dead in the water. Now he was on the verge of a fourth term. It might not even be close.

When the rally was over, they went their separate ways. Cuomo was due in Brooklyn, where Giuliani would be meeting him. Clinton had candidates to see in Iowa and Minnesota and California and Washington State. They were going to lose more races than they won, Clinton recognized. But maybe, just maybe, it wouldn't as bad as it once looked. The embodiment of unreconstructed liberalism was back from the dead in New York. Could it be an omen, a harbinger, a sign

"Don't stop thinking about tomorrow." America's first baby boomer president at a star-studded inaugural ball on the night of January 20, 1993. Along with his daughter, Chelsea, Clinton is joined onstage here by Michael Jackson, with comedian Chevy Chase and Fleetwood Mac's Stevie Nicks just behind him.

Bill, I'm going to be your quarterback and you're going to be my six-hundred-pound gorilla of a fullback." President-elect Clinton with Dan Rostenkowski, chairman of the House Ways and Means Committee, the most powerful of the old bulls whose help Clinton would need.

Gingrich addresses a May 1993 rally against the Clinton budget plan while Bob Michel, who will soon announce his retirement, stands in the background.

By creating a special health-care task force, Bill Clinton manages to skirt nepotism rules and appoint his wife to run it, but the resulting backlash against "HillaryCare" is fierce.

With an estimated twenty million listeners per week, Rush Limbaugh uses his radio show to stir up opposition to President Clinton. After the Republican "revolution" of 1994, he is named an honorary member of the freshmen GOP class.

"**You are my Speaker now.**" House Democratic leader Richard Gephardt hands over the gavel to new Speaker Newt Gingrich on January 5, 1995, officially ending forty years of Democratic House rule.

Dole and Gingrich exit the White House in November 1995, after last-second negotiations fail to avert a government shutdown. Polls would soon show the Speaker's calculation that voters would blame Clinton was misguided.

As the shutdown drags on, Democrats like New York Rep. Charles Schumer seize on Newt Gingrich's complaints about his treatment as a passenger on Air Force One one of numerous media firestorms ignited by the Speaker's penchant for thinking out loud.

Encouraged by Clinton's apparent weakness, a sizable Republican presidential field emerges. Shown here at an October 1995 debate in New Hampshire: Top (from left to right): Former diplomat Alan Keyes; Pennsylvania senator Arlen Specter; Illinois tire manufacturing magnate Morry Taylor; Indiana senator Richard Lugar; Texas senator Phil Gramm. Bottom (left to right): Former Tennessee governor Lamar Alexander; Senate Majority Leader Robert Dole; publisher Malcolm S. "Steve" Forbes Jr.; California Rep. Robert Dornan; political commentator Patrick J. Buchanan.

It is Buchanan who steals the show at Ross Perot's August 1995 United We Stand conference in Dallas, bringing the crowd to its feet with his attacks on free trade, immigration, and multiculturalism.

"**Mount up and ride to the sound o[f] the guns!**" In a jolt to the Republican establishment, Buchanan knocks off front-runner Bob Dole in the February 20, 1996 New Hampshire Republican primar[y]

Ultimately, Dole manages to halt Buchanan's rise, but the process is draining, and by the time he faces Clinton in the fall election, the outcome already seems determined.

A Balanced Budget
That Protects Our Families, Invests in Our People and Cuts Taxes for Middle Class Families

President Clinton signs the Balanced Budget Act of 1997, bipartisan legislation enacted with the support o[f] Speaker Gingrich. The first years of Clinton's second term are marked by political tranquility that proves fleeting.

"Indeed, I did have a relationship with Ms. Lewinsky that was not appropriate." Cornered by prosecutors, the president makes an August 1998 admission that leaves his personal image in tatters—even as his job approval rating remains strong.

"When you're in her presence, you feel a bit of a tingle." The First Lady, catapulted by her husband's scandal to a previously unimaginable level of popularity, campaigns across the country in 1998, but nowhere more intensely than in New York's pivotal Senate race.

Rasputin with a Long Island accent. After likening the Whitewater scandal to Watergate, New York Republican Alfonse D'Amato used his powerful Senate perch to pursue an investigation only to see it redound against him as the politics of his home state shifted.

"Compassionate conservatism." Texas governor George W. Bush supports his home state San Antonio Spurs against Spike Lee's New York Knicks at the 1999 NBA Finals. Bush was rapidly unifying the Republican establishment with a promise to broaden the party's appeal.

Billing himself as a moderate alternative to "Nazi-lover" Pat Buchanan, Donald J. Trump registers with the Reform Party and pursues a potential presidential candidacy in 1999, appearing here on a special edition of *Hardball with Chris Matthews* at the University of Pennsylvania.

that, with not a moment to spare, Democrats were finally turning the lights on?

E lection Night, 1994. The first call came at seven thirty. The polls were closing in Ohio, a big Midwest battleground that had voted for Bush in 1988 and switched to Clinton in 1992. Democratic senator Howard Metzenbaum, a three-term liberal stalwart, was retiring, creating an open-seat contest. Republicans had put up the state's lieutenant governor, Mike DeWine, who ran on the year's familiar GOP themes—antigovernment, anti-Washington, anti-Clinton. Democrats had kept it in the family, nominating Metzenbaum's son-in-law, Joel Hyatt, who in this charged climate had tacked toward the middle, particularly with his support for the death penalty. Once upon a time, this had loomed as the purest of toss-ups, but now it was a certifiable rout for the Republicans. DeWine was on his way to a double-digit victory, and Republicans had their first net Senate gain of the night. They would still need six more to win back the Senate. The Ohio massacre was a good start, but by this point it wasn't much of a surprise. The exit polls were encouraging, but the overall picture was fuzzy. Just what kind of night was this going to be?

Eight o'clock brought a burst of clarity. In Maine, where a frustrated Majority Leader Mitchell was calling it quits after twelve years, Republican congresswoman Olympia Snowe was the winner, easily, over the state's other House member, Democrat Tom Andrews. Another Republican pickup was being called in Oklahoma, where Democrat David Boren had abruptly retired months earlier. Democrats had nominated one of the party's most conservative House members, Dave McCurdy, but his Republican opponent, Representative Jim Inhofe, told voters McCurdy would just empower Clinton's big-government agenda. The voters were siding with Inhofe in another landslide.

Then there was Tennessee, where two Senate races were on the ballot. One was an open contest for what had been Al Gore's seat before he became vice president. There, the Democratic candidate was Jim Cooper, the centrist congressman who had bucked the White House on health care, running against Fred Thompson, who'd parlayed a stint as a Republican lawyer on the Senate Watergate Committee into an acting career. The other seat belonged to Democrat Jim Sasser, who'd been one of Clinton's most unrelenting Senate boosters in the budget fight—so much so that Limbaugh had taken to using clips of Sasser making the case for tax hikes as a recurring bit on his show. He was being challenged by Bill Frist, a heart surgeon and first-time candidate who'd been so detached from politics that he hadn't even voted until the age of thirty-six. Cooper and Sasser: two Democrats, one a very public thorn in Clinton's side, the other a very public ally. Tennessee's voters didn't see much difference, though. Both were getting blown out. So, for that matter, was the Democrat running for the open governorship. Here was a state that Clinton had carried in 1992, a powerful example of his ability to win back turf his party had all but surrendered over the last generation—a state that had launched and incubated his own vice president's political career. "A bitter, bitter defeat for Al Gore," analyst Bill Schneider said on CNN. "He's seen the state of Tennessee elect two Republican senators and now a Republican governor."

Now the net gain was five Senate seats for Republicans. A takeover was within sight. The full picture still wasn't clear, but now there were signs of a wave forming at all levels of the ballot. Republican governors were racking up massive margins in some big places— Ohio, Michigan, Illinois, Massachusetts. There were already House gains in Indiana, Ohio, Kentucky, and Georgia. Democratic incumbents were fighting for their lives all over the place, but it was hard to find a Republican officeholder who was doing much sweating. There were still close races, many states and districts yet to report, but the Permanent Democratic Congress was now facing its gravest threat ever. "If these trends continue," Tim Russert said on NBC, "it's going to be the worst Democratic appearance in fifty years."

Errant scraps of good news sustained Democrats, at least for the moment. Ted Kennedy was called the winner in Massachusetts, earning his seventh term in the Senate. But the story here was that it had ever been competitive in the first place, a consequence of the public airing of Kennedy's dirty laundry a few years earlier. This included, most notably, humiliating details—trumpeted on the cover of *People* magazine—of a night of carousing by the then-fifty-nine-year-old senator that ended with his thirty-year-old nephew, who was staying with him at the family's West Palm Beach compound, charged with rape. (William Kennedy Smith was ultimately acquitted.) Facing the voters for the first time since that night, Kennedy found himself trailing in a September poll against his Republican challenger, a political newcomer named Mitt Romney—the first time a Kennedy had ever trailed in a poll in Massachusetts. But, as Bruce Morton now told CNN's viewers, "Kennedy, pouchy, paunchy, looking so bad, in fact, that one critic referred to him as a manatee in a suit, nevertheless managed to win. He was helped by the debates, which proved—he had set very low expectations—that he could speak a sentence in English." This, so far, was the bright spot for Democrats.

Michigan went next. The Republican Senate candidate, Spencer Abraham, had until recently been building a career as a Beltway power player. He'd served on Dan Quayle's vice presidential staff, organized the GOP's national House campaign effort in 1992, and then sought—and nearly won—the chairmanship of the national party. But when Democratic senator Don Riegle announced his retirement, Abraham jumped into the race and spent the campaign attacking the Democrat, Representative Bob Carr, as a Washington insider. Abraham was the winner by nine points. That was GOP pickup number six. Number seven came soon, when the polls closed in Arizona. Another open seat, another Republican win—Jon Kyl, to replace the retiring Democrat, Dennis DeConcini.

It wasn't official yet. Democrats still had a shot at picking off one Republican seat, but that became moot when the late returns were finally tabulated in Pennsylvania. All night, Harris Wofford had been running ahead of his Republican challenger, a second-term congressman from the blue-collar western part of the state, Rick Santorum. Here was a race that Democrats wanted badly, not just for Senate control, but for its symbolism, with Wofford's name still synonymous with the cause of universal health care. The late precincts broke the other way, though, and when the networks declared Santorum the winner, the Democratic Senate was gone.

On CBS, Dole was being interviewed by Dan Rather. "There are a lot of people out there who do not support the Clinton agenda," he said. A decade earlier, Dole had been elected to replace Howard Baker as the top Senate Republican. For two years, he'd held the title of majority leader, losing it in the Democratic tide of 1986. Now, after eight long years in the wilderness, he would at last reclaim it. Unless—well, unless Dole was looking at this Republican tide and thinking ahead to 1996. He'd already run for president twice before. Would he perhaps forego another stint as majority leader and instead throw himself into one more quest for the White House? "Well, I don't see why you can't do both," he told Rather.

This was momentous by itself. But the picture coming into focus now was far more dramatic than a transfer of power in the Senate. In New York, the Cuomo comeback was turning into a mirage. Trailing slightly with half the votes in, his campaign expressed confidence: the late votes would be from New York City, Cuomo Country. Instead, they were from the Long Island suburbs, and they went Republican. It had been ten years since that magical summer night in San Francisco, when Cuomo opened his mouth and captured millions of imaginations, and now he was done, finished off by the mayor of Peekskill. "You'll recall," CNN's Bernard Shaw told viewers, "that President Clinton campaigned on behalf of Mario Cuomo, but apparently it was not enough." In the suburbs of Philadelphia, meanwhile, Marjorie Margolies-Mezvinsky, the congresswoman who had given in to Clinton's pleadings and cast a vote that saved his budget, was going down, too. She'd outspent her opponent, Republican Jon Fox, who had run a lackluster campaign, but apparently that also wasn't enough.

Even more jarring was the news flash from Chicago: they'd

knocked off Rosty. Even the indictment that forced him to hand in his committee gavel wasn't supposed to stop Dan Rostenkowski from winning a nineteenth term in Congress. This was one of the most Democratic districts in America, machine territory where there was no organized Republican Party to speak of, and he was a local institution. Out of curiosity, national Republicans had taken a poll in the district just a week earlier. The numbers confirmed what was previously unimaginable. The unknown gadfly candidate who had claimed the spot on their line, Michael Patrick Flanagan, could win. They rushed in money, Flanagan went on the air, and now here was old Rosty, with the national networks carrying it live, conceding defeat. "I'm going to go to Washington and clean out my desk," he told the crowd.

It was happening everywhere now, with results pouring in, Democrats fighting for their lives in districts that had never been threatened before. The South, that region they'd dominated in so many presidential races over the last thirty years, was fertile for Republicans, but it didn't stop there. When the West Coast reported, they knocked off four Democrats in Washington State, including the ultimate prize: Tom Foley, now the first House Speaker since Galusha A. Grow in 1862 to be defeated for reelection in his home district.

Now the networks were all declaring it: Republicans were going to hit the magic number in the House, and then some. The Permanent Democratic Congress, born in Eisenhower's first term, had survived Vietnam, the malaise of the Carter years, and national landslides by Nixon, Reagan (twice), and Bush—but not this. It had barely even been threatened in all that time. And now it was gone, felled after two years of unified Democratic control of Washington and unified Republican opposition. There would be a Republican majority in January. The Speaker's gavel that Sam Rayburn, Tip O'Neill, and so many other lions of the institution had once wielded would now be handed over to Newton Leroy Gingrich. "Even the most cockeyed optimist in the Clinton administration may be having some trouble at this hour," Ted Koppel said on ABC. The elder statesman of the network's coverage, David Brinkley, concurred. "There is a revolution in the air."

The leader of the revolution was trying to restrain himself. There wasn't a soul in Washington who didn't know all about him, and plenty of Americans had at least heard his name. But for most of his countrymen, their exposure to Gingrichism had so far been indirect, through the actions of a Republican Party carrying out his strategic vision. They were still waiting to meet Newt the man, and he wanted to make a good first impression. As he made the network rounds now, he tried to project humility. "It feels almost a little bit intimidating," Gingrich told Rather on CBS, "almost in the sense that to go from being whip to being minority leader would have been a pretty big jump, but to go from being minority whip in the Republican Party to potentially being Speaker in one step is an enormous jump." He talked about leaning on Bob Michel for advice. He paid tribute to his soon-to-be-former colleague Rostenkowski. He even said he wanted to find ways to work with Clinton. The words were there, but did he really know any other style, any other speed, than what he'd shown for the last sixteen years—the formula that had just brought him to the very top of a world that not long ago had ridiculed him as a pompous pest? "One of the interesting things to watch," said Tom Brokaw on NBC, "will be whether he can go from being a backbencher and a guerrilla in the House to being a statesman and a leader. He insists that he can."

Back on ABC, anchor Peter Jennings and analyst Jeff Greenfield were looking for the message behind the national landslide. "This shift of historic proportions is taking place in a year when there's no historically explainable reason—no war, no scandal, no recession," Greenfield said. "It's happening for reasons we can speculate. If you don't mind an early one: they don't like Bill Clinton."

"No kidding," Jennings replied.

Clinton waited until the next afternoon to say something. There was no template for an occasion like this. Presidents before him had endured midterm rebukes, but nothing in modern times could rival the political trauma the electorate had just unleashed. Republicans had picked up fifty-two House seats, the biggest jump in a single election since 1948. Their net Senate gain stood at eight at the end of Election Night, and now, on the day after, it had jumped to nine. Richard Shelby, an Alabama Democrat who had been at odds with his party's agenda the last two years, announced he was switching to the GOP. "I had high hopes for Bill Clinton," he explained. "There's not any room for . . . a southern conservative like me in the Democratic Party."

The carnage crossed demographic lines and extended into every region, and not just in congressional races. Before the election, Republicans controlled twenty of the nation's governorships; now they would have thirty. California, Texas, New York, Pennsylvania, Illinois, Ohio, Michigan: seven of America's eight largest states would now be governed by Republicans. Four times in the last generation—in 1972, 1980, 1984, and 1988—Republicans had won landslides in presidential elections. Now, at long last, the trend was extending all the way down the ballot. Gingrich had always said that America was a fundamentally conservative country, and that Republicans would be rewarded with a limitless future if they could just show that they—and they alone—understood its values. Scanning the political landscape now, it appeared his prophecy had been fulfilled.

There wasn't a single state that Clinton could now call a lock for himself in his reelection campaign. The concept of red states and blue states, bastions of tribal loyalists where voters would stay true to their party no matter the national climate, was for the future. In this moment, there were just states, fifty of them, some more friendly to one party, some to the other, but not a single one of them off-limits to the surging Republicans. Their strength was especially intense among evangelical Christians, who made up a fifth of the electorate; gun owners; and white males; but the scope of their rout was broad.

In their shock on this day after, Democrats were pondering the unimaginable, the possibility that they were living out the end stages of their own history. The *Boston Globe* described the mood on Capitol

Hill: "A whole way of life that had developed in the Democrats' four decades of dominance in the House—a set of assumptions about power and comportment—suddenly was rendered irrelevant. Never before had one party ruled the House for so long, and, in an instant, the whole architecture of life in Washington was in a shambles."

Suddenly, a chilling arc seemed plausible: starting with Goldwater in 1964, Republicans had made themselves the party of anti-government conservatism, setting off a steady, three-decade rise that was interrupted briefly by Clinton's accidental 1992 win, which then kicked up a climactic backlash that would now usher in an era of complete, top-to-bottom conservative Republican domination. Was this Republican Revolution the final, irrefutable statement of the American people that they believed Reagan's famous line that "government is not the solution to our problem—government is the problem"? Could the Democratic Party as it had long been constituted continue to exist in a country like this? Was this what extinction looked like? It sounded too dramatic, perhaps, but in the midst of such devastation, who could really say?

Crisscrossing the country just before the vote, Clinton had argued the GOP contract would return the country to the "trickle-down economics" of the Reagan years. Now the eighty-four-year-old Gipper, less than a week after revealing his diagnosis with Alzheimer's disease in an open letter to the public, spoke up one final time. "The American people have sent an unmistakable message about the direction they want our country to take," he said in a statement. "It is clear support for the policies of less government, lower taxes and greater individual liberty." In his syndicated column, conservative George Will wrote that "it is immeasurably satisfying that three days after Ronald Reagan announced his final battle, his countrymen gave him his third national victory."

This was the atmosphere as Clinton appeared in the East Room of the White House on Wednesday afternoon. CNN was airing his remarks live, and so were the broadcast networks. His voice a notch lower than usual, Clinton started by ticking through the same items he'd just touted on the campaign trail. The deficit was down; five million new jobs had been added; trade was expanding; and a stepped-up assault on the "terrible plague of crime and violence" was under way.

"Still," he conceded, "in the course of this work, there has been too much politics as usual in Washington, too much partisan conflict, too little reform of Congress and the political progress, and though we have made progress, not enough people have felt more prosperous and more secure, or believed we were meeting their desires for fundamental change in the role of government in their lives. With the Democrats in control of both the White House and the Congress, we were held accountable yesterday, and I accept my share of the responsibility in the result of the elections."

There was no comeback blueprint for this, if a comeback was even possible. Soon, polls were commissioned testing the 1996 presidential race. Matched against Dole, the closest thing the GOP had to a front-runner at this early moment, the president was running behind—not just nationally, but even in New York. There really wasn't a single state he could count on. Nor was it clear his own party would stand by him. "This election was a severe, sharp, and obvious repudiation of the president," said Bob Kerrey.

"It is still only a question, but a politically portentous one, and it is on the lips and in the minds of prominent Democrats across the country," the *New York Times'* R. W. Apple wrote. "Can Bill Clinton—should Bill Clinton—be the party's presidential nominee in 1996?"

Shortly before Christmas, Clinton made his first major move since the election. It was a prime-time speech to propose what he called "a middle-class bill of rights," a counteroffer of sorts to the Republican contract and an effort to refocus his presidency on the themes he'd run on. The centerpiece was sixty billion dollars in tax cuts for middle-income families. The problem: this was the same Clinton who had campaigned on a middle-class tax cut, then junked it, then claimed he'd never really emphasized it in the first place, then pushed through a tax hike that cost middle-class Americans more money—all in the name of fighting the deficit. Now here he was calling again for a middle-class tax cut and offering few details on how it would be paid for. The Slick Willie taunts started up again.

"Rather than try to persuade the public that the deficit is a corrosive force that must be brought under control to keep it from eating away at the country's economic roots, a view held by most economists but relatively few politicians, Mr. Clinton has opted for the course Republicans have followed for years," the *New York Times* wrote. "For him, as well as for his political opponents, lowering the deficit will play second fiddle to lowering taxes."

Within the Democratic Party, it reopened old wounds. Tsongas, now chairing a deficit watchdog group, characterized the president as "a threat to the well-being of my children." He'd been supportive on Clinton's deficit-reduction agenda the year before, but now Tsongas said he felt duped. "As long as Bill Clinton is president of the United States, you will not have either a balanced budget or be on the path to a balanced budget," he said. "That is a fact, and people who care about what we leave behind ignore it at their peril." Was Tsongas, who'd suffered a cancer relapse after the '92 election, thinking of running against Clinton again? No, he said, but "after [Clinton's speech], I would not support him against a Bob Kerrey, and I would not support him against a Colin Powell."

With his speech, Clinton aimed to reset his presidency and reclaim the initiative, but the maneuver was blowing up in his face. The press was accusing him of flip-flopping yet again, his own party was attacking him, and Republicans were just ignoring him. This was how 1994 ended for the White House. Soon, the new Congress would be sworn in and the first Republican majority in four decades would elect Newt Gingrich as its Speaker. The Senate would reorganize under Republican control for the first time since the mid-eighties, with a majority leader who was now widely seen as Clinton's successor. No president since Nixon at the very end had suffered though such a devastating year—and the next one was promising to be even worse.

SIXTEEN

The first thing they did was stick it to Harvard. In a tradition that went back a generation, the university's John F. Kennedy School of Government was the official host for an orientation program for incoming members of Congress. For a week before their swearing-in, newly elected Democrats and Republicans would be briefed by policy wonks and political practitioners on a range of topics, with some after-hours mingling thrown in. For the pols, it was a chance to bond with each other, brush up on key issues they'd soon be confronting, and build connections to knowledgeable experts. For the Kennedy School, it was a prized mark of its own prestige and stature. It was a creation of the Permanent Democratic Congress, but Republicans had always gone along, and the 1994 session was set for the second week of December.

There'd never been a freshman class like this one, though. They were almost all Republicans—seventy-three of eighty-one of them, the most incoming Republicans since the FDR backlash of 1938—but more than that, they were Gingrich Republicans. When they looked at the Kennedy School, they didn't see a partner; they saw a breeding ground for the elite liberal culture they yearned to dismantle. Sure, the faculty included its share of Republicans and the school's leaders took pains to proclaim the political neutrality of their mission, but the Gingrich Republicans could see right through this. Those Republican faculty members? They included Richard Darman—the same Richard Darman who as Bush's budget director had been instrumental in the 1990 tax hike. The Clinton administration, meanwhile, had dipped deeply into the Kennedy School's roster, and but for a few pockets, the prevailing sensibility all around Harvard skewed unquestionably to the left. The Gingrich Republicans knew the score.

They had a choice, too. A rival orientation program was now on the calendar. Same week, same idea—except this one had no pretense of bipartisanship. Its organizers were all on Team Newt: the Heritage Foundation, a twenty-one-year-old conservative think tank that had provided grist for Gingrich's Contract with America; and Empower America, an activist group launched the previous year by a trio of Gingrich allies. They'd lined up a murderers' row of right-of-center attractions. William Bennett, the former education secretary and drug czar whose *Book of Virtues* had now been on the bestseller list for a year, would be speaking. So would Charles Murray, the conservative social scientist whose new tome, *The Bell Curve*, was stirring controversy with its exploration of the distribution of intelligence in American society. And Jeane Kirkpatrick, the Reagan-era U.N. ambassador who'd famously dubbed Democrats the "Blame America First" party. And Jack Kemp, and Ralph Reed of the Christian Coalition, and Paul Gigot from the *Wall Street Journal*, and on and on. The Republican freshmen were calling themselves revolutionaries, and here was their invitation to strike their first blow. All they had to do was say yes to the rump session—and no to those snobs in Cambridge. It was an easy call.

Harvard pulled the plug just before Thanksgiving. "If only one party participates," the school's spokesman explained, "we are not interested." Those eight freshmen Democrats would have to find another way to get themselves oriented. The revolutionaries, meanwhile, made their way to a Baltimore hotel for three days of speeches, panels, and seminars. "Just think," one of the organizers joked to them, "if you were at Harvard you could be listening to Michael Dukakis." The real treat was saved for the end. The Saturday night closing banquet was headlined by a man who had never sought office but who commanded a constituency vastly larger than any of the politicians in the room. He rose to speak and they hailed him as a hero, a man who had inspired more than a few of them to run and who had made the case for all of them to millions of people.

"I'm in awe of you," Rush Limbaugh told the freshmen. "You took the risks. You are the ones who engaged the opposition."

It was Limbaugh, a radio enthusiast long before he took to politics, who was the biggest success story of conservative talk radio's recent resurgence. Working under the name Jeff Christie, he'd started as a disc jockey in Pittsburgh, returning in failure to his native Missouri, where he linked up with the Kansas City Royals' public relations office. Then it was off to Sacramento and one more shot behind the microphone, this time on the talk side. It was here that his conservative on-air persona took shape, and in the summer of 1988—months after the repeal of the Fairness Doctrine—ABC offered Limbaugh the chance to go national. Soon, hundreds of stations signed up to broadcast his daily taunts of "feminazis," "environmental whackos," and all other creatures of the left. He branched out, too, with a tome, *The Way Things Ought to Be,* that spent fifty-four weeks on the *New York Times* bestseller list.

Introducing him to the Republican freshmen, Vin Weber, Newt's old House ally who'd gone on to help launch Empower America, pointed out that there were more women who listened to Rush just in the state of California than there were members of the entire National Organization for Women, a favorite Limbaugh target. The crowd cheered. Then Weber told them that a poll had found that people who listened to ten or more hours of talk radio each week had voted Republican by a three-to-one margin. "Those are the people that elected the new Congress," he said. "That's why this is the Limbaugh Congress."

When it was his turn, Limbaugh demurred. "What happens on talk radio is real simple," he said. "We validate what's in people's hearts and minds already."

He had advice for the revolutionaries. Together, they had just toppled the Permanent Democratic Congress, and they were all feeling the weight of this moment. "You're coming into the Beltway, inside the Beltway," Limbaugh told them. "And we're all human beings and we all are susceptible to human nature. And we all want to be liked. We all want to be loved. And you all want to live in surroundings which are not hostile.

"But inside the Beltway for people like us, this is not possible. And so sometimes to avoid the hostility we say things and then begin to do things designed to gain the approval of those who are hostile toward us. I want to warn you against it. I want to warn you: you will never, ever be their friends. They don't want to be your friends.

"Some female reporter will come up to one of you and start batting her eyes and ask you to go to lunch. And you'll think, Wow, I'm only a freshman. Cokie Roberts wants to take me to lunch. I've really made it! Seriously, don't fall for this. This is not the time to get moderate. This is not the time to start trying to be liked. This is not the time to start gaining the approval of the people you just defeated."

They didn't give advice like this at the Kennedy School. The freshmen roared. Never before had the Republican Party experienced such a heady moment. Clinton's 1992 victory was now viewed as unquestionably an accident of history, the product of an economy that had been weakened by the betrayal of a Republican president on taxes. But it had been clarifying, too, showing Americans once and for all that the Democratic Party was fundamentally dedicated to big government and the rapid liberalization of cultural norms.

On this night in Baltimore, the Republican future felt entirely without limit. Before leaving the stage, Limbaugh offered this: "I think everybody in this room probably has the idea that you're the beginning of forty years or more of conservative or Republican domination—not just two years of a brief moment in time to fix things."

They certainly did, and so did their leader. On January 5, Gingrich was to be sworn in as the first Republican Speaker of the House since Joe Martin back in 1954. He'd always carried himself as a man of historical import, which once upon a time had made him the butt of jokes around the Capitol. Those days were long gone. A vision that had once seemed grandiose, or just plain delusional, now stood fulfilled. Gingrich had changed the culture of the congressional Republican Party and the nature of the political opposition; and now the Gingrichized GOP had reaped a staggering windfall. "This is what I've worked my whole life for," he said. "It's going to be amazing. It's an open-ended adventure."

It was the morning after the election and he was more sure of himself than ever. His instincts, his reading of history, his strategic sensibility—all had just been validated on a massive scale, and all told him to push ahead now that the climactic battle was approaching. From his Marietta, Georgia, headquarters, he summoned reporters from the two most influential national outlets, Maureen Dowd of the *New York Times* and Dale Russakoff of the *Washington Post*. For them, he sketched a narrative that placed him in the hero's role, empowered by a popular rebellion to dismantle once and for all "the Great Society, counterculture, McGovernick legacy" of the 1960s.

This was the heart of Gingrichism. Strip away the bromides about cutting taxes and slashing regulation and unleashing the free market, and it became cultural and generational. Born in 1943, Gingrich was technically three years too old to be part of the baby boomer generation, but it was with the boomers that he'd come of age politically and it was contempt for their values that drove his political mission. Not all boomers, of course, but a certain type—the ones who'd overwhelmed college campuses with protests, who'd burned draft cards and even American flags, who'd grown their hair long, experimented with drugs (and then some), challenged traditional authority, and rejected the prevailing family structure. A generation later, the boomer elite had now matured into positions of authority, increasingly dominating academia, the media, and—especially—the Democratic Party.

On the outside, they'd long since gone clean, dressing and carrying themselves like polished professionals; but on the inside, Gingrich maintained, they still burned with the values of the counterculture.

"Until the mid-1960s," he explained, "there was an explicit long-term commitment to creating character. It was the work ethic. It was honesty, right and wrong. It was not harming others. It was being vigilant in the defense of liberty."

The world was only beginning to digest what had happened the night before, but to these reporters Gingrich was already setting world-historic stakes for his Speakership. It would be about much more than passing legislation and reining in the federal government. His charge, he declared, would be to "renew American civilization" by vanquishing the counterculture elite once and for all.

Clinton would still be around for the next two years, but Gingrich was already looking past him. The president, he said, would be "very, very dumb" if he tried to derail the agenda that the Republican Congress would soon be pushing. Of course, he fully expected Clinton to fight it. He was counting on it—definition and contrast on the grandest stage. But he was confident in the outcome. In the sixteen years since winning his first election, he'd overthrown a House Speaker, climbed to the top of his party's leadership, and toppled the Permanent Democratic Congress. Now he would finish off Bill Clinton and modern liberalism itself. This time when he said it, no one laughed.

Who was this guy? Even many of the people who'd voted Republican on November 8 didn't have much idea, but now he was all they were hearing about. For a generation, the Republican Party had been defined by its presidential personalities: Reagan's warmth, Bush's affability, Ford's regular-guy blandness. They'd had their critics, Reagan especially, but they'd all shared an instinct for reassurance. Even Nixon, for all his demons, had pursued middle-of-the-road policies meant to pacify the masses. Now here was Newt, calling himself "the most serious, systematic revolutionary of modern times" and proclaiming his desire to "redirect the future of the human race."

"Slash and burn, knife and smear: the Gingrich instincts are unrelenting," charged *New York Times* columnist Anthony Lewis, a pre-eminent liberal voice. The paper's editorial page went further, calling the soon-to-be-Speaker an "authoritarian."

"His language is as revelatory as it is familiar," growled the *Times.* "He describes himself as a battler against McGovernites, liberal elitists and the media. He will restore order and middle-class values. Welcome to Speaker Gingrich's Retro-World. Mr. Gingrich has reinvented the political landscape of his youth—a Sun Belt where politicians communicate in the venerable code words of Barry Goldwater and George Wallace."

Gingrich accused the media of dishonesty. He had also made conciliatory comments since the election, he argued, but was getting no credit for them. Then he reconsidered and reached for the reset button. *Dateline,* the hit NBC newsmagazine that had just expanded to three nights a week, interviewed him before Thanksgiving. The "counterculture McGovernicks" attack came up, but this time Gingrich relented. "There was no point to my saying it. It was a frankly foolish thing to do," he said. He felt the same, he added, about his claim that Clinton threatened "normal Americans."

"I probably need to be 30 percent less pugnacious and 50 percent less negative," Gingrich said.

That sounded easy, but soon he was right back at it, warning on *Meet the Press* that "counterculture people" had infested the Clinton White House. It was clear, he told Tim Russert, that in its early days the administration "had huge problems getting people through security clearance because they kept bringing people in who had a lot of things that weren't very easy to clear." Like what?

"I had a senior law enforcement official tell me that in his judgment, up to a quarter of the White House staff, when they first came in, had used drugs in the last four or five years," Gingrich said.

The uproar was instant. That recent—or even current—drug use was rampant in Bill Clinton's White House was a familiar theme on the talk-radio right, and the D.C. grapevine featured anecdotes about supposed difficulties some staffers had faced with background checks.

But Gingrich now was airing this chatter on national television, with nothing specific to substantiate it. Leon Panetta, Clinton's new chief of staff, exploded. If Gingrich didn't "stop behaving like an out-of-control radio-talk-show host and begin behaving like the Speaker of the House of Representatives," he threatened, the White House would refuse to conduct business with him.

Gingrich had told no one in his camp that he planned to raise the drug use topic. This was how he'd always operated. He sensed no serious blowback potential. A telling detail made it into the *Washington Post*'s write-up: "An aide to Gingrich said that the outspoken congressman has said similar things many times in the last year but that they drew little attention because Gingrich was often not taken seriously by the national news media then. Now that he is about to become one of the most powerful political figures in Washington and is visible on national television, his comments make news, the aide said."

The rules of Newt's world were changing, exposing him to dangers he'd never had to contemplate. Overnight, his stature had swelled. Suddenly, he was every bit as big as his biggest targets—which, of course, made Gingrich himself a very inviting target.

Even his enemies underestimated how slow he was to grasp this. Gingrich had caught the eye of the book publishing industry. For this, he could thank Limbaugh, whose runaway bestsellers had been a revelation. Those millions of devoted conservative radio listeners—they bought books, too. Even before the election, Gingrich had been in talks to pen his own volume, but now his value was skyrocketing. On December 21, he cut a deal with HarperCollins. He would write a book explaining his political philosophy and it would be called *To Renew America*. For this, he would be paid an advance of $4.5 million.

It shattered all precedents. Politicians had written books before, but the custom was for them to be paid based on sales, which were almost always meager. No active officeholder had ever been paid a sum like this upfront. HarperCollins, a source told the *Washington Post*, "is thinking in terms of potential Rush Limbaugh–type sales." Around

Washington, alarm bells sounded immediately. HarperCollins was a major cog in News Corp, Rupert Murdoch's media conglomerate, which counted the Fox broadcast network and numerous newspapers among its other holdings. This meant that Murdoch's business interests were a frequent source of interest to federal regulators. And now the Speaker of the House of Representatives was going to take a check from him for $4.5 million for a book he hadn't even written yet?

Just five years earlier, it had been Gingrich who called Jim Wright corrupt because of the royalties he'd received from bulk sales of his book, nursing the controversy into a full-blown scandal that destroyed the House Speaker's career. The irony was lost on no one. "This is an arrogant act for a man who's about to assume one of the most powerful positions and offices in our land," said David Bonior, about to become the second-ranking House Democrat.

Newt took the hint. He would still write the book, he announced on December 30, but now the advance would be one dollar. "We're about to have the first Republican Congress in forty years," Gingrich explained. "And I did not want to walk in next Wednesday and give the embittered defenders of the old order something that they could run around and yell about." He said he'd make his money from royalties on sales. "He'll make the four-and-a-half [million], I'm sure," his agent predicted. (He almost certainly didn't. The book did sit on the *New York Times* bestseller list for three months, but when sales died down the *Washington Post* estimated that Gingrich made a little over a million dollars from it, a far cry from what the advance would have been.)

The oath of office was administered to the members of the 104th Congress early in the afternoon on January 4. Then, by custom, the vote for Speaker was held. The Democrats nominated Richard A. Gephardt of Missouri, who had slid into the party's top post after Tom Foley's demise. The Republicans put up Gingrich, and besides two conservative Democrats from Mississippi voting "present," the roll call broke on party lines. Newton Leroy Gingrich of Georgia was now Speaker of the House. Wielding a large ceremonial gavel from the rostrum, Gephardt introduced him: "With resignation but with resolve I hereby end forty years of Democratic rule of this House."

Then it was Newt's turn. That morning, Gingrich had told reporters that he was "trying to change my style somewhat," mindful of all the land mines he'd stepped on since the election. In his maiden speech as Speaker, it showed. "I know I'm a very partisan figure, but I really hope today that I can speak for a minute to my friends in the Democratic Party as well as my own colleagues," he said.

He talked about unifying, nonpartisan themes—compassion, humility, even racial reconciliation. "I would say to those Republicans who believe in total privatization, you can't believe in the Good Samaritan and explain that as long as business is making money, we can walk by a fellow American who's hurt and not do something," he said.

Not many Democrats had seen this version of Newt. "It almost seemed as if there was somebody else inside of Newt Gingrich crying out to be heard," mused Maryland's Kweisi Mfume. And not every Democrat was buying it. "He is the most partisan and destructive guy we have ever seen," Barney Frank of Massachusetts said. "I don't think anybody sees any chance of working with him."

Outside the chamber, the image repair effort was already in danger. The night before, CBS News had revealed the contents of an interview that would soon air on Connie Chung's newsmagazine, *Eye to Eye*. The guest was Kathleen Gingrich, Newt's sixty-nine-year-old mother. She was living a quiet life in Pennsylvania with her second husband (Newt's adoptive father), a retired military officer. Both plainspoken and utterly unaccustomed to dealing with the press, Kathleen Gingrich held a lit cigarette while Chung probed her about her famous son. The subject of Bill Clinton came up. What has Newt told you about him? Chung asked. "Nothing," Gingrich replied, "and I can't tell you what he said about Hillary."

"You can't?"

"I can't."

Chung leaned in. "Why don't you just whisper it to me, just between you and me?" In a hushed tone, Gingrich complied. "'She's a bitch.'"

Her voice returning to its normal level, she continued. "About the only thing he ever said about her. I think they had some meeting, you know, and she takes over."

"She does?"

"Oh, yeah. But with Newty there, she can't."

The interview had been taped in Pennsylvania two weeks earlier. Newt Gingrich hadn't been present for it and had no idea what, exactly, his mother had told Chung. Now, on his big day, it was exploding all over the news, another blow to his image. He blasted Chung as "disreputable" and "unprofessional." Indeed, the newswoman was coming in for her share of criticism. Had she been deceptive, tricking Mrs. Gingrich into thinking she was confiding something that wouldn't be aired on national television? Chung countered that Kathleen Gingrich had playfully used the same "stage whisper" earlier in the interview, and that the exchange about Hillary fit with the flow of their conversation.

Newt's rage couldn't change the bottom line. On the very day he hoped to win a fresh look from the country, millions now heard his own mother describe his vulgar opinion of the First Lady. Was Kathleen Gingrich's statement accurate? "You have no idea what I called Mrs. Clinton because I have never commented," Newt said. "And I won't."

Buried in his speech was a brief mention of an obscure House staff position with special significance to the new Speaker, who promised to appoint a new House historian "to teach what the legislative struggle's all about." The next day, he made his choice known: Christina Jeffrey, a college professor from the same Georgia school where Gingrich had briefly taught a course. What Gingrich didn't realize, somehow, was that Christina Jeffrey used to go by the name Christina Price and that while leading a national curriculum review a few years earlier, Christina Price had objected to a Holocaust education program because the "Nazi point of view, however unpopular, is still a point of view and is not presented, nor is that of the Ku Klux Klan." Gingrich fired Jeffrey instantly, but for Americans wondering about this powerful man with a funny name who'd crashed their lives, the episode was more unnerving noise.

It was becoming a pattern, and taking a toll. Most Americans now knew who Newt Gingrich was, and barely one-quarter said they had a positive opinion of him. He was frustrated, and furious with a press that he believed was intent on making a scandal of his every utterance. By week two, he was announcing a month-long moratorium on Sunday-show appearances—time he wanted the media to spend opening its mind: "We are not like any political effort in modern American history. Unless they erase all their political assumptions growing out of Lyndon Johnson and Richard Nixon, and everything that's happened since 1965, they're just not going to get it."

Now this was the Newt that Washington knew. There'd be no image makeover. He would go with the instincts that brought him here—full steam ahead, with the whole world watching, and the White House beginning to wonder if maybe, somehow, this might work out for them after all.

SEVENTEEN

A "conservative opportunity society" or a "liberal welfare state": Newt Gingrich spent years nudging, elbowing, even bullying his own party to embrace that framing and to build around it, and eventually they did. Now, with the '94 election, all of America had made its choice. This was the confidence Gingrich brought to the Speakership. The big questions were settled. He had his mandate to obliterate all that modern liberalism had created, and if Bill Clinton tried to get in his way now, what would it even matter? He'd be gone in two years anyway, replaced by a new Republican president and the top-to-bottom conservative government Americans had always wanted. The country, Gingrich had never been more certain, was with him.

Then again, this was a combination most Americans had never experienced: a Democratic president and a Republican Congress. As Speaker, Gingrich would wield power no Republican on Capitol Hill had enjoyed in generations, and he was vowing to use it to produce abrupt, dramatic change. That meant attention, scrutiny, criticism, all at a level without modern precedent for any congressional figure—

a spotlight befitting a world-historical figure. Already, Gingrich was learning what this could mean when it came to his own behavior and pronouncements. What would happen when he started proposing real legislation?

On this front, he moved lightning fast, and without hesitation. Nothing embodied the "liberal welfare state" like, well, welfare. Bill Clinton the candidate had promised to end it "as we know it," but as president he'd delivered nothing. Now, the new Republican Congress would hold him to his word. Just days after the election, Gingrich made his opening bid, labeling welfare programs "a disaster. They ruined the poor. They created a culture of poverty and a culture of violence which is destructive of this civilization, and they have to be replaced thoroughly from the ground up."

He had every reason to believe the politics were on his side. Technically speaking, "welfare" was a catchall term for an array of programs aimed at helping the impoverished meet their basic needs. As a political issue, though, the action revolved around Aid to Families with Dependent Children, which was conceived during the New Deal and expanded dramatically under LBJ's "war on poverty." AFDC provided monthly cash assistance to households with children in which the principal earner was either absent or jobless. It was the largest single component of the welfare system.

The surge in spending on AFDC since the 1960s coincided with alarming cultural trends. By the early nineties, the rate of violent crime was four and a half times what it had been three decades earlier, and out-of-wedlock births had jumped sixfold. Critics insisted there was a linkage, that the welfare system created backward incentives that discouraged marriage, rewarded indolence, and locked families into poverty. The loaded caricature of a "welfare queen," a term coined by the *Chicago Tribune* in the 1970s to describe a black woman who had lived lavishly while bilking the system, was now entrenched in the political lexicon, popularized in part by Reagan, who made the welfare system central to his narrative of government failure.

In Washington, the consensus now spanned party lines that

change was desperately needed. "The system of Aid to Families with Dependent Children has been in operation for decades," Mary McGrory wrote in her column. "It is a colossal failure, producing generations of dependents." The head of the organization representing officials who administered AFDC programs at the state level told Congress: "There is unquestionable national consensus that the existing welfare system is broken." During the debate over health care, Democratic senator Daniel Patrick Moynihan, who had authored a controversial 1965 report on poverty in black America, announced that "we don't have a health crisis in this country. We have a welfare crisis."

With their Contract with America, Gingrich and the House GOP committed themselves to passing a plan within a hundred days of taking over. It would still need to clear the Senate and be reconciled before actually reaching Clinton's desk, a process that could drag on for months, but they would be taking a clear stand—and daring Clinton to tell them they were wrong. This was the kind of debate Gingrich dreamed of. What he didn't quite see, and what no one yet saw, were the complexities—and contradictions—of public opinion that lurked just beneath the surface.

Broadly speaking, Americans believed the welfare system was a disaster, but they still had their limits, which were tested when Gingrich started talking about orphanages. His idea was to turn AFDC into a block grant program—bundle up the money the federal government was spending and ship it back to the states to tailor their own programs. The block grants would come with stringent guidelines, requiring beneficiaries to work, setting limits on how long they could collect, and banning teenage mothers from receiving aid. That's where orphanages came in: the money saved, Gingrich proposed, could be used to fund them as a home for children born to teenagers unable to care for them.

With that, he gave Democrats an opening they'd never expected. The term "orphanage," for most Americans, conjured grim Dickensian despair. And Newt wanted to send *more* kids to live in them?

The criticism rained down, with Hillary Clinton calling the idea "unbelievable and absurd." Gingrich shot back that she should "go to Blockbuster and rent the Mickey Rooney movie about 'Boys Town,'" a feel-good 1930s fable that critics said whitewashed the realities of orphan life.

"'Boys Town,'" the *Washington Post* wrote, "speaks across generations to the contemporary problems of American juvenilia in the same way that 'The Three Stooges Meet Hercules' provides a cogent sociological exegesis of ancient Greece."

Even Gingrich's allies distanced themselves from the idea. "I don't think government-run orphanages is anything that should cause a conservative heart to beat faster," Gary Bauer, a leader on the Christian right, said. In a poll, nearly 80 percent of Americans said they'd be upset if mothers were forced to send their children to orphanages.

Practically speaking, this was a minor piece of the puzzle, a small, disposable component of Gingrich's massive reform plan—more of a notion, really. But until he finally backed off weeks later, it overshadowed everything. The electorate, it seemed, had a compassion nerve, and Gingrich had pressed on it.

In February, House Republicans formally unveiled their legislation. The headline was their call to end the federal guarantee of benefits to anyone who qualified, block-granting AFDC to the states, just as Gingrich had talked about. There was more, though, another component pressing on that suddenly exposed compassion nerve.

Since 1946, the Department of Agriculture had administered the national school lunch program. Later expanded to include breakfast, it now provided free or low-cost meals to an estimated twenty-five million children each day. Now, under the Republican plan, the federal government would give up its role and ship the money to the states, with a mandate that they spend at least 80 percent of it on meals for "economically disadvantaged children." It would be up to the states to decide which children, exactly, qualified as economically disadvantaged. Republicans said it would save money and promote efficiency. "The states are closer to the needs of the people. They will be totally in charge," Pennsylvania's William Goodling said.

Democrats erupted. The school lunch program was a lifeline for millions, they argued, and it was working just fine. Blowing it up and block-granting it "would really take food out of the mouths of millions of needy schoolchildren," White House Chief of Staff Leon Panetta charged. Words like "unconscionable," "draconian," and "absolutely insane" were tossed around. States, they warned, would be able to abuse the loose guidelines and redirect the money to other budget priorities. There would also be no more federally enforced nutrition standards, and the annual spending increase under the GOP plan would be less than under current law—in other words, Democrats claimed, a cut.

Gingrich was indignant. "It doesn't say anywhere in the Declaration of Independence or the Constitution that anyone is entitled to anything except the right to pursue happiness," he said. "In terms of getting food to children, we ought to find the most effective, practical way to do it. But I don't know that that means you want to keep it as an entitlement for the entire nation."

Just as with orphanages, the furor over school lunches now overwhelmed the coverage, with Republicans left pleading that they weren't trying to deprive hungry kids of breakfast and lunch. Democrats didn't have the votes to the stop the plan, which passed the House at the end of March, but they'd found a surprising toehold with the public.

By now, the first hundred days were nearly over. The Republican Contract had promised up-or-down votes on every item, and the Speaker was determined to deliver on time. The goal was realized on April 5, when by a vote of 246–188 the House approved a package of tax cuts. The centerpiece was a five-hundred-dollar-per-child credit, but it also slashed the capital gains rate and undid many of the hikes Clinton and the Democrats had imposed. "Join us in righting the wrongs of '93!" Tom DeLay, the third-ranking Republican, challenged his colleagues. Democrats blasted it as a giveaway to the rich.

The contract had been fulfilled. Every agenda item had received its promised vote in the House, and save for a constitutional term-limits amendment and the revival of the Reagan-era "Star Wars" missile defense program, each had passed. But so far, that was basically it. Bills to limit unfunded mandates and to force Congress to adhere to federal labor laws had made it through the Senate and been signed by President Clinton, but those were bit pieces. The big, meaty stuff was stuck in the Senate, and would be for a while.

The changed culture was impossible to miss, though. Gingrich and his fellow revolutionaries, the *Washington Post* wrote, "have touched everything from their own rules and children's school lunches to taxes and property rights. They have moved the action from the White House to Capitol Hill. And they have dominated the national debate and pulled it hard to the right."

As if to emphasize that point, Gingrich asked the broadcast networks to air an address to the nation in prime time on Friday, April 7, his summation of the first hundred days and preview of what was next. The networks said yes. It was something Tom Foley, Jim Wright, and Tip O'Neill had never gotten; then again, they'd never thought to ask. Gingrich spoke of a grander purpose behind the House's work.

"While we've done a lot," he said, "this contract has never been about curing all the ills of the nation. One hundred days cannot overturn the neglect of decades. The contract's purpose has been to show that change is possible, that even in Washington you can do what you say you're going to do."

He closed by looking beyond welfare and taxes and to the great, defining battle he planned to kick off soon. For more than a decade, giant deficits had hung over American politics. The two most recent presidents, one a Democrat and one a Republican, had tried to fight the red ink with taxes, and both had failed. Now, the Speaker told the country, the task would fall to a Republican Congress that was prepared to act with boldness. Within seven years, they would balance the budget, and they would do it without a penny more in taxes.

"There are reasons why, as President Franklin Delano Roosevelt said, 'Our generation has a rendezvous with destiny,'" Gingrich concluded. "This is the year we rendezvous with our destiny to establish a clear plan to balance the budget. It can no longer be put off."

The blueprint would be coming soon, he promised, and then the debate would begin.

The president had for these first hundred days been an unusually passive presence, "a barely relevant figure," as the *Washington Post* put it, "compared with the rush of activity on Capitol Hill." But now he'd sized up the new Speaker and saw the makings of a comeback strategy.

Clinton accepted an invitation to address the American Society of Newspaper Editors in Dallas on April 7, the same day as Gingrich's national address. It was an intentional contrast: Clinton reclaiming his spot in the main ring and defining how he would—and wouldn't—deal with the new Congress going forward.

"In the first hundred days, the mission of the House Republicans was to suggest ways in which we should change our government and society," the president said. "In the second hundred days, and beyond, our mission together must be to decide which of these House proposals should be adopted, which should be modified, and which should be stopped."

If there was good news for Clinton, it was that his standing had at least stabilized. His approval rating still hovered in the low forties; hardly great, but also not the total collapse predicted after November 8. Nor had any of the Democrats touted as saviors for their party in 1996 made any moves toward actually challenging Clinton. His overall position remained weak and polls continued to put him behind the most likely GOP candidate for '96, Bob Dole, but he was still in the game. That alone was more than even Democrats had believed possible a few months earlier.

The eruptions over orphanages and school lunches, the Clinton team was coming to believe, had exposed new political space that was both fertile and unoccupied. The close partnership with congressional Democrats that marked his first two years in office had made it only too easy for Republicans to brand Clinton just another big-government liberal, and not the "different kind of Democrat" he'd run as. And now the public had shown that it had definite limits when it came to all-Democratic government. But what about the revolution Gingrich was now trying to implement? Did the public have limits here, too? Conceptually, Americans were all for slashing government, but the skirmishes of early 1995 were hinting at something trickier.

What Clinton was starting to discern was a path between the two poles, one that would bring him back to the turf he'd run on in the first place. He could play the reasonable man, both acknowledging voters' skepticism toward government and standing up to the excesses of the Republican agenda. "Triangulation" was how one of his consultants, Dick Morris, described it—"Take the best from each party's agenda, and come to a solution somewhere above the positions of each party." A character who'd bounced in and out of Clinton's political life, Morris had a résumé littered with Republican clients—Trent Lott and Jesse Helms among them—and was intensely disliked by Clinton's inner circle. But he'd regained the president's ear following the '94 bloodbath, and now Clinton would put his concept to the test.

In his Dallas speech, Clinton declared himself ready to compromise and argued that there was broad common ground: taxes should be cut; the government was spending too much; welfare absolutely did need an overhaul. "I do not want a pile of vetoes," he said. "I want a pile of bills that will move this country into the future. I don't want to see a big fight between the Republicans and Democrats. I want us to surprise everybody in America by rolling up our sleeves and joining hands and working together."

Then he went for where he thought the Gingrich Republicans were vulnerable. Their welfare plan—the one most people knew about through the school lunch kerfuffle—was "weak on work and tough on kids," he said. The tax cut they'd just voted out of the House? "We can't afford it." He wanted a cut, too, he reiterated, but "we have to choose. Do you want a tax cut for the wealthy or for the middle class?" These were themes he'd pressed in '92, themes that had worked.

But the game was different now, transformed by Gingrich's rise. In the Republican Party of yesteryear, there were as many Rockefeller liberals from the Northeast as Taft conservatives from the Midwest. This ascendant iteration, though, was ideologically conservative and found its center of power below the Mason-Dixon line. It had nurtured new alliances on the right side of the spectrum, from freshly mobilized evangelical Christians to increasingly strident gun groups, giving rise to a powerful political coalition. It had been happening slowly, steadily for decades, out of view of many Americans. Now, with Gingrich wielding the Speaker's gavel, it was in everyone's face. And if Clinton was looking to define himself against what this new Republican Party represented, he wasn't limited to Capitol Hill for material.

On the morning of April 19, a rented Ryder pickup truck packed with explosives was detonated outside the Alfred P. Murrah Federal Building in Oklahoma City. The destruction was instant and staggering. One hundred sixty-eight men, women, and children (the building included a day-care facility) were killed on the spot, and more than eight hundred others injured.

Suspicions turned first to international terrorism. Two years earlier, terrorists with roots in the Middle East had set off an explosive device in a parking garage beneath the World Trade Center, a failed effort to topple both towers that left six dead. Quickly, though, it emerged that the devastation in Oklahoma City was an act of domestic terror, conceived and executed by two native-born Americans— Timothy McVeigh and Terry Nichols—and fueled by their deep rage toward the federal government.

There had never been a deadlier terror attack in the United States. Across the country, Americans absorbed images from Oklahoma City with horror and numbness. A memorial service was scheduled for the next Sunday, April 23. Thousands filled the state fair arena, many of them grieving family members of victims. Cameras from all the networks rolled, beaming the proceedings to millions of living rooms. Clinton had spent the day on the ground and now he spoke for the nation, calling the attack an "evil" act and vowing that "justice will prevail."

"Those who are lost now belong to God," he said. "Someday we will be with them. But until that happens, their legacy must be our lives. In the face of death, let us honor life."

The role of national consoler was unique to the presidency, and Clinton was made for it. Even from Republicans, there was praise. "Our president was swift to act," Oklahoma governor Frank Keating said in a national radio address. "He sent us the resources to solve this terrible crime. He offered condolences and heartfelt assistance of a grieving nation." A midterm election drubbing hadn't taken away the presidential bully pulpit, and in this moment of tragedy, Clinton was putting it to use in a way that inspired even his critics.

Then, the next day, he got political. McVeigh and Nichols had emerged from a loose movement of vehemently antigovernment survivalists. They tended to live off the grid, often heavily armed, convinced the forces of the federal government were scheming to deprive them of their freedom.

Two events of the recent past had hardened their resolve. In the summer of 1992, federal marshals attempted to serve a warrant to survivalist Randy Weaver, holed up on his property in Ruby Ridge, Idaho. When he resisted, a firefight broke out, leaving both a deputy marshal and Weaver's son dead. An eleven-day standoff ensued, during which Weaver's wife was killed by an FBI sniper. It finally ended with Weaver's surrender, but to his fellow travelers, it became a rallying cry: if the government could do this to Randy Weaver, imagine what it might do to you.

Less than a year after Ruby Ridge, a similar arc played out near Waco, Texas, this time on a deadlier scale. In a heavily fortified compound, the leader of a cultish sect of Seventh-Day Adventists, David Koresh, was living with scores of followers. In addition to what authorities believed was his cache of illegal weapons, Koresh was suspected of forcing teenage girls into marriages and sexually abusing them.

Agents from the Bureau of Alcohol, Tobacco and Firearms, a Prohibition-era arm of the Treasury Department, tried to serve a warrant, triggering a gunfight. Four agents and five of Koresh's Branch Davidian followers died. For the next seven weeks, a standoff ensued, broken on the order of Attorney General Janet Reno, who authorized the FBI to lead a raid that quickly went awry. As agents stormed the compound, three fires began inside, merging soon into one lethal conflagration. There had been eighty-five living Branch Davidians inside the compound; seventy-six of them perished in the raid, including nearly two dozen children.

The date of the Waco raid, April 19, was the same date as the Oklahoma City massacre, and it wasn't an accident. McVeigh saw his bombing as an act of vengeance. The survivalist world had existed on the edges of public awareness. Now, a spotlight was shining on the paranoia that drove it. Some even pointed to what they claimed was a disturbing overlap between these survivalist beliefs and the antigovernment themes permeating conservative talk radio. It was a connection Clinton himself seemed to draw a day after the memorial service, when he argued that certain voices in American life—he didn't name names—"leave the impression by their very words that violence is accepted."

"We hear so many loud and angry voices in America today whose sole goal seems to be to try to keep some people as paranoid as possible and the rest of us all tore up and upset with each other," Clinton said. "They spread hate."

Talk radio itself was becoming an issue in the national political debate. Its millions of listeners were enraged by Clinton's words, and they dug in further in their opposition to him. "Make no mistake about it," Limbaugh thundered to them. "Liberals intend to use this tragedy for their own political gain." But there were also millions of Americans who'd never listened and who were alarmed by what they were now hearing. To them, the president's words sounded plenty reasonable.

The sudden focus on far-right rhetoric thrust the National Rifle Association into the spotlight. Chartered in the 1870s, it had for much of its existence been a group for recreational sportsmen, but lately, it had taken on a far more political—and conservative—edge. The transformation was completed in 1991, when hardliners took control of the NRA's board and elevated Wayne LaPierre to run the organization.

LaPierre's NRA then suffered two major defeats in Washington, with the passage of the Brady Bill in 1993, which mandated a five-day waiting period for handgun purchases, and an assault weapons ban in 1994. Both were championed and signed by Clinton. LaPierre resolved to push even harder to build the NRA's muscle. He was intent on making Clinton pay and on stopping Congress from ever again passing a single gun control measure. The NRA still had some Democratic allies, but on the whole it proved to be a significant source of political strength for the Republican Revolution.

It was also now mining the same turf where the survivalist right roamed. Before the Oklahoma City attack, LaPierre had sent out a fund-raising letter that decried the overreach of agents from the Bureau of Alcohol, Tobacco and Firearms—"jackbooted government thugs," his missive branded them, clad in "Nazi bucket helmets" and intent on attacking law-abiding citizens. When he first sent it, the letter went unnoticed by the media, but now critics brought it out as a proof of the dark forces that, they said, loomed larger than anyone realized on the political right.

Their case was amplified when one of the NRA's most famous members resigned his lifetime membership. "I am a gun owner and an avid hunter," former president George Bush wrote in a letter to the group's president in early May. "Over the years, I have agreed with most of the NRA's objectives.

"However, your broadside against federal agents deeply offends my own sense of decency and honor; and it offends my concept of service to country. It indirectly slanders a wide array of government law enforcement officials, who are out there, day and night, laying their lives on the line for all of us."

Bush was greeted with wide praise from columnists, editorial boards, and more than a few of the Democratic politicians who'd helped drum him out of office. Clinton himself said he fully agreed with his former opponent's "fine letter," and he used the occasion to note that one of the NRA's top agenda items was also on the new Republican Congress's to-do list.

"I want them to know they can pressure Congress all they want to try to repeal the weapons ban," he said, "but as long as I am president, that ban will be the law of our land."

This was precisely the spot Clinton had set out to claim. He wanted the NRA—this new, intensely conservative, and savagely antigovernment version of the NRA—to be inseparable from the Republican Congress. In the wake of Oklahoma City, GOP leaders quietly backed off their push to junk the assault weapons ban, hoping to ride out the storm. Instead, a January letter from Gingrich to the NRA's chief lobbyist was unearthed. In it, Gingrich had vowed that "as long as I am Speaker of this House, no gun control legislation is going to move in committee or on the floor of the House."

Gingrich had certainly kept that commitment. "You can see who's in control of this Congress," Clinton said.

Just like talk radio, the NRA was facing a brand-new level of exposure and scrutiny, and it was coming against the backdrop of a national tragedy. Not everyone was going to turn against the group; in fact, many supporters would only pull closer now. But among the Americans who'd elected this Republican Congress, there had to be more than a few who'd never thought this would be part of the bargain. This was Bill Clinton's bet.

Through the spring and into the summer, House Republicans provoked flare-ups as they put their revolutionary stamp on annual spending bills.

Republican leaders were fond of talking about their party as a "big tent," united under broad principles but welcoming of diverse viewpoints. For a long time, it had been the avuncular personalities of Republican presidents—Ike, Jerry, the Gipper, Poppy Bush (just a year after his death, the party was still reckoning with the outlier that was Richard Nixon)—that gave the party its identity. Beyond that, your perception of the Grand Old Party probably depended on where you lived, with Republicans generally reflecting the sensibilities of their local communities. For all those decades when the GOP toiled in irrelevance on Capitol Hill, it had never really mattered to anyone who or what drove the congressional wing of the Republican Party.

Now it did. There was no Republican in the White House and for the first time in ages, it was a Republican Congress giving the party its national definition. The Speaker was a self-described "systematic revolutionary." The top three leadership posts were held by southern ideologues, a Georgian and two Texans. The agenda was far to the right, fueled by a freshman class on a mission to wipe out the past three decades of liberal advances.

Plenty of Americans liked this new Republican Party, but how many were seeing it for the first time? Contrast, Newt Gingrich had always told Republicans, was everything—*show the other party for what it is and define ourselves against it and we can't lose.* Now, in this brand-new environment, Bill Clinton was seeing this advice as the answer for his party, too.

EIGHTEEN

The confrontation that would become the legacy of the 104th Congress began with Democrats calling Bill Clinton a sellout.

It was the middle of June 1995 and he was making his opening offer, a ten-year path to a balanced budget that mixed targeted tax cuts and spending reductions. "It's time to clean up this mess," Clinton said in a televised address. The idea was to preempt the Republican Congress, which was about to unveil its own seven-year plan for a balanced budget. Clinton would get ahead of them, assert his own leadership, and demonstrate his openness to challenging his own party's orthodoxy.

That's the part that was infuriating his own allies in Congress. After previously refusing to do so, Clinton was now placing on the chopping block the crown jewel of the Great Society. He would extract, he said, $124 billion in savings from Medicare, which since its enactment in 1965 had provided health insurance to tens of millions of senior citizens. This would not represent a cut to benefits, he stressed; it would be aimed at providers and it would only be a reduction in the rate of growth.

To the left, Clinton's move was as morally dubious as it was strategically self-defeating. Medicare was the work of their party, championed by LBJ and passed over warnings from conservatives of creeping socialism. It had proven to be a lifeline for countless retirees, and a popular one, too. If they didn't stand stalwart in defense of this kind of government program, then who were they?

Months earlier, Clinton himself had seemingly agreed, presenting an initial blueprint that didn't touch Medicare. Moreover, the plan Republicans were about to tee up would be calling for a far deeper rollback. The school lunch skirmish had congressional Democrats convinced the politics were on their side on this sort of thing. Just hold back, they pleaded with Clinton, and let Newt spell out all the ugly details of his plan—then let's pound him for it. Instead, Clinton shirked off his earlier commitment and pronounced himself ready to deal on Medicare.

"I think some of us learned some time ago that if you don't like the president's position on a particular issue, you simply need to wait a few weeks," seethed Representative David Obey, one of the House's leading liberals.

It was triangulation in action. Clinton's goal wasn't so much to represent his own party as it was to make the other one look extreme. He'd concluded there was a compelling case for scaling back spending on Medicare. It was consuming an ever-larger share of overall federal spending, and his own Medicare trustees had just reported that the program's hospital fund was on course for bankruptcy in 2002. The budget deficit, down from its early-nineties peak but still imposing, remained a major issue, too. The Clinton of the previous two years might have reflexively aligned himself with congressional Democrats; this version would show his independence from them— and dare Gingrich to break with his hardliners, too.

That was the key. On Capitol Hill, *USA Today* reported, Democrats were deriding Clinton's plan as "the Dickie Morris budget." But none of it would matter unless Gingrich, who'd spent the past two decades telling Republicans that compromise amounted to surrender, took the bait and met him in the middle.

Medicare's political sensitivity was not exactly lost on the Speaker. In his hundred-day speech, Gingrich had warned that "you will hear screams" when Republicans released their balanced budget plan and asked Americans to "verify the facts on both sides." A GOP pollster had also cautioned him that voters felt particular attachment to the program.

But the Republican Revolution was supposed to be about transformative change, not nibbling around the edges. To Gingrich, it was a test of leadership. Washington had failed to balance its books, the country was bleeding red ink, and exploding Medicare costs were part of the problem. It would take this Republican Congress to confront the crisis directly and make the hard choices that would save Medicare—and America's solvency. "We will not back away from this crossroads and say, 'Let's not balance the budget, it's too scary,'" Gingrich declared.

He wanted it done in seven years—a balanced budget by 2002. It would mean cutting back the increase in Medicare spending by a lot more than the $124 billion Clinton recommended in his June speech. The Republican position hardened a week later when Dole lined up with Gingrich. Their offer: a $270 billion Medicare haircut—and $245 billion worth of tax cuts. Some were surprised Dole signed off. One of his top lieutenants, Senator Pete Domenici, the Budget Committee chairman, had reportedly been urging him to resist any tax cut. But Dole was also running for the Republican presidential nomination. He'd announced his candidacy in April and wanted no public feud with Gingrich and the forces he represented.

With a Republican plan now on the table, Democrats redirected their rage. They'd vented their frustration with Clinton, but what the GOP was offering went much further. There were the cuts themselves, which some experts were now saying could end up affecting benefits. There was also the provision to steer seniors toward

HMO-style coverage—a threat, Democrats charged, to the basic character of Medicare. And there were the tax cuts, like the 50 percent reduction in the capital gains rate, that would disproportionately accrue to the wealthy. To the Democratic eye, it looked like senior citizens would be footing the bill for a giveaway to the rich.

Suddenly, it didn't seem to matter so much that Clinton himself had put Medicare on the table. Congressional Democrats had wanted a unified campaign against the Gingrich plan, and now they would get it—with Clinton leading the way. They organized a rally at the end of July, timed to the thirtieth anniversary of Medicare's passage, and fired away at the Republicans. "Keep your tax-cutting, greedy hands off our Medicare!" Ted Kennedy demanded. "Those who want to gamble on Medicare are risking Americans' lives!" Clinton cried.

Gingrich tried to fire back. "I think to try to scare senior citizens as a reelection technique a year and a half before the election is, frankly, a very despicable strategy," he said. "And I have a simple challenge to the president: you tell the country, before this week is out, what you would do to save Medicare."

He was on the defensive, though, as he'd been on school lunches, trying to quell a growing uproar. Democrats did their best to turn up the heat. A year earlier, when they'd controlled Congress, they'd been in disarray on health care. Now they all read from the same script, condemning the Republican "assault" on Medicare. Republicans returning to their districts for summer recess found themselves hounded by protesters.

Gingrich was scheduled to address a conference on Medicare in Atlanta, but he arrived to the discovery that it had been crashed by a protest being led by one of his colleagues, Representative John Lewis. A Democrat from a neighboring Georgia district, Lewis was a hero of the civil rights movement, beaten to within an inch of his life by police officers on "Bloody Sunday" in 1965.

"Long before I was a member of Congress, I knew how to stand up and fight," Lewis told the impromptu rally. "Don't give up. Don't give in."

The Speaker waited until the protest had cleared to make his re-marks. "My friends in the demonstration industry, who have made it a sort of habit to be obnoxious, had zero interest in a dialogue, and I had no interest in a confrontation," he said.

Strategically, Gingrich was leaning on two messages. His polling told him that Republicans could win the argument if their plan was seen as the only way to preserve Medicare. He also tried to tap into the public's lingering doubts about the president's truthfulness. "I don't detect any panic by senior citizens when a dishonest administration issues another dishonest number," he insisted.

The evidence suggested he wasn't making any headway. An August poll found an outright majority of Americans now disapproved of the Republican Congress. Another showed Clinton with a seventeen-point advantage on the question of whom voters trusted on Medicare. A memo from Senator Domenici's top aide was circulating on Capitol Hill. It urged Republicans to compromise with the president or face "political disaster."

When Congress returned to town after Labor Day, a deadline loomed. The fiscal year would expire at the end of September, and without a budget deal the government would shut down. Clinton was feeling confident now. The Republican plan would "dismantle Medicare as we know it," he said, and he promised to veto it—even if it meant closing the government.

"I am not going to blink at the end," Clinton said. "As awful as it is, it would be better to shut the government down for a few days than to shut the country down a few years from now because we took a radical and unwarranted road here that the American people never voted for."

Gingrich was meeting regularly with the president, and sometimes seemed to be softening. On *Meet the Press* on September 10, he paid Clinton a compliment—"He did the country a service by being blunt and honest about the need to change Medicare"—and said that while Republicans remained adamant about the seven-year window, "how we get to the balanced budget over the next seven years is discussable."

But there was no give beyond that. Gingrich, it turned out, was penned in by the very forces he had incubated. The House Republican conference, and especially its enormous freshman class, was awash in Gingrichism. And under the terms of Gingrichism, this was no moment to go wobbly. Remember 1990? Bush's fatal mistake, Gingrich himself had said, was buckling to the pressure of the establishment and cutting a deal with Democrats instead of standing on principle and creating maximal contrast. These House Republicans had come to Washington to do it the Gingrich way. Rush Limbaugh's plea from their orientation dinner still rang in the freshmen's ears: don't fall for Washington's tricks, don't go native—don't ever give in to *them*.

A letter was drawn up and signed by more than 160 House Republicans, including most of the freshmen. They sent it to Clinton and to congressional leaders, and they said: we won't vote to let the government operate unless we get our budget. "Among those who did not sign," noted the *Los Angeles Times*, "was Gingrich, who warned . . . Clinton at a White House meeting this week that he could not control many of the freshmen's votes."

He did get them to kick the can down the road, a stopgap deal to avoid an October 1 shutdown and keep the government open through November 13. Now there was a new deadline, but the ominous signs were still there for Republicans. Opposition to their plan was growing in the polls. Gingrich and Dole remained committed. This was a showdown their base wanted, badly. They wondered, too, if maybe it would all start to look different to the public when the drop-dead date arrived and the true nature of the stakes came into focus. Clinton, Gingrich advised, "would be wise to think twice about vetoing the balanced budget and jeopardizing long-overdue revolutionary change."

The impasse endured. Then, Gingrich opened his mouth and undermined his own best argument. Republicans, he had been swearing up and down, were only trying to save Medicare from fiscal catastrophe. They wanted to make tough but necessary changes that would "preserve, protect, and strengthen" it, and any claim from Clinton and his crew that they had some ulterior motive was pure demagoguery.

On October 24, though, Gingrich held court before a meeting of the Blue Cross/Blue Shield Association, an insurance group. There were no cameras present and he chose to speak extemporaneously. This was how he liked to communicate, in loosely organized, virtually free-form riffs that mixed his own grand historical vision with whatever was on his mind at the moment. This could be Newt at his most engaging, but it could also be a recipe for trouble—like his musing about "counterculture McGovernicks" the morning after the midterms. On this day, before the insurance group, it was nothing but trouble.

"Let's talk about Medicare," he announced at one point. Eventually, he homed in on the GOP plan's goal of nudging seniors into HMOs and away from the traditional fee-for-service model. "Now, we don't get rid of it in Round 1 because we don't think that that's politically smart and we don't think that's the right way to go through a transition," Gingrich told the group. "But we believe it's going to wither on the vine because we think people are voluntarily going to leave it—voluntarily."

There was no recording and there were no reporters present, but a day later the transcript leaked out, and all hell broke loose. Here was the Speaker of the House saying out loud exactly what Democrats had been accusing him of secretly believing. The Democratic National Committee rushed an ad onto the air highlighting Gingrich's words.

But it was even worse for Republicans. Incredibly, on the same day Gingrich spoke to the insurance group, Dole was addressing the American Conservative Union. He was still the Republican presidential front-runner, but he was also hearing footsteps. At the moment, the loudest were from Texas senator Phil Gramm, whose strategy rested on painting himself as a Gingrich-style revolutionary and Dole as a wishy-washy compromiser.

In his speech to the ACU, Dole tried to shore up his hardliner credentials by reaching into his past. "I was there, fighting the fight, voting against Medicare . . . because we knew it wouldn't work in 1965." As a House member three decades before, Dole had indeed voted against the creation of Medicare, and he hadn't been alone. Almost half of his Republican colleagues in 1965 had done so, along with forty-eight Democrats. Given how popular the program had become, it was not a vote that necessarily held up too well in history—but here was Dole suddenly invoking it with pride.

The White House's spokesman, Mike McCurry, said: "It's now clear that all of the Republican arguments about how they were trying to 'preserve Medicare' are shallow promises when, philosophically, they've now made it clear that what they really want to do is destroy the program."

In the final days before the deadline, here was the Republicans' hope: that no one knew for certain how the public would respond to a shutdown and that the president, for all his bluster, was more nervous than he was letting on. Shutdowns had happened before; seven of them, in fact, in just the past fourteen years. They had almost all been quickies, though, hours-long pauses in the activity of the federal government, many not even requiring employee furloughs. The most recent one, triggered by Gingrich's tax revolt against President Bush in 1990, had been lengthier, but even that one played out over the Columbus Day holiday weekend, when most federal workers were off anyway.

What was shaping up this time around was something very different. The political war between the revolutionary Republican Congress and the besieged Democratic president had riveted the media all year. The looming shutdown promised to be its climactic battle. It would also be televised. CNN, still the lone cable news channel but now available in more homes than ever before, would provide constant, minute-to-minute coverage of any shutdown, adding to the standoff's urgency. And, of course, this one would fall smack in the middle of a normal workweek.

This was where Gingrich hoped there was a chance. The House and Senate hadn't actually passed their comprehensive budget plan yet, and they weren't going to before the deadline. The question was whether there'd be a new stopgap funding bill. The technical term for this was "continuing resolution," and it was a way of buying time: pass a bill that provides money to keep the government open temporarily and use that time to work toward a long-term agreement. Normally, it went without saying that a continuing resolution would just use the existing funding levels for each government program; why haggle over cuts for something that would only be in effect for a few days or weeks?

But this wasn't the time for normal, Newt decided. He would put Clinton to the test—now. A new continuing resolution was drafted, to cover the government for eighteen days, but this one came loaded up with conditions. The Republicans weren't going to settle for existing levels. To keep the government running, even just for eighteen days, they wanted cuts—to education, to environmental regulations, to a bunch of federal programs and services. And the topper: there would have to be a hike—eleven bucks a month—in Medicare premiums. It was a miniature version of their grand blueprint.

This was the continuing resolution they would offer Clinton, and then it would be his choice. If he didn't like it, it would be his veto shutting down the government. Surely, the president would think long and hard. The polls said he had the advantage, but the consequences of a shutdown in this environment were unknown. Was he really ready to take the risk? Or, if he was getting cold feet, he could sign the Republican bill and keep the government running. Of course, by doing so, he'd be accepting the GOP's core framework, which would then shape any long-term deal later agreed to.

By the time the bill landed on Clinton's desk on Monday, November 13, he'd already announced he would veto it. He was scheduled to address the Democratic Leadership Council that afternoon, and it quickly turned into a rally. "As long as they insist on plunging ahead with a budget that violates our values, in a process that is characterized more by pressure than constitutional practice, I will fight it," the president said.

"I am fighting it today. I will fight it tomorrow. I will fight it next week and next month. I will fight it until we get a budget that is fair to all Americans!"

A last-minute meeting with Gingrich and Dole late in the night yielded nothing new, and at 12:01 A.M. the government began shutting down. Later in the morning, about two-thirds of the federal government's civilian workforce—eight hundred thousand men and women—stayed home instead of reporting to their jobs. (Technically, this was a partial shutdown, since Congress had managed to pass bills to fund a few areas of the government.) National parks were shuttered. Shots of national monuments roped off from the public splashed across CNN and network news programs.

Clinton walked over to the press briefing room in the White House, where he delivered a perfectly triangulated message to the country. "The government is partially shutting down because Congress has failed to pass the straightforward legislation that is necessary to keep the government running without imposing sharp hikes in Medicare premiums and deep cuts in education and the environment," he said.

"It is particularly unfortunate that the Republican Congress has brought us to this juncture because, after all, we share a common goal: balancing the federal budget."

At a closed-door meeting of House Republicans, the *Washington Post* reported, Gingrich made remarks that were "so perfunctory that few could remember precisely what he said." Maybe he'd been secretly hoping the reality of a shutdown and the scale of political risk they were now facing would jar the revolutionaries toward compromise. But they were unanimous: Stand firm, they told the Speaker. Politically, the Republican freshmen were Newt Gingrich's children, and one of their leaders now invoked his example.

"If this budget is not balanced in seven years," Kansas representative Sam Brownback told the press, "the same thing will happen here to our leadership that happened to George Bush when he broke his promise of no new taxes. You will have an internal revolt. You have to balance in seven. Period!"

What could Gingrich do? The world he'd created was swallowing him up. An instant poll was conducted that night for CNN and *USA*

Today. Forty-nine percent of Americans blamed Republicans for the shutdown. Only 26 percent said it was Clinton's fault. By a 48–43 percent margin, they approved of Clinton's handling of the budget battle. For Gingrich, the spread was 22–64 percent. The warnings had been accurate. The public was pointing its finger straight at Newt and his party.

It continued like this, in fits and starts, for nearly two months— a brief reopening, then another shutdown, then another stopgap. Republicans did finally pass their full plan, which Clinton promptly vetoed. And Gingrich managed get in the way of his own message again. After flying back aboard Air Force One from the funeral of Israeli prime minister Yitzhak Rabin, Gingrich complained to the press that Clinton hadn't used the opportunity to meet with him about the budget. "Cry Baby Newt," blared the cover of the next day's *New York Daily News,* which depicted the Speaker as a wailing toddler wearing a diaper.

The public's impressions cemented early on and never budged. Slowly, the Republican demand on Medicare came down, eventually to under $200 billion, this after their opening demand of a $270 billion cut. But the 104th Congress never did reach a deal with the president, and eventually it was agreed to punt the whole matter until after the presidential election.

It had been an open question at the start of 1995 whether the Democratic Party would even renominate Bill Clinton. He'd then gone and placed Medicare on the bargaining table, enraging liberals in his party. But they weren't talking about that now. The Medicare debate ended up being driven by something more basic than the mathematical differences between Clinton's proposed cuts and the GOP's. It turned into a referendum on which leader actually meant it when he said he believed in Medicare and wanted to protect it for years to come. That was the fight Clinton had led his party into, and now Democrats stepped forward to shower him with credit. "The president," said Senate Democratic leader Tom Daschle, "stood firm on Medicare."

The Democratic presidential primaries would begin in February, but now they'd be a formality. Clinton would be unopposed, and there was more. A poll pitted him against Dole. At the start of the year, Dole had been well ahead. Now, with 1996 just around the corner, Clinton was up by twelve. "Virtually every finding," the *Times* noted, "showed striking evidence of renewed political strength for the president."

NINETEEN

B ob Dole was the snake-bitten man of national politics. He'd been close before, achingly close, but the big break always managed to elude him. He'd joined Gerald Ford's ticket back in 1976, and together that fall they'd erased Jimmy Carter's thirty-three-point lead, only to lose in a squeaker on Election Night. Then, in 1988, he'd humiliated Bush, the sitting vice president and his longtime rival, in the Iowa caucuses, vaulting overnight into the lead in New Hampshire. Win there and the Republican nomination, and very possibly the presidency, would be his. But Bush assaulted him with a wave of last-minute attack ads that called Dole soft on taxes. Momentum stalled, Bush recovered, and that was that. And now, eight years later, it was happening all over again.

In a way, it was miraculous Dole was even getting another chance. The night of that New Hampshire loss, he'd gone on NBC and was asked by Tom Brokaw if he had any message to convey to Bush. "Tell him to stop lying about my record," Dole snarled. His many friends in the Senate, who knew him as a generous and self-deprecating man, would protest, but the moment cemented an image of Dole as a bitter and angry sore loser. His days as a would-be president, the consensus went, were done.

But the 1992 election left Dole as the senior Republican in Washington. And when Clinton's presidency spiraled into crisis, Dole's stock rose to new heights. As his party's Senate leader, he played the Gingrich game, utilizing every tool he could find to slow, stall, and derail the Clinton agenda. There was no small irony here. In the past, Dole had been an establishment man, not averse to dealing with Democrats, even if it meant higher taxes. He'd been on Bush's side in 1990, when Gingrich launched the revolt that split the party. Dole, it seemed, had taken a lesson from the experience. Gingrich said he was maturing.

A World War II hero whose right arm was shattered in the mountains of Italy, Dole traveled to Europe to commemorate the fiftieth anniversary of D-Day in 1994. When he returned, he began talking about his own future. Maybe, he said, there was "one more mission, one more call." After the GOP revolution that fall, polls made him the favorite for the next Republican nomination, with Clinton looking like a pushover. Dole was old, but probably not too old. He'd be seventy-three in 1996, the same age as Reagan when he'd won his second term. For once, it seemed in that moment, Bob Dole was in the right place at the right time.

But now, as 1996 began, it was looking like a cruel joke. Dole was running behind Clinton, and had been for a while. He could thank the shutdown, but it was more. Gingrich was, by far, the least-liked figure in American politics, and Democrats were intent on tethering Dole to him. And what could Dole do about it? He'd thrown in with Newt during those first two years of Clinton, and anyway Gingrich remained popular with the hard-core party base. Clinton's approval rating kept inching up, now comfortably over 50 percent, each new poll threatening to devalue the Republican nomination further.

But would Dole even win the nomination? As the primaries kicked off, a threat was emerging that no one had foreseen. It was from the same man who'd embarrassed Bush four years earlier, stalking him through the primary season and hijacking the Houston convention. Pat Buchanan was dismissed as a protest candidate back then, profiting off Bush's broken tax promise and little else. His calls for trade protectionism, immigration restriction, and economic nationalism, the line went, were far out of step with the modern Republican Party. But now, Bush was long gone, Buchanan was back, and the ideas were catching fire.

A year earlier, one of the first national polls put Dole at 45 percent, with Buchanan down at just 3 percent, lumped in with Arlen Specter and Richard Lugar, two ultralong shots. Just give me time, Buchanan insisted. "I'm an advocate of causes in which a rising populist majority really believes, from cultural conservatism and traditionalism to a foreign and trade policy that puts our own country and people first."

Catching Dole was daunting enough, but the more pressing threat for Buchanan was Phil Gramm, the Texas senator who was positioning himself to the right of Dole. This was a race Gramm had spent decades preparing for. Originally elected to the House as a Democrat, he switched parties early in the Reagan years and then dramatically resigned his seat. The voters, he declared, had a right to decide if they wanted him representing them as a Republican. He won the special election with ease, then parlayed the publicity into a winning Senate campaign in 1984. Since then, he'd built himself into a Gingrichian senator, embracing partisan combat and testing traditional boundaries. "I was conservative before conservative was cool," he liked to boast in his Texas drawl. He could bring in gobs of campaign cash, too. "I have the most reliable friend that you can have in American politics," he said as he announced his candidacy, "and that is ready money."

The handicappers agreed: Gramm would be Dole's chief conservative rival. It was the same role Buchanan wanted to play, and there were some critical differences between Buchanan's conservatism and Gramm's. A trained economist who'd taught at Texas A&M before entering politics, Gramm was a free-market absolutist, committed to steep tax cuts, deregulation, and unrestricted trade, and suspicious of safety net programs. He saw economic issues as a CEO might, which helped explain his fund-raising prowess.

Buchanan, meanwhile, billed himself as the workingman's conservative. The economy was bouncing back from the early-nineties recession, but not every indicator was positive. Real incomes were in decline and had been since the 1970s. Buchanan blamed trade. Most Republicans, including Gramm, had backed NAFTA, the free-trade deal between the United States, Mexico, and Canada signed by Clinton—but not Buchanan. He was also against GATT, the General Agreement on Trade and Tariffs, another multilateral deal where Clinton and congressional Republicans had found common ground.

"If you force Americans who make ten to fifteen dollars an hour—good wages—to compete with Chinese making twenty-five cents an hour and hardworking Mexican folks making a dollar an hour, you're going to force their wage levels down," Buchanan argued.

There was immigration, too. Gramm approached the issue with a businessman's eye, mindful of the economic value of imported labor. Buchanan preferred to talk about culture. America's values and heritage were at risk, he said, and a moratorium on all immigration—legal and illegal—was needed. "If you go back in American history, you have had periods of very high immigration and then lulls, and the lulls have come about because social tensions have increased and economic problems," he said. "The country has got to regain a measure of social cohesion and assimilation of the twenty-five million who have come in here in the past twenty years." When Buchanan reiterated his longstanding call for a moratorium on all immigration, legal and illegal, Gramm protested: "I'm not ready to tear down the Statue of Liberty."

A constituency was materializing, and Buchanan gained support, soon passing Gramm in New Hampshire. Money began flowing in, small-dollar donations mostly, and a few endorsements, too. Buchanan's "conservatism of the heart" was striking a chord. It was in Dallas that summer that it all jelled.

The setting was a convention called by Ross Perot. By now, Perot was a diminished figure. He'd ended the '92 race on a high note, repairing the damage from his bizarre withdrawal months earlier. There'd been talk after the election of a follow-up bid in 1996, but then came NAFTA. As it moved toward a vote in Congress in the fall of '93, Perot appointed himself leader of the opposition, threatening to deal the White House a devastating blow. To fight back, Vice President Gore proposed a debate, and Perot accepted. They agreed on a venue: Larry King's CNN show, where the Perot phenomenon had started. Instantly, it became the most anticipated nonpresidential debate in history, and a record audience tuned in to the all-news channel—nearly seventeen million viewers, a mark that would stand for more than two decades.

What they watched was a beatdown. Gore was smooth, skillful, and relentless. He interrupted to correct Perot, needled him over his own wealth, and even co-opted one of his trademarks, hauling out a chart of his own to reinforce an argument. Perot had never faced such a concerted and withering assault. His irritation was obvious as he grew flustered and petulant. He repeatedly whined that he wasn't getting enough time to speak and appealed to King for help. He seemed to be unraveling. The polls gave the win to Gore in a romp, and the political ground shifted. Congress passed NAFTA as Perot's image nose-dived. His favorable rating dropped to the same basement level he'd hit before his reentry late in the '92 race. All the image restoration he'd achieved washed away in one night.

And there'd been no comeback this time. Nearly two years after his televised meltdown, Perot remained broadly unpopular, regarded as too flaky and too erratic by too many people. But not everyone had turned on him. His core supporters had taken up his challenge to build a national grassroots organization, United We Stand America. In August 1995, Perot called them to Dallas for a convention, teasing the idea of turning the group into a third political party.

That got the attention of both parties. With his vast treasury, even a weakened Perot could wreak havoc on the 1996 campaign.

Republicans were particularly concerned. Perot's support had come evenly from Bush and Clinton in '92, but the volunteers who'd followed him to United We Stand skewed toward the conservative side of the spectrum. They tended to be older, white, and profoundly distrustful of government. Some of the Republican candidates appealed to Perot to open his convention to them—to give them a chance to show his supporters that there was no need for a new party. Perot liked the idea, and blew up the format. Now it would be a bipartisan cavalcade under the theme "Preparing Our Country for the 21st Century." Invitations went out to every presidential candidate, congressional leaders from both sides of the aisle, and luminaries from both parties, and most of them accepted. President Clinton was invited, too, but sent regrets.

On August 11, the convention began. Thousands of volunteers filed in to the Dallas Convention Center. Every major media outlet was represented. It was the political event of the season, and Perot had to love the optics. One by one over the next three days, he personally introduced each candidate for president, and all of them ladled praise back on him, like subjects courting the king's favor.

No candidate was more eager for his audience than Buchanan. The Perotistas, he was confident, were Buchanan supporters who didn't know it yet. If they tended toward conservatism, it was a flavor that overlapped with his: hostile to free trade and multilateralism, suspicious of foreign intervention, fiercely patriotic. Their only obvious break with Buchananism was on social issues; the Perot crowd was full of traditionalists, but of a more secular stripe. For them, abortion, homosexuality, and school prayer were not animating issues. For Buchanan, it was an opportunity to win over enough converts to change the math of the Republican race. This would be his biggest speech since Houston.

He was the second candidate up on the second day. After thanking his host and introducing his wife ("the individual I intend to nominate to replace Hillary Rodham Clinton"), Buchanan told the crowd that he was unemployed and looking for a job.

"Let me tell you something that I have a difference with the other folks that are looking for that job—that Ross and I have in common. And that is that together, we stood up against NAFTA and GATT and the World Trade Organization and that fifty-billion-dollar bailout of Mexico City!"

They were on their feet right away. The bailout was a reference to Clinton's authorization of a loan months earlier to help the Mexican government stave off a currency crisis. The original price tag had been fifty billion dollars, but when members of Congress balked at that, Clinton went around them to impose a package worth twenty billion. To the Perot army, it was a sellout of America's interests, and Buchanan now stoked the rage.

"Illegal drugs are coming across the border, illegal immigration is soaring," he said. "And what do we get in addition to that? We are required to pay fifty billion dollars to the government of Mexico? For whose benefit was that, my friends? I'll tell you. That was not for the benefit of the working Americans on Main Street. That was for the benefit of the investment bankers on Wall Street, and we all know it."

The crowd had heard from a few candidates already, and they'd hear from a few more later. None of them talked like this, though. He zeroed in on jobs and trade, calling out the names of old factory towns where manufacturing was vanishing, and with it a way of life. "What are we doing to our own people?" Buchanan asked.

He told a story from his first campaign, about a trip to the James River Paper Company in Berlin, New Hampshire. There'd been layoffs the day before. It was just before Christmas. "And I walked through a line," he recalled, "and they were about my age. Middle-age, middle-class guys. And as I walked through and shook hands, one of them looked up at me, and he had tears in his eyes. And he said, 'Save our jobs.'"

The hall was silent. "And I drove down from the North Country for three hours," he continued, "and I couldn't think of a thing to say. Because that guy wasn't going to get another job. He worked at that paper mill his whole life. And then I read in the Manchester paper that the United States Export-Import Bank had just guaranteed a great big loan for a new paper mill in Mexico."

He paused to let that sink in, then returned to his refrain. "What are we doing to our own people, my friends?"

He knew every button to press: multiculturalism and affirmative action ("This was supposed to be a country where men were judged by the content of their character and not the color of their skin"), foreign aid ("If I get there, foreign aid comes to an end and we start thinking about the Americans right here in the United States!"), political correctness ("They took Washington's name off Washington's birthday!"), national sovereignty ("When I raise my hand to take that oath of office, your new world order comes crashing down!").

The cheers were constant, the standing ovations frequent. Chants of "Go, Pat, go!" erupted. The people in front of him were feeling these words as intensely as the man delivering them. He came to immigration and mentioned George W. Bush, just a few months into his first term as Texas governor. Already, Bush was making a name as a champion of the undocumented "who'd come to Texas to provide for their families." Buchanan said he disagreed: unchecked immigration across the southern border was draining America of its resources and stripping it of its identity. "I will build a security fence and we will seal the border of this country cold and we will stop illegal immigration cold in its tracks if I am elected!" They were roaring now. "You have my word on it! I will do it!"

When he delivered his closing line, they all rose to their feet. The commotion nearly drowned out the music. In its own way, this speech was also a declaration of a culture war, but of a different sort. Houston in 1992 had been about morality. Dallas in mid-1995 was about nationalism.

It was territory Buchanan had all to himself. His climb in the polls accelerated through the rest of the summer and into the fall, as Gramm's support waned. Dole remained the runaway leader as the first contest neared, but Buchanan couldn't get at him until he knocked Gramm out.

In late January, he made a surprise play for Alaska, despite the results being nonbinding. With two late campaign appearances and a batch of television ads, he took first place with 33 percent. He pronounced himself "King of Klondike." Gramm took just 8 percent.

Then it was on to Louisiana, where Republicans had scheduled their caucuses for February 6. They wanted to hold the first binding contest in the country, but almost all of the candidates were boycotting out of deference to Iowa and New Hampshire and their traditional leadoff status. Gramm was the exception; he was competing and banking on an overwhelming win. Then Buchanan made his move. He apologized to the Republicans of Iowa and New Hampshire but said he had no choice. Louisiana was his best chance to put Gramm away.

It didn't look like a fair fight. Gramm, from next-door Texas, had flooded the state with money and wired it with organizers. Buchanan was hoping to embarrass him by keeping the margin close, around ten points, if he was lucky. It turned out to be a landslide—for Buchanan. "This wasn't a victory for a man," he told a raucous victory party in Baton Rouge, "this was a victory for a cause."

A stunned Gramm admitted that "it's a setback for us and a small step forward for Pat Buchanan." But the Texan was gasping for breath now, and Buchanan was surging. Six days later, Dole won Iowa. It was no surprise. In his last campaign, back in 1988, his margin in the state had been so large that people took to calling him the president of Iowa. This time, though, Dole netted just 26 percent. Right on his heels, with 23 percent, was Pat Buchanan.

Now the action shifted to New Hampshire, a state that had been as kind to Buchanan in '92 as it had been cruel to Dole in '88. "We shocked them in Alaska," Buchanan thundered. "Stunned them in Louisiana. Stunned them in Iowa. They are in a terminal panic. They hear the shouts of the peasants from over the hill. All the knights and barons will be riding into the castle pulling up the drawbridge in a minute. All the peasants are coming with pitchforks. We're going to take this over the top!"

He was right about the panic. Universally, Republican leaders now considered him electoral poison and were scrambling to check his rise. Feeding their urgency was scrutiny that the media was now applying to Buchanan. Inflammatory passages from his old columns were resurfaced and aired to a wider audience than they'd ever reached before. "The poor homosexuals," he'd written in the early days of the AIDS epidemic. "They have declared war upon nature, and now nature is exacting an awful retribution."

Unsavory characters from far outside the mainstream political world were being drawn to the Buchanan effort by his nationalist rhetoric. David Duke, a former Klansman and professed admirer of Adolf Hitler, declared his support for Buchanan just before the Louisiana caucuses. Duke had run for governor of the state in 1991, startling the nation's political elite by advancing to the runoff, where he was finally defeated.

Buchanan renounced the endorsement, but his candidacy had become a rallying point for other elements of the far right that used newsletters to promote racist and anti-Semitic views. This amounted to a tiny, isolated fringe, Buchanan protested, one he could do little to control. Critics charged him with willful blindness at best, wink-and-nod collusion at worst.

After another debacle in Iowa, Gramm dropped out, then called a press conference with Dole and endorsed him. He reminded reporters: "You don't see David Duke up here."

It was a three-way contest in New Hampshire. Dole had been leading the whole way, but Buchanan was closing. So was Lamar Alexander, the former Tennessee governor who was positioning himself as a soft-edged outsider and running on electability. (To win, he liked to tell Republicans, all they had to do was remember their ABCs: "Alexander Beats Clinton.") Alexander had been the other Iowa surprise, finishing a strong third.

They were tightly bunched in the first wave of returns, with Dole eking into the lead. Then Buchanan edged ahead, then widened the gap, with Dole in second and Alexander in third. More precincts came in, but the result was hardening and there weren't many votes left. The networks finally called it. "A political tsunami," Bernard Shaw said on CNN. Pat Buchanan had won the New Hampshire primary.

He celebrated in the same ballroom he'd used in '92. "I didn't think it would ever be like that again," he told his supporters. "I thought it was the high-water mark of the cause I believed in." But now he was in rarefied company. Since the first modern New Hampshire primary in 1952, every Republican who'd won it had gone on to claim the nomination. He had a warning for the Buchanan brigades: the party establishment would now do everything in its power to keep that from happening. "I'm telling the folks out in the country: They're going to come after this campaign with everything they've got. Do not wait for orders from headquarters. Mount up, everybody, and ride to the sound of the guns!"

At first, when the exit polls showed him running in third place, Dole thought he might have to drop out of the race. But the final result put him squarely in second with 26 percent—a point behind Buchanan, but four points better than Alexander. Republicans, he said, now had a choice between "the mainstream and the extreme."

"We know that we're now engaged in a fight for the heart and soul of the Republican Party," Dole declared.

Buchanan moved to roll his momentum into Arizona, which held its primary a week later, gambling that his hard-line immigration stance would play in a border state. He won the backing of the state's newest political powerhouse, Joe Arpaio, elected a few years earlier as the sheriff in Maricopa County. Arpaio had stirred national controversy by creating a work camp—Tent City—for his prisoners. The race was tight, but Buchanan came up short. His 27 percent was good for third place.

A few days later came South Carolina, where the Republican elite put out the welcome mat for Dole. Senator Strom Thurmond, ninety-three years old and seeking yet another six-year term, enlisted his organization. Carroll Campbell, a popular former governor, returned from his lobbying work in Washington to pitch in. The current governor, David Beasley, was on board, too.

For Buchanan, it was a pivotal test. South Carolina had all the demographic ingredients he was looking for, filled with religious conservatives and working-class traditionalists. Win this one and he'd be in position to sweep through the South and run away with the nomination. He didn't come close. Alexander was fading out, and his supporters were flocking to Dole, who won the state with 45 percent. Buchanan was second with 29 percent—almost the same share he'd gotten in New Hampshire.

From there, the rest of the states fell like dominos for Dole. He cruised through a series of contests the next week, winning all of them by double digits, then flattened Buchanan in Dixie on Super Tuesday. By the middle of March, Dole was far ahead in delegates and safely on his way to the nomination. As in '92, Buchanan stuck around to the end, but his candidacy morphed into a symbolic crusade. His best showing was in Missouri, where he won the lightly attended caucuses with 36 percent. Everywhere else, he was crushed. In total, he grabbed 3.1 million votes during the primaries, only slightly better than his '92 tally.

But this time, no one dismissed it as a protest bloc. These were votes for Buchanan, and for Buchananism. His nationalist conservatism had connected with an audience Republicans didn't even know existed. The nation's immediate economic outlook was improving, but Buchanan had shown that just beneath it were the seeds of a populist rebellion. The decline of American industry, stagnation of wages, and broad trends toward globalization and a more diverse society were stirring unease with a significant segment of Americans. Buchanan had seen it. The rest of his party hadn't.

These were forces threatening to gain potency in the years to come, but in 1996 they only carried Buchanan so far. Dole, on his third try for the Republican nomination, had finally prevailed, but at what cost? He coasted to the San Diego convention on fumes. His campaign had all but run out of money fighting Buchanan off and would need to wait until Dole's formal nomination to collect the giant federal matching funds check that would sustain it through November. Through the primary slog, meanwhile, Dole's positioning against Clinton had only deteriorated. Free from a primary challenge, the president and his party dipped into their flush coffers through the spring and summer to savage Dole. The barrage was relentless, designed to make the Republican candidate indistinguishable from his party's great albatross. One spot promoted Clinton's willingness to stand up to the Republicans, then showed an empty Oval Office. "If Dole wins and Gingrich runs Congress," the narrator said, "there'll be nobody there to stop them." Another seemed to merge the two GOP leaders' names into one: "DoleGingrich tried to cut $270 billion" from Medicare.

Dole was now risking everything, resigning his Senate seat in May in a frantic bid to jar his White House effort to life. "I will seek the presidency with nothing to fall back on but the judgment of the people," he said, "and nowhere to fall back on but the White House or home." It wasn't working, though. By now, Dole was routinely trailing Clinton by double digits. Some polls had the margin over twenty, and for a time in the spring it had even approached thirty. Dole would need the mother of all bounces from his convention, but he got no cooperation from his vanquished primary foe.

For the media, now accustomed to drama-free conventions that functioned more as television ads for the parties, it became the story of San Diego: Would Pitchfork Pat bolt? Dole commanded the overwhelming share of delegates and, remembering what had happened in Houston, used his power to deny Buchanan a speaking slot. All he would offer Buchanan was an appearance in a prerecorded video, where he could issue a brief endorsement of the nominee. Insulted, Buchanan told Dole where he could put his video and then called a rump convention of his own.

In Escondido, California, on the Sunday night before the GOP convention was to open, nearly two thousand Buchanan supporters and an army of media members packed the California Center for the

Arts. There was suspense. Buchanan was refusing to say if he would back Dole and hinting at a third-party effort. A stream of speakers, including Oliver North, the Iran-contra figure turned conservative hero, sang his praises before Buchanan finally appeared. He basked in the adoration of the faithful. There were chants and shouts to bolt the party, but he said he wasn't ready to go that far. Instead, he declared victory on the GOP platform. Hoping to avoid dissension on the convention floor, Dole had permitted hardline language on immigration, affirmative action, and American sovereignty. It was enough now for Buchanan to call for a "truce of San Diego." He never mentioned Dole's name, but he was staying put—at least for this campaign year. "The stone the builders rejected may yet become the cornerstone," Buchanan said.

Four nights later, Dole got his chance to speak. There was speculation Buchanan's delegates might still walk out, and there was no way Dole could ignore it, this force that had been awakened in his party, that was threatening to haunt him all the way to November. "If there's anyone who has mistakenly attached themselves to our party in the belief that we are not open to citizens of every race and religion," he said, "then let me remind you: Tonight this hall belongs to the Party of Lincoln. And the exits, which are clearly marked, are for you to walk out of as I stand this ground without compromise." The delegates roared, the critics applauded, and no one walked out. Tracking polls showed movement; one had Dole cutting the gap to four points. Had he hit his stride? Was he on his way back?

It lasted for maybe two days, and by the next week Clinton's lead ballooned back to double digits.

TWENTY

The president was telling his story of the shutdown to a crowd of thousands in Huntington, West Virginia. The Republicans had threatened him, he said—threatened everyone. They'd demanded dangerous cuts to health care for the elderly, to education, to the future of the country. "We said, 'Have at it,' the president recalled. "We don't stand for blackmail. We stand for America!'"

It was a Sunday afternoon in August. The next day, the Democratic National Convention would begin in Chicago, but Clinton was taking the long route. A special train, meant to evoke the spirit of Harry Truman, was waiting for him. It would shuttle him from West Virginia to Kentucky, then north through Ohio and into Michigan, and back down into Indiana. From there, he'd be escorted to the Second City, his arrival timed to the formal vote Wednesday night that would nominate him for a second term as president.

It was a testament to the mandate Clinton believed was within reach. Two years earlier, it had been conceivable he would lose each one of those states in a reelection campaign—if his party even let him run again. Now, he was favored to win all of them except the old

Republican redoubt of Indiana, and even there he had a puncher's chance. The GOP convention had barely dented his lead, and with the intricately choreographed celebration planned for Chicago, Clinton was poised to build an even bigger advantage over Dole.

Chicago had been the setting for one of the most notorious chapters in the Democratic Party's history, the 1968 national convention. Playing out against the backdrop of Vietnam and the civil unrest of the late sixties, it was marred by chaos and disunity in the hall and riots and cries of police brutality outside it. In one of several infamous episodes from that week, Connecticut senator Abraham Ribicoff used his speech to decry the "Gestapo tactics" of Chicago's cops, prompting the city's ironfisted mayor, Richard J. Daley, to jump from his front-row seat and hurl obscenities back at him.

That had been the tenth—and until now, last—time Democrats chose Chicago for their convention. They had nothing to worry about this time around. The theme of the week mirrored Clinton's strategy since the 1994 revolution: He was a reasonable president and this was a reasonable party, trying to move America forward against an opposition party that was fundamentally unreasonable. The Democrats weren't gathering to ratify sweeping new ideas or to promise dramatic overhauls. The change Clinton was running on this time was smaller in scale, heavy on targeted tax credits and tax cuts, mixed with a celebration of what progress had been made—and lots of warnings that it could all go away if the Republican Party of Newt Gingrich ever got control of the White House.

On Monday night, the star attraction was Ronald Reagan's first press secretary. Fifteen years earlier, James Brady had nearly lost his life when one of six bullets meant for the president instead struck him in the forehead. He emerged partially paralyzed and with difficulty speaking and left his day-to-day life in politics. His wife, Sarah, meanwhile, launched a public crusade for gun control. It was James Brady after whom the 1993 waiting-period law signed by Clinton was named.

Now, as the broadcast networks picked up their one hour of prime-time coverage, Jim Brady, leaning on his wife, Sarah, and walking slowly with a cane, stepped onto the stage in Chicago's United Center and made his way to the podium. "Jim, we must have made a wrong turn," Sarah Brady said. "This isn't San Diego."

"I told you," her husband replied, "this is the Democratic convention."

The delegates ate it up. Also featured on Monday was the widow of one of the six people murdered in a shooting rampage on a Long Island commuter train three years earlier. Like many of her neighbors in suburban Mineola, Carolyn McCarthy had been a Republican, but she no longer recognized the Republican Party she was seeing. Now she stood before the Chicago convention as the Democratic nominee for Congress in New York's Fourth Congressional District, determined to unseat a freshman Republican who had voted to repeal the assault weapons ban. There were, Democrats believed, places like Mineola all around the country, suburbs ready to turn against what the Republican Party now was.

Gingrich's presence in Chicago was constant, even if the Speaker himself was nowhere in sight. At their convention, Republicans had done their best to hide Newt, allowing him to wield the gavel for the ceremonial opening and to deliver a brief speech that was carefully vetted to avoid inflammatory rhetoric. His poll numbers were every bit as toxic now as they'd been during the shutdown, and Democrats were boasting of their plan to make every Republican candidate in America—starting with Dole—indistinguishable from the House Speaker.

The only discordant notes came on Tuesday—safely out of the prime-time window. The week before, Clinton had signed into law a welfare reform plan passed by the Republican Congress. It wasn't what the GOP had initially offered in those early months of 1995, but it wasn't that far off, either, and liberals were irate.

The bill did away with Aid to Families with Dependent Children and the basic commitment that had undergirded it—that everyone who needed it would receive it. Instead, it shifted money and power to individual states, handing them broad new powers to set rules and determine eligibility. And it imposed limits: anyone receiving benefits would have to find a job within two years; and no one could receive benefits for more than five years out of their lifetime.

Two versions of the bill had reached Clinton's desk, and he'd vetoed both of them, but when congressional Republicans took their third pass that summer, Clinton reconsidered. He was on record objecting to many provisions, but he was also on record—dating back to his 1992 campaign—with that promise to "end welfare as we know it." On so many issues, he had turned the tables on the Republicans since the revolution, and he was running well ahead of Dole. But welfare was different. Unlike Medicare, it was not a universally beloved program; in fact, the public was strongly on the side of changing it in major ways.

There were voices telling Clinton not to worry—that he was strong enough politically to take any heat that would come with another veto. But he wasn't nearly so sure. He'd been so low in '94 and he'd defied all predictions with his comeback since then. Why risk giving Dole any kind of an opening now?

On July 31, he announced that he would sign it, even though it had "serious flaws." He promised to fight for changes after it was enacted and argued that the opportunity was too rare to miss. "This is the best chance we will have for a long time to complete the work of ending welfare as we know it." Thirty Democrats in the House voted for it and twenty-three in the Senate, while practically every Republican voted yes, and on August 22, Clinton followed through and made it law.

Denunciations from Clinton's own party had been loud. John Lewis called it "mean." Senator Paul Wellstone, a liberal from Minnesota, said it would "create more poverty and hunger among children in America." Even Moynihan, who had once urged Clinton to tackle welfare, ripped it as "the first step in dismantling the social contract that has been in place in the United States since at least the 1930s." Soon, Peter Edelman, a top official in the Department of Health and

Human Services and the husband of one of Hillary Clinton's best friends, would resign from the administration in protest.

Was this the issue, the media asked, that would send the Chicago convention into chaos? The answer was no. Time was set aside on Tuesday for two of the most eloquent liberal voices. Jesse Jackson and Mario Cuomo had both stolen the show at past conventions. Now they would be free to air their displeasure—and expected by the Clinton campaign to temper it with praise for the president and warnings about the alternative to his continuing rule. They would also do it comfortably out of prime time. It was Evan Bayh, the youthful, moderate Indiana governor chosen by the president to deliver the key-note address, whom the Clinton campaign wanted network television viewers to see—not two aging liberal icons whose political glory days had passed a few years ago.

"In 1968, the tension within our party was over warfare," Jackson told the convention. "In 1996, it's welfare.

"Last week, over the objections of many Democratic Party leaders, and the opposition of millions of Americans, Franklin Roosevelt's six-decade guarantee of support for women and children was abandoned. On this issue, many of us differ with the president."

Then he argued that it wasn't something to abandon Clinton over. "President Clinton," Jackson said, "has been our first line of defense against the Newt Gingrich Contract [with America], America's right-wing assault on elderly, our students, and our civil rights."

It was the same story with Cuomo, who started by acknowledging that "many of us—and I among them—believe that the risk to children was too great to justify the action of signing that bill, no matter what its political benefits." But soon enough, he was sounding the alarm over the "rabid revolutionaries led by Newt Gingrich."

Welfare reform would be a significant component of Clinton's presidential legacy, a source of debate, criticism, and reexamination on the left for years and decades to follow. But in the politics of the moment, it was something that most liberals shook their heads at— and then moved on.

The nominating speech was delivered Wednesday night by Senator Christopher J. Dodd, who also served as the general chairman of the Democratic National Committee. Dodd called back to something few in the party previously dared speak about: the 1993 tax hike. Their votes for it had sent numerous Democrats to their political graves in 1994, but now the ground seemed to be shifting.

Look around, Dodd told his audience. Unemployment, well over 7 percent when Clinton took office, was now barely over 5, and plummeting. The deficit, projected to surge past three hundred billion dollars when Clinton called for action, was falling, too; it would be only a little over one hundred billion dollars in the new fiscal year. Dodd brought up the dire predictions Gingrich and the Republicans had made back in 1993. They'd told Americans the Clinton tax plan would cost jobs, trigger a recession, and make the deficit worse. All of them had voted against it, and none of their warnings had come to pass.

"Tonight," said Dodd, "let me say it plainly: Mr. President, you did the right thing. You did the right thing for America."

Wednesday night was also reserved for Al Gore to accept the nomination for vice president. It was a break with tradition. Typically, the VP would speak on the convention's final night, a warmup of sorts for the main act. But Clinton and his team already had an eye on the next election. They wanted to showcase the man who, they hoped, would affirm Bill Clinton's legacy by running on his record and being elected to succeed him in 2000. During the week, delegates had been polled by the Associated Press about their preference for the next presidential nominee, and 77 percent of them named Gore.

The vice president, the frequent subject of jokes about what could be a wooden public demeanor, opened with a bit of his own, offering to demonstrate his own version of the Macarena, a dance that had recently become an international sensation. Standing motionless and staring straight ahead, he waited several seconds and then said, "Would you like to see it again?"

Laughter filled the United Center. It was that kind of mood in

Chicago. In his speech, Gore was deputized to play the traditional VP role of attack dog, although with a softer edge this time. One of the few areas where Dole enjoyed an advantage with Clinton was on "character." A majority of Americans still said they didn't trust their president, and the memories of the Gennifer Flowers and Vietnam draft scandals remained strong. On late-night television, Clinton was routinely depicted as a hopeless womanizer.

Dole, meanwhile, was a decorated hero of World War II, who'd battled through blood clots and infections for more than three years to recover from the gunfire he took from German artillery. The result of his ordeal was evident for the world to see: the limp right arm that hung by his side, his lifeless hand scrunched into a fist that always held a pencil.

In fact, Dole had made his service—and the service of his generation—the theme of his own speech in San Diego. In a few years, Tom Brokaw would coin the term "the Greatest Generation," and Dole was betting on an untapped market for sentimentality. "Let me be a bridge," he'd said in his speech, "to an America that only the unknowing call myth. Let me be a bridge to a time of tranquility, faith, and confidence in action.

"And to those who say it was never so, that America's never been better, I say you're wrong. And I know because I was there. And I have seen it. And I remember."

It was received well in the moment, but now Gore moved to give Dole's words a different spin. After saluting the Republican nominee as "a good and decent man" worthy of respect, Gore's tone changed. "But make no mistake," he said. "There is a profound difference in outlook between the president and the man who seeks his office. In his speech from San Diego, Senator Dole offered himself as a bridge to the past. Tonight, Bill Clinton and I offer ourselves as a bridge to the future."

Then he brought Newt into it. The two top Republicans in the country, Gore reminded a crowd that knew this story well, had dared the president to accept their budget or shut down the government.

"Let me tell you what Bill Clinton did do," he said. "Bill Clinton took Speaker Gingrich and Senator Bob Dole into the Oval Office. I was there. I remember. And he said—President Clinton said: 'As long as I occupy this office, you will never enact this plan. Because as long as I am president, I won't let you.'

"That's why they want to replace Bill Clinton," Gore concluded. "But we won't let them!"

Clinton's turn finally came the next night, but not until the convention was interrupted by a sex scandal. It involved Dick Morris, the architect of the president's triangulation strategy. The *Star*, the same supermarket tabloid that had splashed Gennifer Flowers's accusations on its cover four years earlier, had caught Morris cavorting with Sherry Rowlands, a two-hundred-dollar-an-hour prostitute, who claimed that Morris would often take calls from the president while they were together and that he would let her listen in.

The story landed late Wednesday night, and by three A.M. Morris had resigned—"so I will not become the issue."

It was a demise no one around Clinton mourned. Morris was seen in the White House as an arrogant self-promoter and by liberals as a threat to many of their core values. His fingerprints were all over Clinton's decision to sign the welfare law and just about every other rightward move Clinton had made since the midterm. Reacting to the news from the campaign trail, Dole said, "Morris has been trying to make President Clinton a Republican. Now maybe he'll revert to the liberal Democrat he really is."

On the Clinton team, there was worry that the story would overwhelm coverage of the president's speech and that the dirt might somehow rub off on him. Then again, this was a president who had already been through his own sex scandal and emerged just fine. The feeding frenzy had calmed by nightfall.

A glitzy twenty-minute video served as the president's introduction, and when he finally strode onto the stage he was treated to a euphoric celebration. Delegates waved baseball pennants with his name on them and renewed one of the oldest traditions at conventions like these. "Four more years!" they exclaimed.

"Mr. Chairman, Mr. Vice President, my fellow Democrats, and my fellow Americans," Clinton began. "Thank you for your nomination. I don't know if I can find a fancy way to say this, but I accept." More rejoicing. More chanting. You could forgive these Democrats for their overflowing jubilation. No one from their party had stood for a second term and won since Franklin D. Roosevelt in 1936, but now Clinton was heavily favored to do so.

He ticked through statistics that showed how far the country had come. Ten million jobs had been created since his inauguration. Interest rates had dropped, inflation remained low. Violent crime was down for the fourth straight year. "Just look at the facts," Clinton said. Dole, he acknowledged, was a man who loved his country and had worked hard to serve it. Then, just as Gore had the night before, he sought to turn his opponent's own campaign theme against him. "Now here is the point," the president said. "I love and revere the rich and proud history of America. And I am determined to take our best traditions into the future.

"But with all respect, we do not need to build a bridge to the past. We need to build a bridge to the future. And that is what I commit to you to do."

He'd come armed with proposals, lots of them. He wanted a tax credit for tuition; a hefty deduction for working-class families paying college expenses; a job-training grant to help the unemployed find work; a corps of volunteers to eradicate illiteracy. There was more, too. The scope of each was narrow, but the effect was supposed to be big—a reasonable president with all sorts of smart, reasonable ideas. There was a sop, as well, for liberals still reeling from his welfare decision. Under the new law, he said, there'd soon be one million beneficiaries required to get work, and "I propose to offer private job placement firms a bonus for every welfare recipient they place in a job who stays in it."

Clinton didn't mention Newt Gingrich's name a single time. He didn't need to. He said he wanted to balance the budget, and to do it "in a way that preserves Medicare, Medicaid, education, the environment, the integrity of our pensions, the strength of our people." Then he told what had become his favorite story.

"Last year, when the Republican Congress sent me a budget that violated those values and principles, I vetoed it," he said. "And I would do it again tomorrow.

"I could never allow cuts that devastate education for our children, that pollute our environment, that end the guarantee of health care for those who are served under Medicaid, that end our duty or violate our duty to our parents through Medicare. I just couldn't do that.

"As long as I'm president, I'll never let it happen, and it doesn't matter—it doesn't matter if they try again, as they did before, to use the blackmail threat of a shutdown of the federal government to force these things on the American people. We didn't let it happen before. We won't let it happen again."

Less than two years earlier, Gingrich and his party had come to power with a simple message for the president: get out of our way or get run over. The Speaker had judged Clinton to be a nuisance, a president in terminal decline, and he hadn't been alone in doing so. The press had written him off as a certain one-termer. So had members of his own party. What no one saw then was what everyone watching Clinton's speech now could. The Republican Revolution that was supposed to destroy his presidency had instead given it purpose.

D ole never led a single general election poll, or even came close. With the deficit well into the teens, his own party cut him loose a few weeks out. "If Clinton is reelected, heaven forbid, the last thing the American people want is for him to have a blank check in the form of a liberal Democrat Congress," Haley Barbour, the Republican National Committee chairman, said. Translation: It was fine with

GOP leaders if candidates in tough races stopped pretending Dole had a chance. In fact, it was encouraged.

This is where the suspense was in the final weeks and days. The House was, stunningly, within reach for Democrats. With a net gain of twenty seats, Gingrich would be out as Speaker and Dick Gephardt would be in. Charlie Cook, a top election analyst, said, "I think this thing is sitting on the edge." Gingrich himself, ever-present in Democratic ads but almost entirely absent from the campaign trail, vehemently defended his tenure but also admitted: "In retrospect, if I were doing it all over again, we would consciously avoid the government shutdown. It was clearly wrong."

Clinton badly wanted to break 50 percent. In his first term, Republicans had taunted him as the president of only 43 percent of the country. Now he wanted a clean majority. It was within his grasp, although tricky, since Perot was running again. This Perot was a shell of the old one, this time locked out of the debates and stuck in single digits. Finally, Dole did catch one break—a last-minute controversy involving the Democratic National Committee's fund-raising practices. In one instance, it was revealed that the party had presented high-end donors with a menu of perks that could be theirs for the right price—including a night in the Lincoln Bedroom. The whiff of scandal, to say nothing of the tawdriness, was potentially enough to make the margin respectable for Dole—and to save the Republican Congress.

But it was not to be. Election Night was quick and brutal. When they called Florida for Clinton, the rout was on. Even in '92, he'd missed out on the Sunshine State, but now he won it by six points. He ended up with 49 percent nationally, to Dole's 41. Clinton would still be a minority president, just barely. In the Electoral College, it was 379 to 159.

The Republican Congress survived, too. Running hard against Gingrich, Democrats knocked out eighteen Republican incumbents, hitting pay dirt in the Northeast, West Coast, and Rust Belt. It put a dent in the GOP majority, but Republicans found some pickups of their own and hung on with a reduced majority of eight seats. It was, Gingrich pointed out, the first time since 1928 that a Republican Congress had been reelected. That was true, but it was also true that this election resulted in nothing like the domination that Republicans had been expecting after their revolution two years before.

If you just looked at those numbers, you might think the 1996 election ended up as more or less a repeat of 1992. The electoral count was almost the same (it had been 370–168 the last time), and Clinton's popular-vote margin wasn't much bigger this time—eight points now, compared to 5.5 in '92. But a closer look told a story of new divisions. Clinton's improvements were not evenly spread out. Overall, his popular-vote margin had grown by two and a half points, but in some places the increase was bigger than others—a lot bigger. In particular, there was the Northeast. In '92, Clinton had carried New York by sixteen points. This time he won it by twenty-nine. New Jersey had gone for Clinton by just two points in '92. Now, his number exploded to eighteen. Connecticut went from seven points in '92 to eighteen in '96. Massachusetts went from eighteen to thirty-three, New Hampshire from two to ten.

The suburbs of the Northeast, packed with culturally moderate white-collar professionals, had traditionally been a source of strength for Republicans. But presented with the Republican Party of Newt Gingrich, they were recoiling. Was this the same party they'd been voting for all those years? It could be seen in other spots on the map, too, and among women, it was especially pronounced. Four years earlier, Clinton had won male voters by three points and females by seven; now he lost men by a point—and won women by seventeen. "Angry White Males" had been the catchall term pundits used to describe the voters at the heart of the Republican Revolution. Now, the same pundits chalked up Clinton's reelection to the rise of "Soccer Moms."

No one had a name for it yet, or knew if it would endure past this election, but Blue America had just been born.

TWENTY-ONE

N ow Newt was in danger. More than anyone or anything else, he was the reason his party had just lost seats, and nearly the House itself. Democrats considered him their most effective political weapon, while the whispers from Republicans were growing louder: Was this guy really worth all the trouble?

Compounding the Speaker's sudden peril was a long-simmering ethics probe now coming to a boil. At issue was his messy mixing of partisan politics, academia, and tax-exempt groups. The heart of the case involved a college course Gingrich taught in Georgia in the early nineties. It was called Renewing American Civilization, and for two hours each week he would stand before a classroom and offer his expansive reading of history, government, and philosophy. Its content was indistinguishable from speeches he'd give on the campaign trail, or in those late-night C-SPAN sessions.

And that was the problem, because the courses themselves were funded by tax-exempt organizations, which were strictly forbidden from engaging in partisan politics. On top of that, Gingrich also arranged for his lectures be taped and fed by satellite to audiences across

the country. Videos were produced and sold. Putting all of this together was a much more expensive endeavor than simply teaching in a
classroom once a week, and to pay for it Gingrich leaned on GOPAC,
the explicitly partisan grassroots group he'd been running since the
eighties.

As with his book deal, the irony was tremendous. In December 1995, Democrats filed a formal complaint with the House Ethics
Committee, the same body Gingrich had turned to when he began
pursuing Jim Wright. "Gingrich's path to power," said Representative David Bonior, the Democratic whip, "was paved with a corrupt
mix of secret contributions, illegal gifts, hidden campaign spending,
and political payoffs."

Gingrich brushed it off as a partisan stunt. "If Newt had a dog,
Bonior would accuse him of kicking it," his spokesman said. But the
Ethics Committee deputized a special counsel to pursue the matter,
and in September '96, the investigation expanded. Now the committee suspected the Speaker of providing information that was not
"accurate, reliable, or complete." The final report was due at the end
of January 1997. It was sure to contain harsh judgments. A formal
punishment for the Speaker was very possible, even likely.

First, though, there'd be a vote on the House floor to elect a
Speaker for the new Congress. Normally, it would be a formality.
All the Republicans would vote for Gingrich and all the Democrats
would support their party's leader, Richard Gephardt. But now there
was the prospect of Republican defections, and it wouldn't take many
to deny Gingrich a majority. "There's no doubt that at least half the
Republican members of the House believe Newt is a problem," New
York representative Peter King said. "Why do we want to reelect the
guy we spent the last six months running away from?"

King, from suburban Long Island, represented one of the many
areas in the Northeast that had swung hard toward Clinton—and
away from the Gingrich GOP. He called for Gingrich to step aside
temporarily, until the ethics report was released. So did Oklahoma's

Steve Largent, one of the revolutionaries from the class of '94, proof of just how precarious Gingrich's position had become.

Just before Christmas, Gingrich admitted to providing misleading information during the investigation, when he'd denied that GOPAC had any connection with the college courses. He called it a technicality and blamed his lawyer—who then quit and said Gingrich had been fully aware of the meaning and significance of every statement he'd signed off on.

The floor vote was set for January 7. On the day before, Iowa representative Jim Leach, a moderate respected by members from both parties, announced he wouldn't support Gingrich. "The party's future and the reputation of the Congress are clearly jeopardized by his continued stewardship of the House." By the end of the day, a total of five Republicans were saying the same thing. Nearly two dozen hadn't announced how they'd vote. The possibility of a stampede was real.

That night, the House Republican Conference held a meeting for all members. It was closed to the press. His fate on the line, Gingrich tried to show contrition and humility. The defectors fired back. California's Tom Campbell, it was reported, accused Gingrich to his face of intentionally deceiving the Ethics Committee. The meeting ran for three hours.

When the vote was called the next day, ten Republicans didn't vote for him. Gingrich survived, but barely. He received 216 votes—just two more than needed.

As he had two years before, Gingrich took the ceremonial gavel and turned to address the House. In 1995, this had been a moment of triumph for him and his party. Now it was something quite different. "Let me say to the entire House that two years ago, when I became the first Republican Speaker in forty years, to the degree I was too brash, too self-confident, or too pushy, I apologize. To whatever degree, in any way, I brought controversy or inappropriate attention to the House, I apologize."

Weeks later, the final report was issued, accusing the Speaker of "reckless conduct" and providing misleading information. Gingrich worked out what amounted to a plea deal. He would admit to wrongdoing and accept the committee's recommended punishment; in exchange, the committee would not charge him with offering the misleading information intentionally. The committee would also not render judgment on whether his arrangement with the college course was a violation of federal tax law. He would simply be charged with bringing "discredit to the House" by not properly vetting the question with legal professionals.

On the first full day of Clinton's second term, January 21, the House voted 395–28 to reprimand Gingrich and to levy a three-hundred-thousand-dollar fine. It was the stiffest penalty ever imposed on a House Speaker. Worse, he didn't have the cash to pay for it. A loan was arranged: fresh off his White House defeat, Bob Dole put up the money and Gingrich promised to pay him back at 10 percent interest.

Some Republicans argued the penalty was too harsh, but for Democrats the floor debate was a delicious opportunity to vent years of pent-up disgust. California's Nancy Pelosi, a prodigious fund-raiser now angling for a spot in her party's leadership, urged Republicans to take the next step and dump Gingrich altogether. "He is technically eligible," she said. "I hope you will make a judgment as to whether he is ethically fit."

A few months later came a coup. At first, it was driven by a restive band of revolutionaries from the '94 class, but soon high-ranking Republicans were in on the scheme too. They wanted Newt out and they might have gotten their way, until Representative Dick Armey tipped off the Speaker, who swung into action and crushed the rebellion. Armey swore he'd never been involved; the plotters said he'd ratted them out when he found out he wasn't their choice to replace Gingrich.

For all the drama, Gingrich was now gaining political strength. So was Clinton, and so was just about everyone in Washington. It was the economy. The bust years of the early nineties had given way first to a creaky recovery and now a full-fledged boom. The Internet, a fuzzy, futuristic concept until recently, was now a mainstream reality, and the source of a tech-sector explosion that was propelling the NASDAQ to one record high after another. The effects went far beyond the dot-com world, though. Unemployment was down to 4.6 percent. The annual rate of GDP growth, which had fallen into negative territory early in the decade, reached a staggering 4.5 percent in 1997. The Dow Jones average had been just over three thousand at Clinton's swearing-in; in the summer of '97, it crossed eight thousand. Everyday Americans were dabbling in the markets and getting rich. Retirement accounts were getting fat. On his watch, Clinton could now boast, fifteen million new jobs had been created, with plenty more on the way.

And then there was this: the deficit had all but vanished. Here was the real key. Three times in the nineties, Washington had descended into political war, and each time red ink was the root cause. Bush in '90 and Clinton in '93 had both raised taxes in the name of taming out-of-control deficits; and it was Gingrich's vow to balance the books within seven years—while cutting taxes—that set the stage for the '95 shutdown.

Now, the economy was on a tear, and revenue was gushing in. Even though the shutdown produced no grand budget deal, the deficit still plummeted from $164 billion in '95 to just $22 billion in '97, the lowest it had been in a generation. A surplus, elusive since the 1960s and unthinkable for most of the nineties, was suddenly within sight. Their effect hadn't been immediate, but it turned out that both tax hikes had done their job. Now that times were good, those higher rates on the wealthy were translating into a windfall for the Treasury.

It meant that there could be peace in Washington if both sides wanted it—and they did. When Clinton and Gingrich now returned to the table, the goal remained a balanced budget, but the hard part was already over, and where there'd been confrontation before, this time there was consensus. Republicans acceded to the kind of

Medicare cut Clinton had first proposed two years earlier—a $115 billion rollback in its rate of growth, concentrated on the provider end, with no structural changes to the program. In return, they got their long-sought five-hundred-dollar child tax credit, along with cuts in the capital gains and inheritance taxes. Clinton also got his college tax credits. Republicans even gave in on some new spending. Earlier in his presidency, Clinton had swung hard and missed on universal health insurance, but he was determined to put some kind of a win on the board before leaving office, and now he got it: twenty-four billion dollars to cover kids from lower-income families that didn't already qualify for Medicaid. It would be called the Children's Health Insurance Program, or CHIP.

These were the contours of what became known as the Balanced Budget Act of 1997. When it was enacted, the projections said, America's red ink days would end within five years. Amazingly, it all went down with little of the drama and none of the brinkmanship that had accompanied the decade's earlier battles. The deal was sealed at the end of July. "After decades of deficits," Clinton declared, "we have put America's fiscal house in order again." With only a few dozen dissenting votes, it passed the House, then rolled through the Senate. As the votes were announced there, *USA Today* reported, "a grinning House Speaker Newt Gingrich watched from the back."

They held a big ceremony on the South Lawn of the White House on August 5—Clinton and Gingrich sharing the same platform, bunting draped from the porch of the South Portico behind them. The president spoke first. He framed the act as the signature feat of his reign: "It's hard to believe now, but it wasn't so very long ago that some people looked at our nation and saw a setting sun. When I became president, I determined that we must believe and make sure that America's best days are still ahead."

The year before, Clinton's team had poured millions of dollars into ads that cast Gingrich as a villainous figure. Now he hailed the Speaker and the top Senate Republican, Trent Lott, as "dedicated partners."

"Today," Clinton said, "it should be clear to all of us without regard to our party or our differences that in common we were able to transform this era of challenge into an era of unparalleled possibility for the American people."

Gingrich returned the flattery. It wasn't long ago that he'd labeled Clinton and his wife "counterculture McGovernicks"; but now the Speaker saluted the First Couple, along with Vice President Gore and his wife, Tipper: "Their willingness this year, coming off their victory, to reach out a hand and say, 'Let's work together,' was the key from which everything else grew. The sincerity of their efforts over the last eight months has made an enormous difference in our capacity to make the system work, and I thank the four of you very much."

The law's title was a misnomer. There were some savings in it, but the country was already on track for a balanced budget even without it.

The crucial actions had come years before. And here, Clinton could rightly take some credit. He had paid a steep political price for raising taxes, but it really had produced more revenue—a lot more revenue, now that the economy was humming. And while they'd resonated with voters in the moment, the dire Republican predictions from the summer of '93—a recession within a year, Gingrich had claimed—sure looked misguided now. The signing ceremony was a bipartisan celebration, but Clinton tried to take a bow, pointing out that on the road to a balanced budget "the first step was taken back in 1993." Of course, the plan Clinton championed in '93 was at odds with the one he'd run on in 1992, when he'd promised a tax cut for the middle class. More to the point, Bush's tax hike in 1990 was at least as responsible for this moment, and probably more so. From retirement, the former president now took his own bow, saying, "I do think history will show that we laid the groundwork for some of this." Then again, Bush hadn't always been this proud of his work; he'd spent the 1992 campaign claiming the tax increase was a mistake that he profoundly regretted and would never repeat.

Not that anyone had much interest in litigating the past right now. No one's hands were entirely clean in this saga, but more than that, everyone was just relieved that the never-ending budget war had actually come to an end. For the first time in nearly two decades, headlines about exploding deficits, warnings about looming bankruptcy, and draining battles over tax rates and spending cuts no longer constituted the background noise of American politics.

There remained occasional cries from the left that Clinton was playing small-ball with his presidency, ducking deeper structural issues for expediency's sake. "Consistently, low-income people come out on the short end of the stick," lamented Minnesota's Paul Wellstone, probably the most liberal voice in the Senate. From the right, too, there remained demands that the Republican Congress push harder for lower taxes and steeper rollback of the welfare state. But between those edges now existed a broad consensus.

There wasn't much to argue about on the world stage, either. In late 1995, Clinton had deployed twenty thousand American troops to Bosnia, to enforce a fragile peace treaty. A civil war had been raging there, triggered by the fall of the Soviet empire and the disintegration of Yugoslavia into a collection of ethnic nation-states. Clinton's decision, on the eve of his reelection campaign, stirred controversy. As always, his own lack of service in Vietnam hung in the air. "It is impossible for this president to explain to the American people why he is going to risk the lives of young men and women," Gingrich said. "The Bosnia policy is quicksand."

It was another dire forecast that never panned out. Now, two years later, the peacekeeping mission was progressing with little drama and Clinton was lightening the American footprint. Just one U.S. service member had been killed. For most Americans, the question of Bosnia was now out of sight, out of mind.

Iraq was a slightly different story. Saddam Hussein, left in power when Bush and his team declined to invade Baghdad after liberating Kuwait, had continued to pester Washington. Twice now, Clinton had ordered air strikes—once after receiving intelligence linking Hussein's government to a plot to assassinate Bush and again when Hussein moved troops toward the autonomous Kurdish region of northern Iraq. Each time, the country rallied around Clinton—and the possibility of a wider confrontation with Hussein loomed.

Here, the shift in American psychology was stark. A few years earlier, in the run-up to Bush's Desert Storm campaign, public opinion had been against a military attack. Back then, Vietnam was the common point of reference, memories of futility seared into Americans' minds. Now, the country's confidence was back. Hussein's forces had been crushed in Desert Storm. The Bosnia intervention had proceeded virtually without incident. The fear that intervention would necessarily lead to a quagmire had been shattered. When Hussein reared his head now, polls showed overwhelming support for a military response. Reservations about the potential cost of war were fading away. A Desert Storm veteran told Newsweek that "when we first got back to the United States, everyone was patting us on the back. And now they say: why didn't you finish the job?"

It was a sea change that would have profound implications a few years down the line. But in this moment, it was enough to say that Americans were again feeling a swagger about their country and its place in the world. Add in the economic boom back home, the end of the budget war, and the new sounds of bipartisan harmony emanating from the capital city, and it was clear that Washington was ready for a truce.

As 1998 arrived, a poll put Clinton's approval rating at 62 percent, the best of his presidency. He still had three years left in office, but the New York Times depicted a president content to preside over harmonious times, and not to wade back into combat with the other party. "Mr. Clinton is far more in command and comfortable in his role. With more than 30 years of elections behind him, Mr. Clinton also believes that many of his goals have been achieved, and he is less urgent and more mindful of the realities of divided government."

The same poll pegged Gingrich's approval at 34 percent, normally a troubling mark, but for this Speaker a giant leap forward. Only

37 percent of Americans now said they disapproved of the job he was doing, making this the closest he'd come to breaking even since the start of his Speakership. For the first time in memory, Newt was headed in the right direction. The news was even better for his fellow Republicans, who'd been so bogged down by their leader's unpopularity in 1996. Now, the Republican Congress's approval rating stood at 47 percent, with just 32 percent disapproving. As with Gingrich, it was the best they'd fared in a long time.

Gingrich and his party were already looking to the November midterm elections with anticipation. History said the White House party was all but certain to lose seats. Gingrich embarked on a rehabilitation tour of the country. Within the Republican Party, he was regaining his power, and now he would reclaim the role he truly coveted, as the preeminent national voice of Republicanism. On January 13, he spoke to the state legislature in Olympia, Washington. On this day, a reporter tagging along wrote, he was "ebullient."

"This country's a romance," Gingrich told the crowd.

Peace was at hand and both parties were profiting. Not one week later, it would all be gone, and nothing would be the same again.

TWENTY-TWO

For more than three years, Ken Starr's investigation had been the background noise of American politics. Every now and then, it would crackle to life with the possibility of a breakthrough that would lead all the way to the top. But try as he might, Starr never had gotten there. Politically, Whitewater had long since devolved into a stalemate. The right remained convinced the Clintons were guilty of *something*, a conviction reinforced by the Clintons themselves, who'd not exactly been models of forthrightness throughout the investigation. Of course, their supporters fired back that the First Couple's defensiveness was understandable; wouldn't you be on guard in the face of an open-ended investigation your loudest opponents were cheering on?

No doubt, it had all taken a toll on Bill and Hillary. Even if most Americans ignored the details of news about Whitewater, its unending presence in the media served as a reminder of all the doubts about honesty, integrity, and character that had dogged Bill Clinton from the start. But by the start of 1998, the president could also look to the future with new confidence. He was more popular than ever, the economy was roaring, the deficit was vanishing, and Starr was spinning his wheels. In private, Clinton fumed at the mention of the independent counsel's name. Publicly, he was poised to get the last laugh.

What no one knew was that Starr did have something—and it had nothing to do with arcane real estate transactions. It was something every American would have no trouble understanding, and on January 21, 1998, they found out what it was:

> Independent counsel Kenneth W. Starr has expanded his investigation of President Clinton to examine whether Clinton and his close friend Vernon E. Jordan Jr. encouraged a 24-year-old former White House intern to lie to lawyers for Paula Jones about whether the intern had an affair with the president, sources close to the investigation said yesterday.

So read the front-page story in the *Washington Post* that instantly threatened to bring down Bill Clinton's presidency. Technically, it was old news. Days earlier, a crude, graphics-free site on the World Wide Web called *The Drudge Report* had published an item claiming that *Newsweek*'s Michael Isikoff had learned of a possible extramarital affair involving Clinton and a White House intern and that the magazine was refusing to publish it. Drudge's site was largely unknown, though, and without social media to spread it, his scoop had percolated just below the radar—until now.

Newsweek threw the Isikoff story onto its America Online site, generating traffic that before this moment had been unthinkable for a news outlet on the Internet. The rest of the press had been on the story, too, and now rushed to reveal its findings and unearth more.

The intern in question was named Monica Lewinsky. She'd graduated from college in 1995, then come to the White House as an intern in Chief of Staff Leon Panetta's office. Eventually, she moved to a full-time post at the Pentagon, where she'd struck up a friendship with a coworker, Linda Tripp. Lewinsky, at least according to the reporting, confided to Tripp that she and the president had begun an affair while she was an intern—when Lewinsky was twenty-one years old and the president forty-nine.

All of this was becoming public knowledge now because of a different scandal involving Clinton, a sexual harassment allegation lodged in 1994 by Paula Jones, a former Arkansas state employee. Jones contended that Clinton had, while governor, enlisted a state trooper to lure her to a hotel room, where Clinton made unwanted advances, then dropped his pants and asked her to perform oral sex. She filed a lawsuit; Clinton denied the story, then claimed that executive privilege shielded him from any civil lawsuit while he served as president. That set off a legal dispute that bounced from one courtroom to another until finally the Supreme Court ruled that the case could go forward. Now, opening arguments in *Jones v. Clinton* were set to begin in May, and depositions were already being taken.

Lawyers for Jones were trying to establish a "pattern of conduct," to show that Clinton had behaved toward other women in a way that would lend credibility to Jones's accusation, and they'd been scavenging for names. Before they'd ever heard of Lewinsky, Linda Tripp was on their list, because of her role in yet another accusation against Clinton. This one had come from Kathleen Willey, who claimed that while she was a volunteer aide in the White House in 1993 the president had kissed and groped her in his private study. Tripp had been working in the White House at the time and said she'd encountered Willey just after the episode. But where Willey was saying the advance was unwelcome and that she'd been in distress, Tripp reported that she'd seemed "happy and joyful." Clinton was denying anything had happened at all.

When it came to Lewinsky's confessions, Tripp hadn't been discreet. She'd spilled the beans to a friend, Lucianne Goldberg, who just so happened to be a book agent well connected in conservative circles in D.C. Goldberg encouraged Tripp to begin taping her conversations with Lewinsky and looped in Isikoff, the *Newsweek* reporter. Soon enough, the Jones lawyers caught wind of rumors about

Lewinsky and Clinton, and by the end of 1997, both Tripp and Lewinsky were under subpoena to provide depositions in the Jones case,
triggering a rapid series of events that would send it all crashing into
the open.

Lewinsky told Tripp about pressure from the president and from
Vernon Jordan, his confidant and an accomplished Washington fixer.
She said they wanted her to deny the affair and that Jordan was using
his clout to land her a private-sector job. Deposed by Jones's lawyers
on January 7, Lewinsky denied any involvement with the president.
Then she pleaded with Tripp to back her up when it was her turn to
be deposed. But Tripp, through Goldberg, had already gone to Ken
Starr, who listened to her recordings and slapped a wire on her—
which she then wore for one more conversation with a frantic Lewinsky, who shared even more details.

Unaware of any of this, Clinton provided his own deposition on
January 17. Like Lewinsky, he denied the affair. But later that same
night, the *Drudge Report* story went online, and now the stories Lewinsky had been telling Linda Tripp were at the center of a full-blown
scandal.

Already, it was taking the country into uncharted waters. A
presidential sex scandal was new enough, but this one came with very
specific—and potentially dire—legal implications. Clinton had issued a denial under oath. But now Starr had tapes and leverage over
Lewinsky. Had there actually been an affair? That would mean perjury by the president. If he'd also urged Lewinsky to lie? That would
be suborning perjury. If he'd tried to keep her quiet by using Jordan
to procure a job? Obstruction of justice. Or maybe abuse of power,
or both. The Constitution provides no instructions on how to define "high crimes and misdemeanors," the threshold it sets for the
impeachment of a president, but the I-word was already finding a
prominent place in the coverage.

Clinton's first public comments poured gasoline on the fire. Before the *Post* story hit, he'd agreed to a January 21 interview with Jim
Lehrer, the PBS *NewsHour* anchor, and he didn't want to back out

now. "The news of this day," Lehrer began, "is that Kenneth Starr, independent counsel, is investigating allegations that you suborned perjury by encouraging a twenty-four-year-old woman, former White House intern, to lie under oath in a civil deposition about her having had an affair with you. Mr. President, is that true?"

"That is not true," Clinton replied. "That is not true. I did not ask anyone to tell anything other than the truth. There is no improper relationship. And I intend to cooperate with this inquiry. But that is not true."

Lehrer burrowed in. "You had no sexual relationship with this young woman?"

"There is no sexual relationship," the president responded. "That is accurate."

Three times in the interview, Clinton shifted the tense in which was speaking midsentence, in each instance placing his Lewinsky remarks in the present tense.

"The nation, recalling past brushes with disclosure from that man who famously did not inhale, seized on his use of the present-tense 'is,'" the *New York Times* wrote. "Was this, Americans instantly wondered, Mr. Clinton's linguistic escape hatch from the tale of the intern, Monica S. Lewinsky, and her reported tape-recorded confessions alleging that the President advised her to lie under oath to protect him?"

The president tried again, a few days later. At the end of a normally scheduled event in the White House, he looked up at the cameras. "I want to say one thing to the American people," Clinton said. "I want you to listen to me. I'm going to say this again. I did not have sexual relations with that woman, Miss Lewinsky. I never told anybody to lie, not a single time—never. These allegations are false."

Now he wasn't using the present tense. Just as notably, the First Lady was standing with him as he made this new denial. Speculation about her reaction had become its own dramatic storyline. Did this mean she believed him? Yes, Hillary Clinton emphatically said the next morning on NBC's *Today*. It was her first interview since the scandal broke, and she played off as absurd the mere idea of an affair between her husband and an intern. The first she'd heard of Lewinsky, she told Matt Lauer, had been the morning the story had broken in the press, when her husband woke her up and said, "You're not going to believe this, but . . ."

By now, more details had trickled out about what Lewinsky had told Tripp. It was, specifically, oral sex that she said she'd engaged in with the president. Gifts had been exchanged, too, and there'd been lurid phone calls and private conversations in which, Lewinsky felt, Clinton had raised the possibility of one day leaving Hillary for her. Lauer said to the First Lady: "So where there's smoke . . ."

"There's no fire," Hillary interjected.

The accusations would all be proven untrue, she said, and were the product of a vendetta by Starr. "We've got a politically motivated prosecutor who is allied with the right-wing opponents of my husband who has literally spent four years . . . looking at every telephone call we've made, every check we've ever written, scratching for dirt, intimidating witnesses, doing everything possible to try to make some accusation against my husband."

She pointed to the wild claims that had been lodged against her and her husband—like the murder charge on the video Jerry Falwell hawked—and the behind-the-scenes connections between many of the people who'd made allegations against the First Couple. "The great story here for anybody willing to find it and write about it and explain it is this vast right-wing conspiracy that has been conspiring against my husband since the day he announced for president."

That line became an instant joke. It's not that she didn't have a point. There was indeed a well-funded conservative network that had been working to unearth and promote bombshell accusations against the Clintons. The problem here was that few thought you needed a conspiracy theory to explain how Bill Clinton might have gotten caught up with an intern. On *The Tonight Show,* Jay Leno cracked: "And cheating husbands everywhere are now saying, 'Honey, it wasn't me, it was those right-wing people who put that lap dancer in my lap!'"

Clinton's denial—the present-tense one—at least bought his party breathing room. Some expressed faith in his words. "I believe him completely and I know he's telling the truth," said Paul Begala, his old campaign aide and now a media commentator. Others were more cautious. "We don't know whether it took place or not," Nancy Pelosi said. "But if it did, it comes as a great big surprise to a great deal of people. I don't think that for one minute there's an atmosphere of sexual harassment in the White House."

Either way, they didn't have to say anything else for now. Lewinsky's accusations were on tape, Clinton's denial was on the record, and Starr was on the case. For Gingrich and his fellow Republican leaders, that last part was key. They believed Clinton was lying and that Starr was on his way to proving it. In fact, there were already reports that the FBI might have seized an article of clothing from Lewinsky that might have incriminating evidence on it. So why say anything now? "Every citizen ought to slow down, relax, and wait for the facts to develop," Gingrich said. "When we know, then is the time to comment."

The first polling numbers started coming in. A majority of Americans, according to Gallup, believed it was likely the president had engaged in a sexual relationship with Lewinsky and that he'd lied about it under oath. For that matter, 62 percent said he'd probably had other affairs as president, and 57 percent rejected the notion that the Lewinsky story was part of a right-wing conspiracy.

And yet, the bottom line was encouraging for Clinton. His approval rating hadn't dropped; in fact, it was up slightly. A majority believed the Starr investigation wasn't fair, 59 percent said most past presidents had affairs, too, and 75 percent said Clinton hadn't done anything worse than his predecessors. Was it time to for the House to lay the groundwork for impeachment proceedings? Eighty-five percent said no.

But even Clinton's loyalists weren't ready to put much stock in this. There was no telling what Starr was going to come up with, and if he did corner Clinton, opinions could shift fast. This was a scandal that wasn't just going to fade out. Did the public have a breaking point with this president? If it did, it was likely to be tested in the near future. One of the president's advisers, Doug Sosnik, said: "I don't sit at home at night popping the champagne corks. I don't believe that these poll numbers will hold up over time."

Starr cut the deal with Lewinsky in July. He would get her cooperation, and she would get immunity. She also handed over phone recordings with the president's voice and a dress, said to contain physical evidence of one of their encounters. Starr's team would now crosscheck it with a DNA sample from the president.

Lewinsky testified before the grand jury on August 6. The press was barred from the courtroom, as if that would stop the proceedings from leaking out. She recanted her denial from her Jones case deposition and provided a detailed account of a sexual relationship with Clinton that, she said, lasted for eighteen months. Now her story was at odds with the president's. She testified that Clinton hadn't explicitly instructed her to lie, but that the two of them had discussed "cover stories."

Eleven days later, on August 17, it was the president's turn. Clinton desperately wanted to avoid the spectacle of giving his testimony in court, and after some haggling, Starr spared him that. The president would be deposed by the independent counsel's team in the Map Room of the White House. Bill Clinton had been called a liar for years, but he'd never confessed to being one. "The degree of comfort the president can take in public opinion as he testifies is not so clear," wrote one pollster.

His testimony lasted four hours, then he addressed the nation in prime time. "Indeed, I did have a relationship with Ms. Lewinsky that was not appropriate," he said. "In fact, it was wrong. It constituted a critical lapse in judgment and a personal failure on my part for which I am solely and completely responsible."

But Clinton's tone belied those words of contrition. His testimony about Lewinsky in January had been "legally accurate," he insisted, and he'd failed only to "volunteer information." He mentioned the pain inflicted on his wife and daughter, but there was no apology to Lewinsky, no acknowledgment of the devastating consequences all of this would have on her life. The rest of his remarks, which clocked in at just under five minutes, amounted to a defiant critique of Starr and an investigation that had "gone on too long, cost too much, and hurt too many innocent people."

"Even presidents have private lives," Clinton said. "It is time to stop the pursuit of personal destruction and the prying into private lives and get on with our national life."

It was now on the record. The president of the United States had engaged in a sexual relationship with a onetime intern less than half his age, and he'd lied about it to the country, and maybe under oath as well. It was a momentous admission, though by this point hardly a shocking one. By overwhelming margins, polls had shown all year that Americans believed Clinton and Lewinsky were involved sexually, and his job approval still hadn't taken a hit.

It raised the question: Was there anything about this story that actually would budge public opinion? There were the legal issues involving perjury and obstruction of justice that Starr was pursuing; if he could establish either, it might become the basis for impeachment proceedings in the House. But voters were also telling pollsters that they assumed Clinton had lied in his deposition—that he'd committed perjury. And the majority of them said they didn't think it should disqualify him from continuing as president.

What about his behavior itself? Did Americans expect more from their president than revelations of Oval Office trysts? Would they demand his exit for cheapening his office? For dishonoring it? For setting such a poor example? There was also the power dynamic— the commander in chief being pleasured by a young woman with no stature or authority of her own. Was "affair" too anodyne a word for this? Was it something more depraved—lecherous, maybe? Even predatory? Many major corporations defined a relationship like Clinton and Lewinsky's as sexual harassment. A CEO conducting himself as Clinton did might be fired. Why shouldn't the president face a similar penalty?

There were voices raising these points, but so far they were being drowned out by a consensus in popular culture that this was all a matter of privacy. Lewinsky was young, yes, but she was nonetheless of legal age and she'd engaged in the relationship willingly. On this point, Clinton was already receiving cover from many feminists, starting long before his admission. In the *New York Times*, Gloria Steinem wrote that she saw no issue with a relationship between Clinton and Lewinsky because "welcome sexual behavior is about as relevant to sexual harassment as borrowing a car is to stealing one." Another feminist writer said: "The only person who minds that Bill Clinton's having sex without being in love is Ken Starr."

There was a sense that the underlying question had already been litigated in the 1992 campaign, when plenty of people who'd doubted Clinton's denials of Gennifer Flowers's claims decided it still didn't matter—that whatever had or hadn't happened was between Bill and Hillary. (Ironically, it now emerged that Clinton had also been asked in his deposition about Flowers, and that he'd acknowledged a sexual relationship with her.) The Flowers scandal, in effect, had conditioned Americans to assume the worst about Clinton when it came to rumors of extramarital activities. The notion of him as a man unable to control his libido had long ago been baked into perceptions of him, a caricature that comedians had been mining for yuks for his entire presidency. When the news about Lewinsky first broke, it could feel like little more than the next punch line. "What have we learned?" Craig Kilborn, host of *The Daily Show* on Comedy Central, asked. "We learned the president's taste is improving."

An overnight poll after his confession had 65 percent of Americans saying they were satisfied with what Clinton said and that it should conclude the matter. That was the good news for the White House. The bad news: the media and political classes weren't buying it. After all, he'd been lying to their faces for months now. They listened for remorse in Clinton's remarks, a sense that he was genuinely troubled by his own behavior and dissembling, and they heard nothing.

The *New York Times* issued an editorial call for Starr's investigation to continue. The president's habit of "stonewalling, of misleading by omission or concealment or fabrication or failure of memory," the paper wrote, made it impossible to accept his simple assertion that he'd committed no underlying crimes.

"The outcome is hardly satisfactory for those who had hoped he would meet this challenge in a less characteristic way," the editorial read. "Too much of a too-short speech was devoted to another blast of the familiar dichotomous blarney. His touching admission of lying to his wife was coupled with the insulting contention that his earlier denial, under oath, of a sexual relationship was 'legally accurate.'"

Democrats who'd vouched for Clinton now felt used, betrayed. California senator Dianne Feinstein was one of them. Clinton's remarks, she told the press, "leave me with a deep sense of sadness in that my trust in his credibility has been badly shattered." In the *Washington Post*, Dan Balz described the speech as "a political failure that has unleashed a torrent of anger among some of the president's most loyal supporters and created problems no one at the White House anticipated."

But would any Democrat take the next step and call for Clinton to go? In the immediate aftermath, the answer was no. They were all waiting to see what happened, measuring their own anger and disgust with the reality that—so far—their own voters were squarely behind the president. The next days did bring one change in the polls, a twenty-point plunge in Clinton's personal favorable rating. Americans were registering their disapproval with his behavior, but his job-approval numbers remained sky high. There was still no sign they wanted him to go.

If there was a Democrat who might turn on Clinton and bring others along, it was Joe Lieberman. The Connecticut senator enjoyed a reputation in Washington as his party's most credible moral voice.

An Orthodox Jew, he'd partnered with William Bennett, the conservative author of *The Book of Virtues*, in a campaign to shame the entertainment industry for its emphasis on sex and violence. He also went way back with the president, who as a Yale Law student in 1970 had volunteered on Lieberman's maiden campaign for the Connecticut State Senate. They were fellow centrists, DLC men both, and when Clinton announced his presidential campaign, Lieberman had been one of his first supporters. Now Lieberman was being tested. He spoke frequently of the importance of morality in public life. Would he now call for Clinton to lead by example and resign his office?

Lieberman took the Senate floor on September 3, and was unsparing in his critique of Clinton's conduct. "In this case," Lieberman said, "the president apparently had extramarital relations with an employee half his age and did so in the workplace in the vicinity of the Oval Office. Such behavior is not just inappropriate. It is immoral. And it is harmful, for it sends a message of what is acceptable behavior to the larger American family—particularly to our children—which is as influential as the negative messages communicated by the entertainment culture."

But he went no further. He made no demand for resignation and urged his colleagues to hold off on any talk of impeachment at least until they received a report from Starr. This, Lieberman said, would create space for Clinton to find "additional opportunities to accept personal responsibility, to rebuild public trust in his leadership, to really commit himself to the values of opportunity, responsibility, and community that brought him to office, and to act to heal the wounds in our national character." He was warning Clinton, but he wasn't defecting.

In the early days of the scandal, Gingrich had led the Republican charge, confident that Clinton's deceit would be exposed and public opinion would turn. At one point in the spring, he vowed that "never again, as long as I am Speaker" would he make a speech without mentioning the Lewinsky story. And after Clinton's speech, he told an audience in Atlanta that big gains would be coming for Republicans in the November midterms, forty seats perhaps. Now, he was quieting down, aware of what would be coming: Starr's report. If the independent counsel found impeachable offenses, it would be up to the House to initiate the proceedings, and Gingrich wanted to project an image of fairness. "My deepest belief," he said, "is that all of this is part of a much bigger story, and we have to wait until the report is in before we make judgment."

In the Republican grass roots, meanwhile, impeachment fever was already running wild. On talk radio and among activists, where assumptions of criminal wrongdoing by the president preceded the Lewinsky revelations, Clinton's acknowledgment was the opening they'd been waiting for. Finally, he'd been revealed for what he'd always been. On his radio show, Limbaugh treated his audience to clips of Democrats from the months before Clinton's admission professing their faith in his word. "I think the president as a president, as a governing leader, is toast," he said.

Congressional Republicans were talking the same way, and not just in the backbenches. "It is bad enough that our president is guilty of having an extramarital sexual relationship with one of his young interns," said Tom DeLay. "But it is much more damaging that this president looked the American people in the eye and knowingly lied to us."

"If Republicans already felt good about this fall's congressional and gubernatorial campaigns before this week," *USA Today* wrote, "they can barely contain their glee over their prospects now that President Clinton has confessed to an 'inappropriate' relationship with former White House intern Monica Lewinsky."

Starr delivered his report to the House on September 10. He said the president had committed perjury, tampered with and intimidated witnesses, obstructed justice, and abused his power. The main report ran more than four hundred pages. There were boxes of evidence, too. Absorbing Starr's write-up, members of Congress expressed surprise at the graphic descriptions and sordid details he'd chosen to include. Then they voted to make the report available to all Americans. It would be posted online the next day.

The lurid content was Clinton's fault, Starr said, because the president was still denying he'd committed perjury during his initial deposition. How could that even be possible? Clinton had denied "sexual relations" with Lewinsky, but was now claiming that oral sex did not constitute such relations. So Starr opted to document with exacting precision exactly what Clinton and Lewinsky had done.

"The president's defense to many of the allegations is based on a close parsing of the definitions that were used to describe his conduct," Starr wrote. "We have, after careful review, identified no manner of providing the information that reveals the falsity of the president's statements other than to describe his conduct with precision."

To critics, the Starr report was cheap smut. "Unconscionable over-reaching," the Clinton legal team charged. The public seemed to agree. In one poll, two-thirds said it wasn't appropriate for Starr to include so much detail, and 59 percent said he did it to embarrass Clinton—not to prove perjury. But Clinton's approval rating remained robust, and it even went up the next week, when the videotape of his August grand jury testimony was released and played unedited on the broadcast networks.

The consensus was evident. Americans believed Bill Clinton had lied under oath, that he'd encouraged Lewinsky to do the same, and that he didn't share their moral values. But they also saw this as a private matter, one where Washington's best course of action would be to just drop it.

Democrats remained angry at their party's leader. It wasn't just the lying. His scandal had also undermined their agenda. No major legislative achievement was possible in the environment his misconduct had created. They could read the polls, which still had Clinton's approval rating at well over 60 percent. Plus, the president was ready to give them some help. Lieberman had asked for more contrition, and Clinton now delivered it. Speaking to a group of ministers, he expanded the list of those hurt by his actions to include "my friends, my staff, my cabinet, Monica Lewinsky and her family, and the American people."

"I don't think there is a fancy way to say that I have sinned," Clinton said.

On Capitol Hill, a new idea was being floated. Congress could put itself on the record censuring Clinton, an alternative to impeachment that seemed to align more with public opinion. Media voices on the left were already rallying behind the idea. The *New York Times* declared that it would "give the country room for a pragmatic judgment that does not require a societal endorsement of his pathology" and endorsed the idea—provided that Clinton accepted such a reprimand without quarrel. The White House signaled he would do just that, and it became the default Democratic position: yes on censure; no on impeachment. Activists mobilized online with a petition demanding that Congress "censure and move on." More than 250,000 names signed on—the foundation for what would soon become a new group: MoveOn.org.

Republicans didn't even nibble. "I don't understand how people can rush to a solution before they finish the investigation," Gingrich said of the censure idea. "I just think there's an awful lot we don't know yet and there's an awful lot of evidence that hasn't been gathered yet." DeLay distilled it further: "Either we impeach him or we do not impeach him."

And on that question, the conviction of rank-and-file Republicans was only hardening. Two weeks after the Starr report's release, the Christian Coalition held its annual convention. Under Clinton, Pat Robertson told the activists, the White House had been turned into "the playpen for the sexual freedom of the poster child of the 1960s. . . .

"I say to you today: We will be silent no longer!"

The prominence of the Christian right was one of the impeachment drive's vulnerabilities, though, reinforcing notions that Starr was persecuting immorality, not investigating illegality. Democratic congressman John Conyers called him "a federally paid sex policeman spending millions of dollars to trap an unfaithful spouse." Starr had long been attacked as a partisan hack; now he was pilloried as a prude and a religious extremist. Feminist leaders staged a rally to defend Clinton. "Women will not stand by," Patricia Ireland, the head of the National Organization for Women, said, "as the Republicans use their power in Congress to overturn a president they elected." Eleanor Smeal from the Feminist Majority Foundation said: "We do risk ushering into power a puritanical or fundamentalist sex police who speak of freedom but allow government to destroy the right of privacy."

Republicans had their own dilemma. The polls were telling them to stop, but how could they? Hadn't they been saying—for years—that just a little more digging would bring to light the scandal that would end the Clinton presidency? And now the independent counsel was telling Congress he had identified eleven grounds for the impeachment of Bill Clinton. Wasn't this just what Republicans had been priming their base for?

Gingrich was presiding over the chamber on October 8 as the House took up the question of whether to launch a formal impeachment inquiry. The debate was heated, and personal. Nancy Pelosi invoked Gingrich's own censure the year before. Hadn't he been guilty of providing misleading information during an investigation, too? "I believe that it is a double standard that absolves the Speaker and seeks to impeach the president," she said.

The main Democratic argument was that Clinton's wrongdoing just wasn't what the Founders had in mind when they created the impeachment mechanism. "We're talking about Monica Lewinsky," Florida's Robert Wexler said. "God help this nation if we trivialize the Constitution of the United States. While the president betrayed his wife, he did not betray the country."

From the Republican side, seventy-four-year-old Henry Hyde rose to speak. As the chairman of the Judiciary Committee, the Illinois Republican would lead the impeachment investigation if the House authorized it. He was smarting from a new report from the liberal online magazine *Salon* that he'd engaged in an extramarital affair in the late 1960s. This was another aspect of the scandal. With so many Americans believing the Clinton-Lewinsky relationship was strictly a private matter, attention was shifting to the Republicans pushing for impeachment. What was in their closets? The pornographer Larry Flynt, whose legal fight against obscenity charges had been celebrated in a recent Woody Harrelson movie, offered a one-million-dollar bounty for anyone who could provide "documentary evidence of illicit sexual relations with a congressman, senator, or member of the executive or judiciary branches." The *Salon* story about Hyde was not a result of Flynt's offer, but how many other Republicans now had reason to fear?

On the House floor, Hyde emphasized what was now the standard Republican line. This was not about sex. It was about specific allegations of lawbreaking. "The question asks itself," he said. "Shall we look further, or shall we look away?"

The last time the House had opened an impeachment investigation had been in 1974, when Richard Nixon was president. Then the vote to do so had been nearly unanimous. Now it was 256 to 178—close to a party-line vote, though thirty-one Democrats did cross over to support the inquiry. Hyde said it would start after the election, now less than a month away, and that his panel would report back to the House by the end of the year. For any voter who wanted to know, though, Republicans were now on record: they were the party driving impeachment.

As the debate played out on the House floor, Charlie Rangel, a long-serving Democrat from New York, had gestured toward his colleagues across the aisle: "Voters will be judging you on November the third."

TWENTY-THREE

The instructions from the First Lady were clear. Her schedule for Monday, November 2, the day before the midterms, was to remain empty. Her work was done and she wanted to catch her breath. All fall, requests from Democratic candidates had been streaming in, pleading for a visit, even just a quick one, from the woman who was suddenly the most popular political figure in America. She'd accepted plenty of them, growing quickly into a role that would have been unimaginable just months before: public face of the Democratic Party.

Six years earlier, Americans had been introduced to Hillary Rodham Clinton, a whip-smart career woman with Wellesley and Yale degrees who promised to be an equal partner in her husband's administration—"two for the price of one," as Bill Clinton put it. It was a brand-new proposition, a First Lady assuming a formal policy-shaping role. To feminists, her ascent was an inspiration; to conservatives, an affront to traditional values.

"Polarizing" was the word invariably used to describe her, and it remained so well into her husband's presidency. When Bill put her in charge of health care, critics cried nepotism. If she did a poor job, was her husband really going to fire her? She had her admirers, plenty of them, and they saw sexism at work. Detractors said she was overstepping her role: If she wanted to set policy, why didn't she run for office herself?

In fact, she had thought about it, fleetingly, in Arkansas, back when Bill seemed to be wearing out his welcome. He'd floated the idea publicly, and they even took a poll: What if Hillary ran for governor instead of him? Then they got the results, and that was the end of that. What had been true in Little Rock seemed just as true when they moved to Washington: Hillary stirred very different reactions in different people.

The Lewinsky mess turned it all around—not because of her initial response, when she told Matt Lauer about a "vast right-wing conspiracy"; that only fed the critics. The turnaround came later, when the world watched how she conducted herself in the face of a public humiliation no other First Lady had known. When Bill Clinton admitted to cheating, the press demanded details from her. When did she find out the truth? What did she tell him? Would she leave him? How distraught was she?

They got a brief statement from her spokeswoman, issued as the First Family headed off for a private trip to Martha's Vineyard: "Clearly, this is not the best day in Mrs. Clinton's life. She's committed to her marriage and loves her husband and daughter very much and believes in the president, and her love for him is compassionate and steadfast. She clearly is uncomfortable with her personal life being made so public but is looking forward to going on vacation with her family and having some family time together."

She said nothing about the scandal in public, and somehow that said everything. There was no lashing out, no tearful breakdown, no juicy tell-all interview; she returned to Washington with her husband, kept her head up, and went about her business as she always had. Everyone was watching her now, and they imagined the pain she must be feeling, the suffering, the embarrassment. And then they recognized how strong she must be, how much poise she had to have, to carry herself this way. Where they'd seen hunger for power before, they now saw grace. Her favorable rating shot through the roof. Nearly three out of four Americans said they admired her courage. She was more popular than Bill—more popular than anyone.

It made her the solution to her party's dilemma. The nation was still absorbing the Starr report as the midterms heated up in September, and Bill Clinton was finding himself an unwelcome guest on the campaign trail. His approval rating was high—very high—but Democrats were still apprehensive. His personal ratings had crashed, and they feared his sleazy image might rub off on them.

When Clinton went to Cincinnati for a rally, the mayor, a fellow Democrat, said she only showed up because "the president is still the president. I will extend to him the courtesy due his office." Representative Adam Smith, a Democrat in Washington State, pronounced himself "incredibly repulsed by what I've seen the president do over the last year. It goes beyond betrayal." Pat Schroeder, a liberal stalwart who'd left Congress the previous year, counseled Democrats to seek other assistance: "They have to deal with the fact that this is not the role model we want for the Democratic Party."

And so it happened that Hillary became the Clinton that Democratic candidates clamored to showcase in the autumn of 1998. "The attraction of Mrs. Clinton," Jay Inslee, running for Congress in Washington, said, "is that she focuses on the issues rather than the distractions swirling around Washington, D.C. The president is a distraction."

She hit the road: New Jersey, Minnesota, Pennsylvania, California, Rhode Island, Illinois, to name just a few stops—"a one-woman campaign machine," the *Boston Globe* called her. She never brought up impeachment, but no one had trouble reading between the lines when she said of Republicans: "They'd rather spend their time dividing our country, diverting our resources, doing anything but focusing on the real problems of America." In any other context, those words might have sounded formulaic, but now they carried an extra punch.

In this moment, she was invulnerable. There were Republicans—and even some Democrats—who thought her new image a sham. In the Clintons, they saw a cold political partnership, and little else. But they didn't dare say it in public. "Attacking Mrs. Clinton is off limits in about 85 percent of the country," said Republican pollster Frank Luntz.

Her national tour ended in Tampa on the Sunday before the election. For twenty minutes, she spoke at a rally for Buddy MacKay, the Democrat running an uphill race for Florida governor against Jeb Bush, one of the former president's sons. Then she flew back to the White House, intent on taking it easy on Monday and preparing for the nation's verdict.

But then came one final plea, a personal one from a candidate she'd already appeared with three times. Standing next to her less than a week before, the candidate had said of Hillary: "When you're in her presence, you feel a little bit of a tingle. I do." Now, he asked her to blow up her schedule and to do one final event—a rally with him on Election Eve. He was close to winning, he could feel it, but he needed a closer, and there was no one better than her.

The candidate's name was Charles Schumer, and he was taking a big risk. He'd represented a House district in Brooklyn for eighteen years, but he'd put it on the line to run for the U.S. Senate. Already, he'd won a Democratic primary no one had given him much chance in, and now he was on the verge of unseating a Republican who'd proven maddeningly elusive in the past. Hillary heard his request and told him she'd be there. Maybe she liked Schumer's pluck, or maybe it was his flattery, or just the pull of party loyalty. But probably it was something a little more human: she despised his opponent.

The way the Clintons saw it, there was no separating the Starr investigation from Alfonse Marcello D'Amato. The chairman of the Senate Banking Committee had latched on to Whitewater early and refused to let go. So many of the developments that had prolonged their ordeal and eroded their credibility could be tied to his handi-work.

Now, D'Amato was on the ropes back home in New York. At least it seemed that way. He was trailing by a few points, but with him Democrats could never be sure. The sixty-one-year-old D'Amato had been defying political gravity his whole career.

A Republican from suburban Long Island, he crashed the po-litical scene in 1980, challenging four-term senator Jacob Javits in the Republican primary. Javits was an icon of liberal Republicanism and a veritable Senate institution. D'Amato was a town supervisor in Hempstead. D'Amato called Javits too liberal and ran an ad suggest-ing that Javits, "at age seventy-six and in failing health," wouldn't live through another term. D'Amato won the primary going away, then squeaked out a win in the fall.

Democrats took him for a dim bulb. "Senator Pothole," they started calling him, a glorified local hack who'd accidentally wound up in the corridors of national power. He turned it around on them: all it meant was that if you were his constituent, there was no issue too small for him to roll up his sleeves and handle. He won a second term in 1986.

Then scandal struck. D'Amato stood accused of helping friends cash in on federal housing programs. The Ethics Committee rebuked him, his poll numbers collapsed, and Democrats pronounced him dead. But D'Amato got lucky in his next campaign. He drew an opponent who proved an ideal foil, state Attorney General Robert Abrams, an impeccably credentialed policy wonk ill at ease with the street fight politics practiced by D'Amato. Baited and hectored, Abrams finally snapped just before the election, when he branded D'Amato a "fascist" at a public event. It was all the senator could have hoped for and more. D'Amato accused Abrams of anti-Italian bigotry, then trotted out his mother, 77-year-old Antoinette "Mama" D'Amato, who looked into the cameras and asked of Abrams: "Does he talk that way at home?" It was a comeback for the ages. On Election Day, Clinton carried the state by 16 points, a big Democratic margin at the top of the ticket. But those same New Yorkers also returned Republican Al D'Amato to the Senate for six more years.

It put him on a collision course with the Clintons. Like many of his fellow Republicans, D'Amato thought there might be something to Whitewater, maybe a lot. As the ranking Republican on Banking, he had leeway to pursue it. D'Amato, remember, was the one who had cornered Roger Altman at that hearing in February 1994, forcing him to admit the White House and the RTC had been talking about the Whitewater probe.

What a fateful moment that turned out to be. For the media, it became a tipping point, the revelation that convinced them there really might be something to all of this. *If they don't have anything to hide, why do they keep acting like they do?* Momentum then built quickly for an independent counsel, and soon enough the three judges were installing Starr.

And Whitewater became D'Amato's passion. Speeches, press conferences, TV shows: he threw himself into the cause of convincing America there was something major to be found. At one point, his friend Bob Dole presented him with a gag gift, a bottle of cologne. It was called White Water for Men. With the Republican takeover of the Senate in 1994, D'Amato grabbed the Banking gavel and convened his own Whitewater hearings. "To me," he said, "this could be the culmination of all the political endeavors I've ever undertaken."

D'Amato's final report in the summer of 1996 took dead aim at Hillary. It depicted her as serially deceitful, destroying inconvenient documents and conspiring with her inner circle to hide information from the committee. Democrats objected, but Republicans had the votes to adopt it anyway. "History will judge these hearings as a revealing insight into the workings of an American presidency that misused its power, circumvented the limits on its authority, and attempted to manipulate the truth," D'Amato said. "We have witnessed a pattern of deception and arrogance that undermines the fundamental core of the American democratic system."

It was "a remarkable document," the *New York Times* wrote, because "never before in modern history has a congressional committee so aggressively challenged a First Lady." The Clintons' lawyer called it "the politically preordained verdict of a partisan kangaroo court." There was no smoking gun to threaten the White House with instant immolation, but the years of scrutiny—from D'Amato, from federal investigators, and from the media—had eroded Hillary's image. She'd originally claimed to be a passive bystander to the travel office firings in early '93; it turned out she'd been instrumental to them. She'd once invested one thousand dollars in cattle futures and turned a profit of nearly a hundred thousand dollars in less than year; she said she was a hobbyist who got lucky, but experts were skeptical. For two years, she failed to provide billing records from her old law firm, then claimed they'd turned up in an area of the White House that no staffer had access to. Even if Americans still weren't sure what Whitewater was about, the First Lady's behavior fed the impression that she had something to hide. By the time of the report's release, a majority of Americans said they believed Hillary was lying about Whitewater and that her actions were either unethical or illegal.

D'Amato didn't walk away clean, either. War with the Clintons played well in the administration's early days, but New York had since rallied hard around the president—and against Gingrich and the Republican Congress. Even if New Yorkers shared those doubts about Hillary Clinton's honesty, they'd decided that Whitewater was more of a partisan obsession than a genuine scandal. And whatever doubts they had about the First Lady's character seemed to vanish in the face of her stoic response to her husband's affair.

Once again, D'Amato found himself an endangered incumbent. Everyone figured his opponent in '98 would be Geraldine Ferraro, Mondale's old running mate, attempting one more political comeback after cohosting CNN's *Crossfire,* but Schumer outhustled her in the September primary.

That's when Hillary made her first trip to New York. Introduced by Schumer as "an American hero," the First Lady wasted no time going after "Chuck's opponent." She called D'Amato a "Jesse Helms clone" whose Senate votes would "keep women down and back." Hillary's attack, the *Daily News* noted, was "far more pointed and personal than the stump speeches she normally delivers." Not that anyone wondered why.

For Clinton, the reversal was surely sweet. In the past, when D'Amato was launching accusations against her, she'd had no choice but to take it. Lashing out herself would only have advanced the notion that she was hiding something. Now, she was free to hit him as hard as she could—and because she was so popular, especially in New York, he couldn't say much back.

A few weeks later, she was back for more: "Al D'Amato does not deserve the votes of the women in New York. He has consistently voted against so many issues that women care about, that we think about." At the end of October, a third trip: "I have a good feeling about this race," she said, "because I believe the people of New York are ready for a change and they are ready for someone they can trust and someone they can be proud of." Now she'd campaigned more in New York than anywhere else.

Behind the scenes, she maneuvered as well. The nation's leading gay rights group, the Human Rights Campaign, was in a bind. Typically, it backed Democrats, and Schumer checked all the right boxes. But D'Amato was an enigma in his own party. In the fight over gays in the military, he'd been for ending the ban, and he'd voted to outlaw employment discrimination, too. Part of the Human Rights Campaign's strategy was to have the back of any Republicans willing to stand with them; an endorsement of D'Amato was under consideration. Then Hillary made some calls, and the idea was scotched.

Schumer was ahead in the Election Eve polling, but D'Amato was in the game. He'd revved up his old campaign machine, dumped a record-shattering twenty million dollars into it, and dusted off his favorite attack lines, shredding Schumer as "hopelessly liberal" and ready to tax "everything that moves and some things that don't." In the closing days, he brought back his secret weapon from campaigns past, Mama D'Amato, now a feisty eighty-three (and the author of a new book, *Cooking and Canning with Mamma D'Amato*).

"He will win!" Mama D'Amato told the crowds. "Why? Because he is a good senator. He helps everyone in this state!"

Sentiment in New York was squarely against impeachment, and D'Amato knew enough to stay away from the subject. Now you could barely get him to say one word about the Clintons.

Schumer's rally was just off Times Square, at the headquarters of Local 1199, New York's most powerful union. Back in Washington, Bill Clinton was calling in to radio shows aimed at nonwhite audiences, quietly trying to goose turnout among reliable Democratic constituencies. In the home stretch of this midterm, he'd traveled to just four states, mostly to raise money. It was the other Clinton who now took the stage with Chuck Schumer, basked in the roar of a crowd that was smelling victory, and wound up to deliver what she badly hoped would be the knockout blow.

"I know you have the best baseball team," she told the New York crowd. "You've got the best hot dogs, bagels, and delis in the world. Wouldn't it be nice to have the best U.S. senator?"

They whooped and hollered. Was it actually possible that the scandal that was supposed to topple Bill Clinton would end up taking down Al D'Amato instead?

By now, the scandal, or at least their party's reaction to it, was threatening to take down a lot of Republicans. So many times, they'd felt it was coming: the moment when the public would decide it had had enough and turn on Clinton once and for all. It never happened.

After voting to move forward on impeachment, Republicans had said as little as possible about it on the campaign trail. Gravity, they believed, was still on their side. This was a midterm election, when the opposition party was all but guaranteed to gain seats. The last time the White House party gained in one was back in 1934, when

FDR's Democrats picked up seats amid the trauma of the Depression. And this was the sixth year of Clinton's presidency; the last time the White House had added seats in a sixth-year midterm was under James Monroe in 1822.

The signs in the home stretch were getting ominous, though. There was Clinton's staggering approval rating, but there was also this: the Republican Congress's numbers were falling back to shutdown levels. Just 34 percent now liked the job they were doing. October also brought news of a budget surplus, an estimated seventy billion dollars. Clinton had been telling Americans to separate their personal feelings about him from his job performance; now he could show them he was delivering. A White House ceremony was held commemorating the first year of black ink since 1969.

In the final days, Gingrich's desperation showed. Fearful of a voter revolt, he'd been talking about everything but impeachment, but the issue was still dominating media coverage. And why wouldn't it? For the third time in American history, Congress was moving to charge a president with high crimes. It was still, overwhelmingly, the main topic of political conversation. Democratic candidates had even started running ads against impeachment.

And yet, conservatives had been growing frustrated that Republicans weren't saying *enough* about it. Suddenly worried his party's base might stay home, Gingrich commissioned a last-minute wave of ads—ten million dollars' worth of them—playing on the Clinton scandal. "That is the question of this election," one of them went. "Reward Bill Clinton. Or, vote Republican." Reporters asked the president for his response, and he relished the chance: "How can I object to them exercising their free-speech rights in saying what they think the election is about?"

In the exit poll, 62 percent of voters said Clinton shouldn't be impeached. Fifty-seven percent said Congress should drop the entire matter without doing anything else. Gingrich was back in his Georgia district, and as the East Coast results came in, he stepped forward to assure the crowd that "we will gain seats as the evening goes on."

They were actually losing ground. In Pennsylvania, Jon Fox was going down. Four years before, he'd unseated Marjorie Margolies-Mezvinsky, running against her pivotal vote for Clinton's tax hike. Now, his suburban Philly district was rising up against impeachment and the party behind it.

Next door in New Jersey, Mike Pappas, a conservative freshman, was meeting the same fate. Overcome with zeal over Clinton's woes, Pappas that summer had sung the independent counsel's praises on the House floor—literally:

Twinkle, twinkle, Kenneth Starr,
now we see how brave you are!

Pappas's opponent, Rush Holt, took that ditty and turned it into an ad. Now voters were showing their Republican congressman the door.

They called the New York Senate race before 10 P.M. There were no miracles left for D'Amato, who was losing by double digits. The First Lady, her spokeswoman would claim, was watching a movie in the White House theater when it went official. "She really feels that that's a real plus for the state of New York."

There was almost as much delight in the White House at the result from North Carolina. It was Lauch Faircloth, remember, who'd met with David Sentelle just before the three-judge panel made its surprise decision to appoint Ken Starr—and no one in Clinton world had ever seen this as a coincidence. Now, Faircloth's Senate career was over. Voters had ousted him in favor of his Democratic challenger, a trial lawyer named John Edwards.

The West Coast finally reported, and Newt's predicted gains remained illusory. Back in the summer, when Clinton made his admission, Republicans pegged California as one of their top Senate targets. Now, Democrat Barbara Boxer was returned to office in a landslide. The governor's race was lopsided, too. Dan Lungren, once a Gingrich ally in the House and now the state's attorney general, had flamed out against Gray Davis. In Washington State, Democratic senator Patty Murray was holding on. In Nevada, the counting took a while, but Democrat Harry Reid escaped what not long before had seemed like certain defeat by an ambitious congressman, John Ensign.

Over the summer, Gingrich had talked of netting forty new Republican seats. Instead, the GOP lost five. For a moment as the returns came in, there even seemed a chance Democrats would take back the House altogether. As it was, the Republicans would control the chamber by the slimmest of margins going forward—223 seats to 212. In the Senate, where Republicans had once had their eyes on cracking the sixty-seat mark, they gained nothing. It was historic, the first time in 176 years that the White House party had won seats in a sixth-year election.

"Democrats who once thought they were facing the firing squad in yesterday's midterm elections instead woke up today with a new lease on life," David Broder wrote the next morning, "a gift from voters who simultaneously registered their strong disapproval of impeaching President Clinton."

Gingrich strained for a positive frame. "This will be the first time in seventy years Republicans kept control of the House for a third term," he pointed out. But his own party wasn't buying it. "This was a stunning victory for Democrats and a major defeat for Republicans," said Representative Chris Shays, a moderate Connecticut Republican. From Oklahoma representative J. C. Watts: "I personally believe that we got whipped, and when that happens, I think you need to evaluate everything from the CEO to the janitor."

The president was subdued. Gloating wouldn't be a good look, not when he was still trying to show contrition. "Astonishing results," he said at a cabinet meeting the next day. "I am very proud of what our party did yesterday in the face of the tide of history and an enormous financial disadvantage."

A few weeks later, on a Friday afternoon, his lawyer would announce that Clinton had settled the Paula Jones case. He agreed to pay eight hundred thousand dollars, with no admission of wrong-doing.

There was no end zone dance from the First Lady, either. But for the first time in their partnership, she was receiving as much credit for one of her husband's triumphs as he was. As *Time* wrote in its next issue: "We all thought a woman who has loved Bill Clinton would dramatically influence the midterms, and we were right. It just wasn't Monica.

"Most people might have gone into therapy or hiding after what Hillary suffered this year. She tore up the campaign trail instead. The operative analogy best describing her ceased to be Tammy Wynette. It became something more like Jackie Joyner-Kersee."

Three days after the election, New York's other senator called a press conference. Daniel Patrick Moynihan was seventy-one years old and had been in office for three terms. He was going to hang it up in 2000, he now announced. Political junkies perked up. Open Senate seats were rare in New York, and the jockeying to fill this one figured to be fierce. The press speculated on the political titans who might run: Rudy Giuliani, Colin Powell, Jack Kemp, and even D'Amato on the Republican side; for the Democrats, maybe one of the Cuomos—Mario or his son Andrew? Or how about a Kennedy? Carolyn, John John, or RFK Jr.—take your pick.

A few days later, in its Bold Face Names column, the *New York Times* threw one more possibility out there, seemingly for the hell of it: "The announcement by **SENATOR DANIEL PATRICK MOYNIHAN** that he would retire in 2000 has some New York Democrats buzzing about a potential candidate who definitely knows the ins and outs of campaigning: **HILLARY RODHAM CLINTON.**"

Back in Washington, meanwhile, the Republican recriminations intensified. The election results were making it open season on Newt, and not just over his midterm strategy. There were moderates still seething over the shutdown; there were conservatives who still thought he'd sold them out on the shutdown; there were pragmatists who thought he was too erratic, too inflammatory, too undisciplined. Four years of pent-up hostilities were coming to the surface.

He'd survived a coup a year earlier, in part because the plotters failed to agree on a successor. Now a challenger stepped forward. It was one of Newt's closest allies. At least that was Bob Livingston's reputation. The Louisianan chaired the powerful Appropriations Committee, a perch he owed to Newt's assistance. When Livingston had toyed with retirement months before, Gingrich told him to stay. Now, on the Friday afternoon after the election, Livingston saw his moment and claimed it. "Revolutionizing takes some talents—many talents," he said at a press conference. "My friend Newt Gingrich brought those talents to bear and put the Republicans in the majority. Day-to-day governing takes others. I believe I have those talents."

It didn't take more than that. Gingrich was still in Georgia, scrambling to keep the pieces together, but the defections were swift and sizable. Within hours, he gave up and put out a statement: "Today I have reached a difficult personal decision. The Republican conference needs to be unified, and it is time for me to move forward where I believe I still have a significant role to play for our country and our party."

In twenty years, he'd climbed from the backbenches of a supposedly permanent minority party all the way to the Speaker's chair, and now it was all over. The majority he'd built had lost faith in him. On a conference call with his soon-to-be former Republican colleagues, it was reported, Gingrich closed by saying, "I love you all. Take care." A few days later, he was back in Washington to speak at a dinner for GOPAC, an organization he'd once run, one that had been instrumental to his rise. They shouted his name and treated him to a champion's entrance. "For a few brief moments," the *Washington Post* wrote, "it was almost exactly the way it was four years ago, in the weeks after the Republicans took control of the House for the first time in 40 years."

But times had changed. When they won that majority in 1994, Newt had told Republicans it was just the start. They lived in a conservative country that wanted conservative government, and they were going to deliver it and reap the rewards for decades to come. By now, Clinton was supposed to be long gone, replaced by a Republican president, and the welfare state was supposed to be in terminal retraction, with Americans cheering on its demise, welcoming a new era of decentralization and self-reliance. Instead, Republicans were facing a deeply uncertain future. They'd been licked by Clinton three times— the shutdown, the 1996 election, and now the midterms—and their image was deteriorating. Even this crowd, Newt's crowd, knew he could take them no farther. He told them he agreed, that staying on would only mean "divisiveness and factionalism."

"The ideas are too big," Gingrich said, "the issues are too important for one person to put their office above the good of the party and the country." He wasn't just stepping down as Speaker. He was resigning his seat in the House. He was fifty-five years old, young enough to author a new act of his life. He wouldn't be running for office, at least not for now, not with his toxic poll numbers, but he promised to stay in the arena.

Even with Gingrich gone, the Republicans pressed forward on impeachment. It was by now obvious they wouldn't knock Clinton out of office. In the House, they had the votes to impeach him. But it would take a two-thirds supermajority to get a conviction in the Senate. That meant at least twelve Democrats would have to turn on Clinton, and not a single one was about to. They were trapped, though. They'd set the wheels in motion before the election, and their base expected them to follow through.

It got ugly. On December 19, as the House debated the articles of impeachment, Livingston went to the well. Republicans had officially tapped him as their next leader; the formal vote to make him Speaker was a few weeks away. But Livingston had just received some startling news. Someone had responded to Larry Flynt's bounty offer and given him the goods on Livingston. It was all about to become public. To gasps from his colleagues, many of whom had no idea of the circumstances, the Speaker-in-Waiting resigned on the spot.

"I believe I had it in me to do a fine job," he said. "But I cannot do that job or be the kind of leader that I would like to be under current circumstances."

Eventually, Republicans turned to a sixth-term Illinois Republican with little appetite for the media spotlight. Far in the future, long after his retirement, J. Dennis Hastert would be implicated in a stunning sex crime of his own and would admit to molesting students in his days as a high school wrestling coach. In this moment, though, "Denny" seemed to his colleagues an honorably modest man—a welcome antidote to the leadership they'd lived under for the past four years.

The House impeachment vote broke on party lines that day. A few Republicans did break rank, while no Democrat voted for any article. In the Senate, it was the same story, and the president was acquitted on February 12, 1999. There was never any suspense. As the Senate votes were cast, Gallup surveyed the country. Clinton's approval rating was 68 percent.

Days later, a new story emerged, this one much darker. An Arkansas woman named Juanita Broaddrick was accusing the president of having raped her. She said it happened back in 1978, when Clinton was first seeking the governorship. She ran a nursing home and met him when he stopped by for an event, and he encouraged her to reach out if she was ever in Little Rock. He would give her a tour of his campaign office. When she found herself in Little Rock for a conference a few weeks later, she called his headquarters and reached him. He suggested meeting for coffee in her hotel lobby; then, when he arrived, he called up to her room, told her the lobby was too crowded, and asked if they could meet in her room instead. Not long after she opened the door, she said, it happened. "It was twenty years ago and I let a man in my room and I had to take my lumps," Broaddrick told the *Washington Post*. "It was a horrible, horrible experience and I just wanted it to go away."

It was, obviously, an explosive charge—"sensational yet ancient and unproven," is how the *Post* characterized it. She'd told other people about it when it happened and rumors had circulated in Arkansas ever since. Reporters had picked up on them in 1992. So had Paula Jones's lawyers. Broaddrick had provided a deposition for that case, denying anything had happened. Now she was telling a very different story. It was her young granddaughter, she told the *Post*, who triggered a change of heart. "When they ask me about this in a few years, I want them to say, 'That was a neat thing you did.' I didn't want them asking me, 'Why didn't you come forward?'"

While the impeachment trial was under way, Broaddrick had taped an interview with NBC News, but it didn't air until after the vote. Clinton refused to address the allegation himself, referring all questions to his lawyer, who said: "Any allegation that the president assaulted Ms. Broaddrick more than twenty years ago is absolutely false. Beyond that we are not going to comment."

Was a brand-new, even more serious scandal now erupting? The answer was no. The press reported Broaddrick's claims, but as a stand-alone story, not the gateway to another round-the-clock political and media soap opera. Republicans called on Clinton to make a full accounting of whether and how he'd known Broaddrick, but he ignored them. A personal denial from the president, the *Washington Post* editorialized, would carry no weight anyway. "Mr. Clinton's word in this realm by now has no value," the paper concluded. But beyond acknowledging that Broaddrick's story was both plausible and ultimately unprovable, the commentary class said little. A poll found that a majority of Americans doubted Broaddrick's story. Democrats had no interest in looking further. Asked if Clinton should answer questions about the accusation, Tom Daschle, the Senate Democratic leader, said: "I think he's addressed it the way you would expect the president to address it, and I don't think you're going to hear anything more from him, nor do I think it's probably going to lend any additional insight or new information. Let's move on."

And so everyone did.

TWENTY-FOUR

I t was happening again. The fiscal year was expiring. An emergency bill to fund the government for a few more days was rushed through the Congress and signed by the president, who now warned that if House Republicans insisted on a budget that would "turn its back" on the vulnerable, he would veto it and there would be another shutdown.

Newt was gone, but at the end of September 1999, his old House crew was up to familiar tricks. At issue this time: how to manage America's first surplus in a generation. Republicans wanted big tax cuts, and they didn't want Democrats to be able to accuse them of risking a return to red ink. They needed some savings to offset the tax cuts.

What they'd come up with was an accounting trick. They would take the earned income tax credit, now a lump-sum refund paid annually to twenty million Americans, and parcel it out in twelve monthly installments. This way, some of the monthly payments would technically fall outside the next fiscal year, and—poof! Savings!

They were walking right into a trap, again. The earned income tax credit was popular, easy to defend, and widely enjoyed. It targeted families just above the poverty line—those with too much money to qualify for assistance who were still struggling to get by. It was worth nearly two thousand dollars a year to the average qualifying family. As with Medicare, this was a benefit that few saw as a handout. You had to work to get it.

By switching to monthly payments, Republicans were effectively penalizing those who saved or invested their tax credit money, cutting into the interest payments they'd otherwise receive. More simply, they were inviting suspicion. Why should anyone believe this would be the limit of their tinkering?

"Let me be clear," Clinton said. "I will not sign a bill that turns its back on these hardworking families. They're doing all they can to lift themselves out of poverty, to raise their children with dignity. I don't think we should be putting roadblocks in their way."

Then, from three thousand miles away, on a campaign swing in California, a surprise voice piped up. "I'm concerned about the earned income tax credit," George W. Bush announced. "I'm concerned for someone who is moving from near poverty to middle class. I don't think they ought to balance their budget on the backs of the poor."

Bush, the governor of Texas and son of the former president, was seeking the Republican presidential nomination. Now he was telling his own party's leaders in the House to knock it off, and they weren't happy about it. "It's certainly not what we wanted to hear," Speaker Hastert's press secretary told reporters. Tom DeLay said, "It's obvious Mr. Bush needs a little education on how Congress works. I don't think he knew what he was talking about."

But they backed off anyway. Three days later, the plan to switch to monthly payments was dropped. "There's too much flak and it's too hard to explain," Ways and Means chairman Bill Archer said. The government stayed open, and Bush was treated to a round of flattering news coverage. He had, the *Washington Post*'s Dan Balz wrote, "stolen a page from President Clinton's political playbook, distancing himself from the unpopular congressional wing of the Republican Party in the same way Clinton played off congressional Democrats as he mounted his recovery after the GOP landslide of 1994."

It looked more courageous than it was. Bush was perched at the helm of one of the most formidable campaign machines ever built by a nonincumbent candidate for president. He'd locked down endorsements from nearly every Republican governor, the majority of national committee members and state party chairs, and more than a hundred House members and senators. His lead over his nearest rival was approaching fifty points.

Picking this fight with House leaders? It didn't threaten this outpouring of support. It was the reason he had it in the first place.

They'd decided he was the solution to an urgent crisis. The Republican Party that emerged from impeachment in the winter of 1999 was facing its worst image with the public since Watergate. Just 38 percent of voters now viewed the GOP favorably, compared to 56 percent for the Democrats. As alarming for Republicans was the generic ballot. Asked which party's candidate they'd support in the next year's congressional elections, voters now picked the Democrats by eleven points. Clinging to control of the House by just six seats— the slimmest majority in forty-five years—Republicans now had to contend with the real possibility of losing their crown jewel.

Charging ahead with impeachment despite strong public opposition—and especially after a rebuke at the polls—was the proximate cause, but the damage went back further. The '94 revolution introduced most Americans to what the Republican Party had become. It was increasingly dominated by southerners, flavored with evangelical Christianity, rigidly conservative, and intensely combative. It was the party of Newt Gingrich.

In 1995, as the new GOP Congress aligned itself with the NRA and moved to pare back environmental rules and Medicare spending, political analyst Kevin Phillips had a warning. "Two decades ago," he wrote, "the Republicans nailed the Democrats—and not inaccurately—as the party of acid, amnesty and abortion. Now, they themselves could be just as validly indicted as the party of gunk, gun clubs and granny-bashing."

Just a few years earlier, Republicans seemed on the cusp of top-to-bottom control of government. Now it was possible, even likely, that the 2000 election would put Democrats back in charge of everything—the House, the Senate, and the presidency. Without a makeover, and fast, Republicans were heading for a wipeout.

The field of would-be saviors was enormous, at least at first. For the past three decades, GOP nominating races had been orderly affairs. In 1980, 1988, and 1996, the party simply tabbed the runner-up from its last open contest. In one case, George H. W. Bush in 1988, that runner-up was also the sitting vice president.

This time, though, there was no heir apparent. The second-place candidate from 1996 was already running again, but Pat Buchanan's views were seen as too extreme by the party establishment, whose opposition checked his ascent. The rest of the field included two other '96 vets (Lamar Alexander and Steve Forbes), three senators (John McCain, Orrin Hatch, and Bob Smith), a former vice president (Dan Quayle), a governor (Bush), a former cabinet secretary (Elizabeth Dole), a House member (John Kasich), and two religious conservative leaders (Alan Keyes and Gary Bauer).

By dint of pedigree, Bush and Dole claimed the early front-runner spots, but no one was putting much stock in that. Republicans hadn't endured a White House drought this long since the sixties, and the man who'd been their public face in the Clinton era—Newt—was off the stage. Surely, this was the kind of leadership vacuum that would trigger a volatile, chaotic, and plain unpredictable nomination battle—"probably the most open fight since 1948," predicted the Republican chairman from the first-in-the-nation primary state of New Hampshire.

Then Bush, or "W," as the branding went, got to work. It started with the governors. They were the party's bright spot. Republicans now ran thirty-two states. They tended to be a lot more popular than their party mates in Washington—"the only Republicans who got away with their shirts after the 1998 election," as *Time* put it—and they were tired of their own political fortunes being threatened by the GOP Congress's image. They decided: it should be one of their own, not a D.C. Republican, leading the national ticket in 2000.

A few of them wanted it, but only one could inspire consensus. "It became pretty clear pretty early that the one among us on whom fate seemed to be settling was George W. Bush," Utah's Mike Leavitt would later say.

The family name gave him instant credibility, and the large built-in donor network didn't hurt, but what the governors mainly saw in W was electability.

Personality was key. It was the Gingrich blend of pomposity and inflammatory partisanship with which the electorate now identified the Republican Party. Bush, a reformed drinker who retained the cheerful, wisecracking style of his life-of-the-party days, was Newt's polar opposite. He smiled readily, laughed at himself instinctively, and put audiences at ease instantly.

Then there was the packaging. Clinton had effectively branded the Gingrich Republicans as extremists, practitioners of "the politics of personal destruction" who were hell-bent on shredding basic protections for the vulnerable. To date, the GOP's rebuttal had been feeble, but now Bush was offering his party a framework.

He would present himself as "a uniter, not a divider" and argue that government wasn't always the problem; that it could be the solution to problems. "Government should do a few things and do them well," Bush said. Clinton could expertly infuse his policy arguments with an empathetic sensibility, and it had helped him make the Gingrich Republicans look coldhearted and mean. Bush was promising Republicans what they'd been missing: a frontman who could do the same thing. He would fuse conservatism with compassion.

"It's conservative to cut taxes," he explained. "It's compassionate to give people more money to spend. It's conservative to insist upon local control of schools and set high standards and insist upon results.

It is compassionate to make sure that not one single child gets left behind."

Not everyone on the right was ready to buy it. In his column, George Will cautioned that "a compassionate government's work is never done." But Bush could point to evidence that it worked with voters.

Texas was not yet the Republican bastion it would become, and Bush's 1994 victory over Democrat Ann Richards had been an upset. And his reelection margin the previous November, thirty-seven points, was the best performance ever by a Republican candidate in Texas. To a party that had received just 38 and 41 percent of the vote in the last two presidential elections, Bush had the markings of a winner—especially when the early trial heats put him ahead of Al Gore, the likely Democratic candidate, by double digits.

He was, in effect, the Republicans' own Bill Clinton, a likable southern governor pulling the party's image back from the extreme edge. The big-state governors climbed aboard fast. Michigan's John Engler led the way. Tom Ridge from Pennsylvania was there, too; and Jim Gilmore in Virginia; George Ryan in Illinois; Paul Cellucci in Massachusetts; George Pataki in New York; Bob Taft in Ohio; eventually Wisconsin's Tommy Thompson, too. And, of course, Jeb Bush, just elected in Florida on his second try, was there from the start.

Momentum was spreading. Scores of House members signed on. They'd lived the trauma of running in Newt's shadow; now they imagined sharing the ticket in 2000 with someone who could actually beat the Democrats.

Bush might be the antidote to other vulnerabilities, too. California was the biggest prize on the electoral map and it had once been a Republican redoubt. Lately, though, it had been slipping away. After siding with the GOP six straight times, the state broke hard for Clinton in 1992 and again in 1996. There were a number of factors, including the revolt of culturally liberal professional-class white voters against the party of Newt.

But what Republican leaders focused on was something else: a

1994 state ballot initiative—Proposition 187—that proposed denying non-emergency public services to illegal immigrants. Then-governor Pete Wilson championed it, pulling off a come-from-behind victory, and voters passed it with 59 percent support. The debate, though, was heated and divisive, with liberals denouncing it as racist and even some national Republican leaders disavowing it. Moreover, California was rapidly diversifying; long-term projections showed Hispanics surpassing non-Hispanic whites as the state's largest ethnic bloc sometime in the next generation. Would the taint of 187 limit the GOP's ability to expand its support with this emerging plurality of voters?

Behind California, Bush's Texas had the nation's second-highest Hispanic population. But instead of seizing immigration as a wedge issue, he'd tried to make it a showcase of his compassion. In his reelection race, Bush secured more than 40 percent of the Hispanic vote, becoming the first-ever Republican to carry the heavily Hispanic city of El Paso. As a presidential candidate, he pledged to contest California aggressively. "We wouldn't have had a 187 in my state," he said. Soon, most of California's twenty-four Republican members of Congress put their names on a letter urging Bush to formally enter the race.

He had skeletons. That was the chatter, anyway. Bush acknowledged drinking to excess in his early years, but stressed that he'd quit at forty. He was a religious man now, with disciplined personal habits, he said. But rumors of wild behavior extending well into adulthood, particularly drug use, were making their way into the press. He refused to address them directly, issuing only a denial that he'd used any illegal drugs within the last twenty-five years. "When I was young and irresponsible," he said, "I was young and irresponsible."

It didn't bother his party a lick. The endorsements kept rolling in, and so did the cash. Maybe it was a consequence of Clinton's scandals, which had tested the bounds of behavior Americans would accept from their leaders. Or maybe it was just Bush's charm, or his poll numbers against Gore. The stampede only accelerated. In *Newsweek*, columnist David Brooks wrote: "Right now, the GOP is so scared of total meltdown that it's going to do whatever it takes to win. Even if it means imitating Bill Clinton."

At the start of July, the Bush team told reporters to gather around. The candidate had news he wanted to deliver himself. The second quarter of the year had just ended, and campaign finance reports from the campaigns were due. Bush now announced his total: thirty-six million dollars. Jaws dropped. Before this, the most any presidential candidate had ever raised in a three-month period was thirteen million. Bush had basically tripled that. None of his rivals were even in the same ballpark. The closest, John McCain, had brought in six million. Even Gore, the sitting vice president, had only managed eighteen million dollars. If you added every single Republican candidate together, their combined haul was still less than Bush's.

"I am humbled," Bush said. "It's clear that people are hungry for something."

No one had ever seen a financial show of force like this. It spoke to the deep-pocketed friends Bush and his chief strategist, Karl Rove, had cultivated, but also to the overpowering sense of inevitability now fueling his campaign. People were signing on and chipping in for fear of being left out if—or when?—Bush won.

The dispiriting effect on his opponents was just as strong, and they began dropping one by one. Kasich went first, a few days after the fund-raising numbers came in. "George Bush's term of 'compassionate conservative' really kind of defines what John Kasich is all about," he said. "I feel as though I have a soul brother." Then it was Smith's turn; the New Hampshire senator at first vowed to wage a third-party bid instead, then thought better of it and simply exited the stage. Alexander and Quayle both followed. Now, as Bush took his shot at the House Republican budget, Dole looked just about finished, too. The most wide-open race in a generation was turning out to be a snooze.

Bush would soon go further. "Too often," he told a New York audience in mid-October, "on social issues, my party has painted an image of America slouching toward Gomorrah. Too often, my party has focused on the national economy, to the exclusion of all else, speaking a sterile language of rates and numbers, of CBO and GNP. Too often my party has confused the need for limited government with a disdain for government itself."

Slouching Towards Gomorrah was also the title of a recent polemic by Robert Bork, whose 1987 Supreme Court nomination was scotched after Democrats branded him a dangerous ideologue. To the right, Bork would forever be a hero, and it was customary for Republican office seekers to pay homage to his martyrdom. Bush's signaling was clear. He was looking for ways to defy the Puritan caricature voters were now attaching to Republican politicians.

It was the bargain his bandwagon signed up for. They would give him space and he would return them to the White House. So far, they were holding up their end.

Dole pulled the plug on October 21. At the beginning, she'd been running close to Bush, but money had proved a struggle and few meaningful endorsements ever materialized. Now, the field of a dozen was down to seven, and there was news that it was about to shrink by one more.

But this looming withdrawal was different from the others. Instead of easing Bush's path to the nomination, this one had the potential to derail him in the general election. It wasn't just the Republican race that Pat Buchanan was leaving. He was quitting the Republican Party—and teaming up with Ross Perot.

T his was the first time that Buchanan came to a presidential campaign as a powerhouse. In 1992 and 1996, the surprise had been his strength against the establishment. When he announced this time around, it was assumed he'd give Bush a serious run. "With this campaign," Buchanan declared in March 1999, "I intend to redefine what it means to be a conservative in America—to reshape our party into

the natural home for the men and women who work in this country."
But Bush was putting something together that was much stronger than
what Buchanan had run against before. The game, he became convinced, was being rigged against him.

He had three options. He could quit, like the others. He could
stick around and wage a doomed but principled battle against Bush. He
Or he could join up with America's newest third party. That would
be the Reform Party. Perot had followed through with his threat and
launched it after his Dallas convention, using the new party as the
vehicle for his 1996 presidential campaign, which proved a bitterly
disappointing experience for him. Shut out of the debates and given
short shrift by the press, Perot won just 8 percent of the vote. Still,
that was enough to guarantee the Reform Party ballot access for 2000
in a bunch of states. More importantly, based on the Federal Election Commission's formula, it would mean $12.6 million in campaign
funds for the next Reform nominee—enough to run a real campaign.

Perot wasn't running again, at least it didn't seem that way, and
Buchanan found himself tempted. The nomination would be winnable, and he knew the language of the Perot voter. The free money
was alluring, too. No one had yet harnessed the Internet to finance
a campaign. As the Reform nominee, he'd be a player in the general
election. Polls already showed him creeping into double digits in a
race with Bush and Gore. If he could get into the fall debates, who
knows what might happen?

That's what was making the Bush campaign so nervous. Buchanan's appeal wasn't confined to the Republican Party, but it seemed
unquestionable he'd draw more votes from Bush than from Gore—
a lot more. Buchanan's dance away from the GOP played out in slow
motion, and through it all, Bush treated him with kid gloves, hoping
he might yet reconsider. The effort was in vain, and on October 25,
Buchanan bolted.

"The day of the outsider is over in the Beltway parties," he said.
"The moneymen have seen to that. Never again will our political establishment permit a dissident to come as close to capturing a nomination

as we did in 1996. They have rearranged the primary schedules and rigged the game to protect the party favorites."

Within the Bush family, it was an article of faith that Perot had cost the father his reelection; was Perot's new party now going to thwart the son?

First, Buchanan would have to secure the Reform nomination. On paper, he was a good fit, but the party was in turmoil. It lacked a coherent infrastructure and faced an identity crisis at the very top. Perot regarded the party as his baby and expected to keep calling the shots, even from behind the scenes. But now he had competition, from the Reform Party's biggest—and only—success story.

The year before, Jesse Ventura, known as "The Body" in his former career as a flamboyant professional wrestler, had run under the party's banner for governor of Minnesota. He caught fire late in the race and won in an upset. Now he was the only Reform Party member in the nation holding a major office. His platform was vaguely libertarian and his blunt blow-up-the-system rhetoric made him a hit with the Perotistas, who urged him to run for president. He didn't want to do that, but he did want to control the party.

The Ventura and Perot forces collided in July 1999, at the convention to elect a new party chairman. Perot put up a candidate, but Ventura objected. "It is time for him to step aside," he said of the founder. Ventura had his own candidate: Jack Gargan, the retiree who'd helped ignite the Perot phenomenon all those years earlier with his Throw the Hypocritical Rascals Out (THRO) campaign. Gargan prevailed in the vote.

With his chairman in place, Ventura moved to find a presidential candidate. He wanted Lowell Weicker, who'd been a liberal Republican senator before winning election as Connecticut's governor as an independent in 1990. Weicker, now sixty-eight, had left office after a single term but remained a vocal participant in national politics. He began exploring a campaign.

That possibility seemed to provoke an old Weicker nemesis. Six years earlier, Donald J. Trump had sought to expand his casino empire from Atlantic City into Connecticut, where the Mashantucket Pequot Indian tribe had capitalized on a Supreme Court ruling to build a gaming resort on its reservation land. The Foxwoods casino was an instant hit and Trump wanted in on the action, appealing to Connecticut lawmakers to authorize casinos on nonreservation land. Governor Weicker opposed him and it got personal. When Trump claimed the leaders of the Pequots "don't look like Indians to me," Weicker called him a bigot and a "dirtbag."

"My opposition to casinos isn't just casinos," he explained. "It's opposition to Donald Trump." Trump replied that Weicker was "a fat slob" who should "concentrate on losing 125 pounds." Weicker ended up getting his way, but now, as he flirted with a presidential bid, Trump began intimating his own interest. He issued a statement: "If the Reform Party nominated me, I would probably run and probably win."

At the July convention, some delegates distributed homemade literature promoting Trump. A straw poll tested possible presidential candidates. Perot came in first with 22 percent. Trump was second with 17. Weicker was back in single digits. "I don't know anything about what's happening," Trump responded. "But I'm nevertheless greatly honored by all the hoopla for Trump."

Weicker backed away and Ventura grew smitten with Trump. There was common ground in their styles and personalities ("He and I think a lot alike"), but Ventura was also alarmed by Buchanan, who now looked likely to jump in the race. "He has social issues that he puts very strongly on the front burner, and we in the Reform Party tend to leave social issues aside," he said.

Ventura began wooing Trump, who sounded like he was warming to the idea. "When he asks me to look at something, I'm going to consider it," Trump said. Like Ventura, he also pronounced himself opposed to Buchanan: "I think his views are prehistoric."

Fleetingly, Trump had flirted with a presidential campaign once before, delivering a high-profile speech in New Hampshire in the run-up to the 1988 campaign. His remarks played on the theme of America being kicked around and "ripped off" by foreign governments. He proposed attacking Iran, a menace to Washington since the Islamic Revolution of 1979, and seizing its oil fields.

He'd been an icon of the go-go eighties back then, an audacious Manhattan developer with an abundance of self-regard and a playboy lifestyle that made him a tabloid fixture. The nineties had brought hard luck, though. Trump overextended himself and watched his fortune crumble into bankruptcy, only to refashion his name into a personal brand and bounce back. He'd topped the bestseller list in 1987 with *The Art of the Deal*. A decade later, he made it back again with *The Art of the Comeback*.

Nominally, he was a Republican, but politics had nothing to do with the Trump image. He'd donated almost equally to Democrats and Republicans and expressed sympathy for Bill Clinton during the impeachment drama. Sensing a publicity stunt, the media treated the Trump rumblings skeptically. Then, he saw a chance to raise the stakes.

Buchanan's new book was stirring up a hornet's nest. *A Republic, Not an Empire* amounted to a treatise against foreign intervention. He made his case by reviewing past military campaigns and challenging their necessity—including World War II.

Hitler, Buchanan wrote, even after gobbling up large chunks of Europe in the late 1930s, had made "no overt move to threaten U.S. vital interests" and posed "no physical threat" to America. In essence, he was arguing that America allowed itself to be pulled into a confrontational posture, prompting Hitler to respond in kind and setting off a war that could have been avoided.

This was, needless to say, white-hot stuff. Many of the Republicans who'd been treating Buchanan delicately now called for him to go ahead and leave their party. John McCain called his analysis "so far outside of the philosophy of what America is all about that it's unacceptable."

On Sunday, September 19, Buchanan was a guest on CBS's *Face the Nation*. An hour before the show, Trump faxed a statement in to the producers:

> Pat Buchanan's stated view that we should not have stopped Adolph [sic] Hitler is repugnant. Hitler was a monster, and it was essential for the Allies to crush Nazism. Buchanan denigrates the memory of those Americans who gave their lives in the Second World War in the effort to stop Hitler. I am proud of the vital role that the United States played in stopping Hitler. I think it is essential that someone challenge these extreme and outrageous views by Pat Buchanan.

On the air, Gloria Borger, who was joining host Bob Schieffer in the questioning, read some of the statement to Buchanan. "Well, that's a silly and false caricature of my position," he replied. "Hitler declared war on the United States. We had no alternative but to fight him."

"So you're saying he's mischaracterizing you?" Borger asked.

"Well, sure. Of course. Hitler declared war on the United States December 11th, 1941. What alternative does Donald Trump suggest?"

Now Trump was part of the story. "Pat says Hitler had no malicious intent toward the United States," he told the *New York Times*. "Well, Hitler killed six million Jews and millions of others. Don't you think it was only a question of time before he got to us? He tackled Europe first and we were next. Pat's amazing."

"Asked whether he had read the book," the paper noted, "Mr. Trump said, 'I've seen the phrases we're dealing with.'"

There was at least some resistance to Buchanan within the Reform Party, but who would give it voice? Trump seemed to relish the role. Now the press was getting interested. To the *New York Daily News,* he pronounced himself "very serious" about running.

In 1999, CNN remained the undisputed cable news leader. It still had the biggest audience, and even for other media professionals, it was the go-to source for breaking developments throughout the day. Trump recognized this, and on October 7, he demonstrated he knew how to exploit it, too.

He started the day by taping an interview with Larry King, to air on his show at nine o'clock that night. Trump came equipped with a nugget of news. He was forming a presidential exploratory committee. "The polls have been unbelievable," he told King. "So I am going to form a presidential exploratory committee. I might as well announce that on your show. Everyone else does."

Practically speaking, the move had little significance, but it sounded very official. Now, CNN had "breaking news" that it could build its programming around for the rest of the day.

It wasn't just CNN. Trump was turning himself into a content machine for every outlet. On NBC's *Today*, he rebutted the suggestion that he lagged in the polls by pointing to a survey from the *National Enquirer*. He told the *Washington Times* he'd love to have Oprah Winfrey as a running mate: "If she'd do it, she'd be fantastic. I mean, she's popular, she's brilliant, she's a wonderful woman." On the newsmagazine *Dateline*, he said he'd spend "as much as necessary for me to at least have a shot. If it took twenty or thirty or forty million dollars, I'd be willing to spend it." The *Wall Street Journal* offered him space for an op-ed. "I believe non-politicians represent the wave of the future," he wrote. *Saturday Night Live* imagined a strategy meeting with Perot, Buchanan, and Trump.

His strength was hard to measure. Forty-seven percent of Americans had a negative view of Trump, according to one poll, although that was actually *better* than the numbers for Buchanan and Perot. In matchups with Gore and Bush, he was sitting around 10 percent, essentially the same level as Buchanan.

Within the Reform Party, things were even blurrier. The party was promoting what would be the first-ever online vote for a nominee, but who would comprise the electorate was unknown. Each candidate was supposed to mount a petition campaign to help the party qualify for the ballot in the states where it didn't have automatic access.

Anyone signing a petition accepted by state officials would then be eligible to vote in the nomination race, along with every voter already registered with the Reform Party.

But fault lines were coming into focus. Perot's running mate from 1996, Pat Choate, endorsed Buchanan. A sign of where the founder was casting his lot? Perot refused to say, but his loyalists were now maneuvering to regain control of the party. They controlled a majority of the executive committee, a check on the power of Gargan, the Ventura-aligned chairman. Disputes were breaking out; the schism at the top was deepening.

On October 24, Trump went on *Meet the Press* with more news to make. He was leaving the Republican Party and filing paperwork to join the Reform Party. Host Tim Russert asked why. "I really believe the Republicans are just too crazy right," Trump replied. The same description, he went on to say, applied to Pat Buchanan: "He's a Hitler lover. I guess he's an anti-Semite. He doesn't like the blacks, he doesn't like the gays."

"It's just incredible," Trump said, "that anybody could embrace this guy. And maybe he'll get four or five percent of the vote, and it'll be a really staunch right wacko vote. I'm not even sure if it's right. It's just a wacko vote. And I just can't imagine that anybody can take him seriously."

The bluster and bravado; easy disparagement of critics and rivals; constant references to dubious polls: so many of the traits Americans would confront a generation later were on display as Trump explored a 2000 bid. But the content of his message differed dramatically from the Trumpism that would later emerge.

He presented himself as a social liberal and economic conservative. On abortion, he was "totally for choice—I think you have no alternative." He said he wanted deep tax cuts, but also called for a onetime 14.25 percent tax on the assets of anyone worth at least ten million dollars. It drove conservatives crazy; they said it would crash the stock market. He said it would pay off the national debt once and for all. He was skeptical of gun control, but also critical of the NRA.

On health care, he called himself "liberal" and advocated expanding Medicare to cover the cost of prescription drugs.

There was some overlap with Buchanan. Trump, too, was against NAFTA and spoke of global trade deals as a drain on American jobs. And he was for a strict immigration policy. "We have to take care of the people who are here," he said. But he drew a bright line when it came to Buchanan's tone ("He seems to be a racist") and accused him of cultivating support from the bigoted fringes.

"On slow days," Trump wrote in an op-ed, "he attacks gays, immigrants, welfare recipients, even Zulus. When cornered, he says he's misunderstood."

On a trip to California, between a meeting with Reform activists, a paid speech, and a taping of *The Tonight Show with Jay Leno*, Trump visited the Simon Wiesenthal Center's Museum of Tolerance, which sought to shine a light on racism and injustice around the world. After his tour, Trump told reporters that Buchanan should "come here and have a talk with Rabbi Cooper and his staff and talk things out a bit."

He added: "We must recognize bigotry and prejudice and defeat it wherever it appears."

Catching up with him during the California swing, a *Los Angeles Times* reporter took note of one of the would-be candidate's fixations:

> "I got a call from one of the biggest politicians in New York," he says during a half-hour interview. "He couldn't believe he was watching CNN for a whole hour and they kept announcing 'Donald Trump and the other presidential candidates coming up at 7 o'clock.'
>
> "They use me to announce they're going to discuss presidential politics," he brags before conceding that "whether or not TV ratings can transfer into votes is a very interesting question."

He kept it up into the new year. By now, he was also touting a new book, *The America We Deserve*, adding to the debate over whether this was all one big publicity stunt. Trump swore he was serious, but if he was, there were new obstacles now.

The clash at the top of the party was worsening. Perot's forces were moving to oust Gargan as chairman. A court fight loomed, and it threatened to drag on for months and maybe even tie up the big pile of cash—those $12.6 million in federal matching funds—promised to the winner of the Reform Party nomination.

The new party's decentralized nature was also making it a magnet for all sorts of niche interests. Lenora Fulani, self-described "black-nationalist Marxist" who'd previously run for president under the New Alliance Party banner, teamed up with Buchanan and was bidding to gain control of the party's levers in New York. Fulani had been accused of anti-Semitism—"Jews had to sell their souls to acquire Israel," she'd once said—and Trump condemned her.

David Duke popped up again, too, announcing his interest in joining the party and supporting Buchanan. John Hagelin, previously the candidate of the obscure Natural Law Party, which emphasized "harnessing the body's natural healing mechanisms," was now talking about mobilizing his troops to compete for the Reform nomination for himself.

In January, Trump canceled a speech to the Reform Party's state convention, citing the party's increasing instability. He wrote a letter to Perot and Ventura encouraging them to bury the hatchet and prevent a "fratricide." But the Reform Party was plunging into chaos now. Perot's executive committee voted to expel Gargan, who called the move illegitimate. Then Ventura, enraged by Perot's maneuvering, threw his hands up and quit the party. Trump said he supported the decision and that it was a "devastating blow" to the Reform Party.

On February 14, Trump himself walked away, claiming the internal discord was rendering the nomination worthless. "The Reform Party now includes a Klansman, Mr. Duke, a neo-Nazi, Mr. Buchanan, and a communist, Ms. Fulani," he said. "This is not company I wish to keep." He elaborated in an op-ed for the *New York Times*.

"When I held a reception for Reform Party leaders in California, the room was crowded with Elvis look-alikes, resplendent in various campaign buttons and anxious to give me a pamphlet explaining the Swiss-Zionist conspiracy to control America," Trump wrote.

Lest anyone think the experience had soured him on politics, he closed the piece by noting that he'd seen Gore on CNN recently with an "obvious look of drudgery" on his face as he campaigned.

"My experience was quite different," he wrote. "I had enormous fun thinking about a presidential candidacy and count it as one of my great life experiences. Although I must admit that it still doesn't compare with completing one of the great skyscrapers of Manhattan, I cannot rule out another bid for the presidency in 2004."

No one put much stock in that. Pat Choate, Perot's '96 running mate, said Trump's campaign had been nothing more than a sham, "a complete hustle of the media, and I think the media should send him a massive bill on it." On that count, there was wide agreement within the media world.

With Trump out, the chaos turned to farce. Buchanan amassed a sizable number of delegates, but the factional infighting of the party was overwhelming, with endless leadership battles and court fights. A year earlier, he'd been threatening to wreak havoc in the general election, at Bush's expense. Now he was barely given a thought by the Republican campaign, or anyone else. Pitchfork Pat was a has-been.

It turned out that the biggest threat to George W. Bush was his own party and its image with Americans.

After cruising through 1999 and winning the leadoff Iowa caucuses with ease, Bush was walloped in the New Hampshire primary by John McCain. No one had seen the scale of it coming. McCain won by eighteen points, a landslide powered by independent voters, who sided with him by a 61 to 19 percent margin. Like that, the Bush machine's inevitability was thrown into doubt.

McCain's appeal was rooted in some stark contrasts with the front-runner. Bush had tapped the country's richest donors to amass the heftiest bankroll of any candidate in history; McCain, with a fraction of Bush's cash, was running on a vow to curb the power of big money through campaign finance reform. Bush had steered clear of combat duty during the Vietnam War, serving instead in the Texas Air National Guard; McCain spent five and a half years in a North Vietnamese prison camp, an ordeal he recalled in a book—*Faith of My Fathers*—that hit the bestseller list as his campaign took off. Bush was tightly managed as a candidate; McCain would cheerfully hold court with reporters in nothing's-off-the-record bull sessions on his campaign bus, the Straight Talk Express.

McCain was selling the idea of independence and authenticity, and after New Hampshire there were buyers lining up everywhere. His support shot up in the next key state, South Carolina, and small-dollar donations adding up to more than a million dollars a day were gushing in over the Internet. There hadn't been an overnight sensation like this in a presidential race since Gary Hart in 1984.

Bush did manage to beat back the wildfire, but at a heavy cost. He ran hard to the right, playing up his staunch opposition to abortion and gun control and challenging McCain's loyalty to the party. Bush was the vehicle for conservative activists and interest group leaders who wanted to win back the White House, and they rallied fiercely behind him now. Pat Robertson and Jerry Falwell vouched for him and went after McCain. McCain fired back, calling the televangelists "agents of intolerance." Rush Limbaugh used his show to rail against McCain as a tool of Democrats.

Bush accepted an invitation to speak at South Carolina's Bob Jones University, a Christian school where interracial dating was banned and the founder's son had called Roman Catholicism a "Satanic cult." When he mentioned none of this in his remarks, he was showered with disgust by the national press. There were stories about "push polls" and underground efforts by Bush-friendly groups to smear

McCain, including anonymous fliers suggesting he'd fathered a biracial child out of wedlock.

The theory of the Bush candidacy was that a winning personality and "compassionate" rebranding of conservatism could sweep away the electorally polarizing baggage the GOP had accumulated in the nineties. Instead, Bush found himself leaning on many of the forces that had helped to create that baggage. The pattern became glaring. In primary after primary, Bush would win overwhelmingly with registered Republicans and self-described conservatives, while McCain cleaned up with moderates and independents.

The numbers were on Bush's side, at least in the primaries; after a few more scares, he put McCain away and wrapped up the nomination. But now, as the fall battleground map took shape, his appeal faced clear limits. Except for New Hampshire, New England was off-limits to the Republican campaign. In New York, a state Reagan carried twice in the eighties, Bush was so far behind that he was now likely to drag down Rick Lazio, who was waging a spirited Senate race against Hillary Clinton. The Bush campaign wasn't even bothering to contest New Jersey, a bastion of suburban Republicanism only a few years earlier. Illinois looked gone, too, and all the early buzz about a Bush-led Republican revival in California now felt like a joke.

The problem wasn't Bush himself. Voters tended to like him personally, polls showed. But that could only get you so far anymore. Partisan warfare had been the permanent condition of the 1990s. Both parties were now more clearly defined, and polarization was rising. Bush was both fortified and trapped by the image of a Republican Party that had welcomed a government shutdown, impeached a president, and made alliances with the Christian right, conservative radio, and an increasingly radical National Rifle Association.

Barely a decade earlier, his father had won forty states, and Reagan had taken forty-nine before that. Now, it was becoming clear, there were whole sections of the country where Bush just wouldn't be competitive.

His saving grace: the same was true for the other party.

Gore marched through the Democratic primaries undefeated. His opponent, former New Jersey senator Bill Bradley, ran to the left, promising an end to "old politics" and implicitly accusing the Clinton administration of ducking thorny issues out of political expedience. Bradley's best showing was in New Hampshire, where he lost by six points. He was trounced everywhere else. Gore's path to the nomination was among the smoothest ever.

He brought his own liabilities to the general election. There was tarnish from the fund-raising scandal that erupted in the final days of the 1996 campaign; the controversy had deepened after the election and Gore became a central player, earning the nickname "solicitor in chief" for his heavy-handed role in raising money from questionable sources.

He lacked Bush's natural charm and was dogged by charges that he had a penchant for embellishment. Eventually, Bush opened a wide lead on the question of honesty.

There was the Clinton factor, too. The public's approval of the president's job performance remained as high as its regard for his character was still low. Clinton was itching to play a major role in his number two's campaign, but Gore opted go it alone, and for a running mate chose the most prominent Democratic critic of Clinton's conduct, Joe Lieberman.

As with Bush, big chunks of the map were out of bounds for Gore and his party. In much of rural America, the gun control laws enacted under Clinton smacked of an attack on a way of life. To millions of churchgoing Americans, the criticism that resonated was not of the Republican Party for being too religious but of the Democrats for being too secular. Gore's fervent environmentalism, celebrated in liberal circles as a sign of moral enlightenment, could be its own liability among blue-collar voters, who worried it might cost jobs.

Bush's intellect became its own issue, fueled by his frequent malapropisms ("Rarely is the question asked: Is our children learning?"). On late-night television, he was cast as a dimwit. A broader impression

took hold in popular culture and the media of Bush as lacking curiosity and being insufficiently worldly. For many voters, it was cause for
alarm. Was the country poised to elect someone with a proudly anti-
intellectual streak? For just as many voters, though, the entire line of
criticism reeked of condescension. The sophisticates calling Bush a
rube might just as well have been sneering at them.

It was the kind of cultural divide Clinton had skillfully navigated,
but Gore was a different candidate. He shared his boss's southern
roots, but not his good ol' boy charm. It made him far more vulnerable to the Republican charge that he'd forsaken his roots and become
just another national liberal politician. Late in the campaign, Bush
delivered one of his most potent jabs in Knoxville, Tennessee. "I feel
good about carrying Texas," he said. "My opponent says that he wants
to carry his home, too. Well, he may win Washington, D.C., but he's
not going to win the state of Tennessee!"

It was close, but Bush appeared to have the edge when a last-
minute bombshell struck: a report just four days before the election
that he'd been arrested for drunk driving in 1976. He'd been thirty
years old at the time, had pleaded guilty, and never publicly revealed
the infraction. Did it cast new doubt on his honesty? Reawaken old
questions about his maturity and seriousness? The effect was hard to
pinpoint, but when the returns started coming in on Election Night,
it was clear the contest was tighter than it had looked just a week
earlier.

In the not-too-distant past, there'd been no standard election color
scheme on television. One network might assign blue for the Republicans while another would use red; then, for seemingly arbitrary reasons, they might each switch the next time around. But on the night
of November 7, 2000, they were all synced up. The result was a brand-
new shorthand that would define American political culture in the new
century.

In the Northeast, Gore was swamping Bush: by sixteen points in
New Jersey, twenty-five in New York, pushing thirty in Massachusetts. On their screens, viewers saw a virtually uninterrupted blue blob

from Washington, D.C., north through Maine; another one in the upper Midwest, where Gore claimed Illinois easily and Michigan, Iowa, Minnesota, and Wisconsin with great effort; and one more streak of blue along the Pacific Coast, stretching from Seattle to San Diego.

The states in these blue areas shared some characteristics. Most had large cities with diverse populations. The political clout of nonwhites was growing; they now accounted for almost one in five voters, and they were breaking for Gore at a 76 percent clip.

There were densely packed suburbs and metropolitan areas, where college degrees, white-collar jobs, and liberal cultural values were now the norm. These were voters who'd reacted with such hostility to the Gingrich Congress, and even though Newt was gone now, they weren't buying the GOP's Bush makeover.

And especially in the Midwest, there were blue-collar whites, many of them unionized, staying just loyal enough to the party of their birth to push Gore over the top.

The blue blobs contained a lot of voters, enough to give Gore 267 electoral votes—the precipice of victory. But they also represented a shrinking geographic footprint for the Democrats. Of the nation's more than three thousand counties, Gore was winning just 659 of them. That was fewer than Dukakis had won even as he barely cracked one hundred electoral votes in 1988.

Almost everything else on viewers' screens was red: the entire mountain West (minus New Mexico, where Bush was 366 votes short); the Plains states, Appalachia, and the South. All of the incursions Clinton had made into rural America and the Old Confederacy were now wiped away.

West Virginia, where working-class whites had rallied around the party of the New Deal for generations, now switched to Bush and the Republicans. Kentucky, won twice by Clinton, wasn't even competitive. Missouri switched back; Ohio and Louisiana, too. And the topper: Tennessee, turning its back on its own native son. Bush had been on to something in that speech. A lot of folks in the Volunteer State didn't think of the boy from Carthage as one of their own anymore.

The swaths of red dominated the screen but only added up to 246 electoral votes. The tiebreaker was Florida, unresolved until early December and still disputed for years to come. When a Supreme Court ruling put a stop to the recount, Bush was declared the winner of the state and, thus, the presidency. Out of almost six million votes counted in Florida, his margin was 537 votes.

I t was the closest thing to a perfect tie American presidential politics had ever seen. Any factor could credibly be singled out as the difference maker. Democrats called it a stolen election, accusing the five Supreme Court justices who ended the recount—all of them Republicans—of doing Bush's bidding, and casting suspicion on Florida's top election official, Republican Katherine Harris. Their fingers also pointed at Ralph Nader, the consumer activist who ran as the left-wing Green Party's candidate and gobbled up almost three million votes nationally—and ninety-seven thousand in Florida.

It was at least as plausible, though, that the third-party candidate who damaged Gore the most was Pat Buchanan. The culprit was the bizarre "butterfly ballot" used in Palm Beach County. Candidates were listed on both sides of the page and voters were instructed to punch a hole in the middle. The result was rampant confusion—and 3,704 votes for Buchanan, by far his best showing in any county in the state.

Palm Beach was a Democratic bastion, heavily populated with older Jewish voters; hardly Buchanan's natural constituency. The evidence that most of his votes were intended for Gore and cast in error was anecdotal, but even Buchanan admitted it was compelling. In the end, the candidate who'd entered the race poised to ruin Bush may have actually saved him from defeat.

Take a step back, though, and the story wasn't the intricacies of the Florida process or the personalities of the candidates. It was how politically divided the nation had become; how its voters were now so polarized—tribalized, really.

The election of 2000 was the product of the decade-long partisan war that preceded it. When the 1990s began, Democrats seemed incapable of winning the presidency, just as Republicans were seen as a permanent minority party in Congress. The collision of Bill Clinton, the first Democratic president in a dozen years, and a Gingrichized Republican opposition unleashed a series of convulsive events that gave new definition to both parties and compelled Americans to pick sides once and for all.

The electoral map spoke to the new divide. It was regional. It was demographic. It was cultural. Above all, it was deep. Red states and blue states. The terms had never existed before; now they told the story of what America had become: a nation of two political tribes, each with its own value system, its own grudges and resentments, its own worldview.

"The candidates have worked out a compromise," David Letterman joked as the recount dragged on, "and, thank God, not a minute too soon. Here's how it goes: George W. Bush will be president for the red states and Al Gore will be president for the blue states." The audience laughed at a joke that never before in American history would have made any sense.

Definition and contrast, Newt Gingrich had insisted long ago, would propel the Republican Party to a limitless future. It turned out he was half right. There were bedrock red states now, where voters would check off every Republican name on the ballot, from president down to dogcatcher. But every action brings a reaction, and with the rise of Red America had also emerged a Blue America.

Its capital, arguably, was New York, where Gore collected 60 percent of the vote. But it was another contest on the New York ballot, just one spot below the presidential race, where the citizens of Blue America would find their solace in this moment of disappointment. By a ten-point margin, Hillary Rodham Clinton had won the race for U.S. Senate, and already the question was when—not if—she and her husband would seek a return to the White House.

Whenever the time came, they would do so in a country whose

politics had been transformed under Bill Clinton. A decade before, he'd been a small-state governor pleading with Democrats to move to the middle and promising to win back the South for them. Now, the South was gone; every single state south of the Mason-Dixon line had sided with Bush. The Clintons weren't going back to Little Rock. Chappaqua, a tony suburb of New York City, would be their new home. Like their party, they could see where their future was.

ACKNOWLEDGMENTS

This project has been in the works so long that I think it might actually have started in the '90s. I owe thanks first and foremost to Zack Wagman, my editor at Ecco, who took me to lunch many, many years ago and told me I had a book in me. Both of us have changed jobs since then, a few times in my case, but after a few dumb ideas (all mine), dead ends, and false starts, somehow we've ended up here. I am grateful for his confidence, his deft editing touch, and his super-human patience with my appalling disregard for deadlines. Oh, and also his willingness to humor my obsession with Al D'Amato.

The entire team at Ecco, led by Dan Halpern, has been nothing but supportive and enthusiastic. I hope the final product is everything they imagined it could be—or at least something close enough. Thanks as well to Will Lippincott for seeing potential in me and showing me the ropes in an industry that still baffles me; Sloan Harris for helping to guide this to the finish line; Jean Sage for assuring me it was possible to learn television and book-writing at the same time; James Andrew Miller, a source of unfailingly clear-eyed wisdom; and Olivia Metzger for her invaluable advice and professional savvy.

This is a book that was inspired in part by my first-hand recollections of the '90s, when I was a teenage enthusiast absorbing political theater for first time. For me, the research process represented a joyful excavation of formative events and also a jarring one, as my youthful perceptions were upended and replaced by a much deeper and more nuanced understanding. Particularly valuable to this process were John Marino, Tom Davis, and Vin Weber. My thanks as well to Joe Grandmaison, who held the paperwork that almost changed history on December 20, 1991. And while I've never met him and don't know anything about him, I feel compelled to thank Michael Renn, who has performed a priceless service in uploading the broadcast networks' coverage of various election nights onto YouTube. My appreciation also extends to Olivia Paschal, who aided me with research, and Emma Janaskie, whose efforts to help me secure photo rights have been nothing short of heroic.

At MSNBC, I owe gratitude to Phil Griffin, who sensed value in this book and offered me encouragement and latitude. It is not lost on me that my association with this network has introduced me to viewers who—God willing—will now become readers of this book. I would therefore be remiss if I didn't thank the first person ever to book me as a guest on MSNBC, back when I was toiling away at the New York Observer during the 2008 campaign, Lisa Nelson. Along those lines, I also want to express thanks to John Reiss, who took the time to talk with me when he didn't have to and who put me on the radar of some of the marquee shows here, and to Izzy Povich, without whose kindness and wisdom I wouldn't have lasted a year in this business.

I am lucky as well that my introduction to daily television came on a show produced by the legendary Steve Friedman, who taught me more than I ever realized at the time. And I will forever appreciate the generosity that Chris Matthews, Rachel Maddow, and Lawrence O'Donnell showed in welcoming me onto their programs in the early days and in supporting me off the air.

The task of balancing my television workload with researching and writing this book was always difficult, especially as the 2016 race played out. Any success I had in meeting that challenge is a credit to the supremely talented producers I've been fortunate to work with here at MSNBC. In particular, I offer thanks to Casey Schaeffer for running a tidy ship, shielding me from distraction and chaos, and always finding just the right way to channel my peculiar fascinations; Adam Noboa for his wizardry with the ever-uncooperative big board (and his readiness to kick and yell at it when I've run out of energy myself); Eric Greenberg for his peerless dependability and infectious creativity; Jack Bohrer for his eagerness, no matter the circumstance, to go the extra mile in the name of doing it right; Betsy Korona, a source in equal measure of prudent counsel and great humor; Eelin Reily, who admirably endeavored to build a show around my voice when I was far less sure of what that voice was; and Cal Perry, who I still think of as the mayor of 30 Rock even though he now does his reporting from London. I am indebted as well to Margaret Menefee, Patrick Spence, Katie Brinn, Mary Murphy, Sarah Bernstein, Nikki Egan, Elizabeth Sedran, and the one and only "Wex"—David Wexler.

I only made my way to MSNBC because of the people who believed in me earlier in my career, including Josh Kurtz, Joan Walsh, and Kerry Lauerman. I am especially grateful to Josh Benson, the very best of the good guys in the media world, who has provided me with more support and guidance than I could ever repay him for; he's also one of those rare people who's up for sitting around and sharing Sharpe James stories, which I appreciate (almost) as much.

Speaking of New Jersey, I also want to thank David Wildstein, who hired me to write about politics in the Garden State when I was a twenty-two-year-old Massachusetts kid with only a vague sense of what he wanted out of life. The three years I spent roaming Jersey remain among the best of my life. I'm convinced I learned more about politics—and how the world really works—by watching Steve Adubato Sr.'s North Ward machine in action on one single election day than any student could ever get out of four years in an Ivy League program. So many New Jerseyans made meaningful and lasting impressions on me, in particular: Tom Barrett, Nick Acocella, Linda Cavanaugh, Ingrid Reed, Tom O'Neil, Michael Aron, and Jim McQueeny, who gave me my very first television gig, on News 12

New Jersey. Nor can I write about those years without including one of my all-time favorite people, the late David Rebovich of Rider University. I wish he were here to see this book and to bust my chops over it.

This book's dedication page honors my terrific parents, Anne and Steve Kornacki, who must be relieved to know that the troubling amount of time their young son young spent watching C-Span ended up being of at least some measurable use. I'm thankful as well to my sister Katie, the actual brains of our clan, for making me a lot sharper than I otherwise would have been. I am blessed in life with an amazingly fun and close-knit extended family; to all the Ramonases and Kornackis, my endless thanks and love.

The same goes for friends who have been there since way back, including Jeff Burns, Matt Kaplan, Dave Kempski, and Paul Hanlon. For more than half of our lives, I've been boring them to death with stories about many of the characters in these pages. The least I can do is thank them here. "Michigan Jeff" Eldridge is a good friend and a talented writer whose feedback was blunt but also beneficial. Finally, as I wrote this book, there were many occasions when I felt on the verge of failure. I had never written anything like this before and I came closer to quitting than I'd care to admit. It was in those moments of doubt and self-pity that I invariably turned to and found the encouragement I needed from Brian Duffy. Without him, there wouldn't be a book to read, and for that I can never be grateful enough.

NOTES

PROLOGUE

1 "Mike, excuse me": "CBS News Election Night (7:00 PM ht ET)," CBS News transcripts, November 7, 2000.

2 "very, very, very": "Election Night 2000: Bush vs. Gore" (1 of 2); YouTube video, August 29, 2015, https://www.youtube.com/watch?v=Ucdj8qHqqdQ.

2 "a roadblock": "Presidential Race to Be Decided by Florida Results," CNN transcript, November 7, 2000.

2 "The networks called": "CBS News Election Night (7:00 PM ht ET)," CBS News transcript, November 7, 2000.

2 "if we say": Peter Marks, "A Flawed Call Adds to High Drama," *New York Times,* November 8, 2000.

2 "It gets a lot": "Election Night 2000: Bush vs. Gore" (1 of 2); YouTube video, August 29, 2015, https://www.youtube.com/watch?v=Ucdj8qHqqdQ.

2 "CNN right now": "Presidential Race to Be Decided by Florida Results," CNN transcript, November 7, 2000.

3 "What the networks": "Election Night 2000: Bush vs. Gore" (1 of 2); YouTube video, August 29, 2015, https://www.youtube.com/watch?v=Ucdj8qHqqdQ.

3 "Something to report": "Presidential Election Still in Doubt," CNN transcript, November 8, 2000.

4 "we're officially saying": "Election Night 2000 Bush vs. Gore" (2 of 2)," YouTube video, August 29, 2015, https://www.youtube.com/watch?v=Sn06pE1Ryek.

4 "This is astonishing": "ABC 2000: The Vote, 4 A.M.", ABC News transcript, November 8, 2000.

5 "just the right": Kurt Andersen, "The Best Decade Ever? The 1990s, Obviously," *New York Times,* February 6, 2015.

ONE

11 "His eloquence": Bill Simmons, Associated Press, December 5, 1978.

11 Georgetown University wanted: Henry Mitchell, "Any Day: Mocking-birds' Songs and Rites of Passage," *Washington Post*, May 30, 1980.

11 "He may be president": Ibid.

13 Her husband: Bill Simmons, Associated Press, December 5, 1978.

14 "I made": Dan Balz. "Wunderkind Seeks 2nd Chance," *Washington Post*, May 23, 1982.

14 Checking in: James Dickenson and David Broder, "Changing All Those Changes," *Washington Post*, February 28, 1982.

14 Clinton's appearance: David Broder. "Zigzagging in Search of Identity; The Democratic Convention," *Washington Post*, July 15, 1984.

15 "If Truman were": "Give 'em hell, Demos," United Press International, July 16, 1984.

17 Then, the opening: "Mario Cuomo 1984 Democratic National Convention Keynote Speech," C-Span Video Library, July 16, 1984, https://www.c-span.org/video/?323534-1/mario-cuomo-1984-democratic-national-convention-keynote-speech.

19 "New York Gov. Mario M. Cuomo": Haynes Johnson, "Cuomo Won Convention's Heart and Nearly Changed Its Mind; Convention Journal," *Washington Post*, July 18, 1984.

19 Gushed one delegate: Lou Cannon and Helen Dewar, "Gov. Cuomo Rouses Dispirited Delegates," *Washington Post*, July 17, 1984.

19 Said another: Ibid.

19 And a third: Haynes Johnson, "Cuomo Won Convention's Heart and Nearly Changed Its Mind," *Washington Post*, July 18, 1984.

19 We've got to: Lou Cannon and Helen Dewar, "Gov. Cuomo Rouses Dispirited Delegates," *Washington Post*, July 17, 1984.

19 Those watching: Tom Shales, "For Openers, Cuomo's Roof-Raiser Meets Agenda Gap," *Washington Post*, July 17, 1984.

21 "last liberal hope": Mary McGrory, "Democrats' Last Liberal Hope," *Washington Post*, November 11, 1984.

TWO

27 He was twenty-seven: Elizabeth Williamson, "Newt Gingrich's College Records Show a Professor Hatching Big Plans," *Wall Street Journal*, January 18, 2012.

28 This stuff's: Michael Barone, "Who Is This Newt Gingrich," *Washington Post*, August 26, 1984.

31 "A revolt": Art Pine, "Revolt Against Taxes . . . and Performance," *Washington Post*, June 11, 1978.

32 Ads of mysterious origin: Dale Russakoff and Dan Balz, "After Polit-
    ical Victory, a Personal Revolution," *Washington Post*, December 19,
    1994.

34 "Mr. Nice Guy": Margot Hornblower and Richard L. Lyons, "House
    GOP Picks Michel As Leader," *Washington Post*, December 9, 1980.

34 "He doesn't have": Associated Press, "Hard-Nosed 'Nice Guy' Wins
    Post," Henderson (NC) *Daily Dispatch*, December 9, 1980.

36 He and: Mary McGrory, "Reagan's Pals, the Democrats," *Washington
    Post*, May 20, 1984.

37 They took turns: T. R. Reid, "Outburst," *Washington Post*, May 16, 1984.

37 The operator complied: Peter Carlson, "Is Bob Walker the Most Obnox-
    ious Man in Congress," *Washington Post*, September 7, 1986.

38 Presiding over: "The Speaker's Words Are Ordered Taken Down," C-Span
    Video Library, May 15, 1984, https://www.c-span.org/video/?93662-1
    /speakers-words-ordered-taken.

40 "What a great show": T. R. Reid, "Tip's Greatest Hits," *St. Petersburg
    Times*, May 30, 1984.

40 Lott acknowledged: "O'Neill Assails a Republican and Is Rebuked by
    the Chair," *New York Times*, May 16, 1984.

41 "I am just": T. R. Reid, "Its 'Tip's Greatest Hits,' Electrifying a Closed
    House GOP Circuit," *Washington Post*, May 29, 1984.

41 "Veteran Republicans": Steven V. Roberts, "Congress; New Conflict a
    Threat to Old Ways," *New York Times*, May 19, 1984.

42 "They are trying": Steven V. Roberts, "The Indiana Imbroglio," *New York
    Times*, January 3, 1985.

42 "On any point": Margaret Scherf, "Panel Splits 2–1 On Ballot-Count-
    ing Rules," Associated Press, March 11, 1985.

43 "What happened": Dan Balz, "McCloskey's Victory Lights Partisan
    Fury; House Republicans Seek New Vote in Indiana," *Washington Post*,
    April 20, 1985.

43 "I think": Dan Balz, "House GOP Sets Protest on Seating; Disputed
    Election in Indiana District Prompts Disruptions," *Washington Post*,
    April 23, 1985.

43 "Would the gentleman": "House seats Indiana Democrat; GOP walks,"
    Arizona Republic, May 2, 1985.

43 Michel called it: Associated Press, "GOP Plans Boycott in Indiana Dis-
    pute," *Chicago Tribune*, May 2, 1985.

44 "I try": Steven V. Roberts, "Working Profile: Representative Robert
    H. Michel; Forging Alliances to Get Minority's Plans Passed," *New
    York Times*, May 13, 1985.

44 "We trust": Ibid.

44 "I think you're": Associated Press, "GOP Plans Boycott in Indiana Dispute," *Chicago Tribune*, May 2, 1985.

THREE

46 "The perception is": Dan Balz, "Southern and Western Democrats Launch New Leadership Council," *Washington Post*, May 1, 1985.
46 "People are saying": William E. Schmidt, "Teachers Up in Arms Over Arkansas's Skills Test," *New York Times*, January 17, 1984.
46 He'd praise integration: Bill Simmons, Associated Press, December 5, 1978.
47 Robb, observed David Broder: David S. Broder, "Chuck Robb's 'Instant Credibility,'" *Washington Post*, October 26, 1986.
47 "When he strode": Phil Gailey, "Sam Nunn's Rising Star," *New York Times*, January 4, 1987.
48 "It would mean": Paul Taylor, "Bumpers Decides Against Presidential Bid; Arkansas Senator's Surprise Announcement Says Campaign Would Disrupt," *Washington Post*, March 21, 1987.
49 At his announcement: E. J. Dionne, Jr., "Gore, Entering 1988 Race, Projects 'New South' Image," *New York Times*, June 30, 1987.
49 At a gathering: Thomas B. Edsall, "Conference Showcases Democrats' Dilemma," *Washington Post*, June 18, 1987.
51 She had a list: David Maraniss, "On Brink of Running, Clinton Called It Off," *Washington Post*, February 7, 1995.
51 The *Arkansas Democrat-Gazette*: Bob Wells, *Arkansas Democrat-Gazette*, July 15, 1987.
51 "Frankly," he said: Bob Wells, "Clinton: Time Will Come," *Arkansas Democrat-Gazette*, July 16, 1987.
51 "I want": Associated Press, "White House Remains Part of Clinton's Plan," *New York Times*, July 16, 1987.
51 "I know that": Bob Wells, *Arkansas Democrat-Gazette*, July 15, 1987.
52 There was, Clinton insisted: Bob Wells, "Clinton: Time Will Come," *Arkansas Democrat-Gazette*, July 16, 1987.
52 "For what it's worth": Associated Press, "White House Remains Part of Clinton's Plan," *New York Times*, July 16, 1987.
52 "Words and phrases": Jeffrey Schmalz, "Cuomo, in a Visit to South, Offers First Statement on Issues for '88," *New York Times*, February 17, 1987.
53 "In my opinion": Paul Taylor, "Cuomo Says He Won't Run for President; N.Y. Governor Surprises Aides," *Washington Post*, February 20, 1987.
53 "Cheering could be heard": Marc Humbert, "Cuomo Bows Out of Possible 1988 Presidential Race," Associated Press, February 20, 1987.
53 "It's silly": Ibid.

53 "No, no, no": Jeffrey Schmalz, "11 Weeks Later, Cuomo Still Says He Won't Run for President," *New York Times,* May 10, 1987.

54 "The Democratic Party": Lois Romano, "Jesse Jackson: His Charismatic Crusade for the Voters at the End of the Rainbow Coalition," *Washington Post,* July 31, 1983.

55 A poll asked: Jay Mathews, "Jackson Gains Top Ranking as Spokesman for Blacks; Poll Sought to Assess Farrakhan Influence," *Washington Post,* December 21, 1985.

55 "I respect him": Milton Coleman, "Jackson Rebukes Hecklers; It's Embarrassing, He Tells Delegates," *Washington Post,* July 19, 1984.

56 He defended: Rick Atkinson, "Peace with American Jews Eludes Jackson," *Washington Post,* February 13, 1984.

56 "If you harm": Dorothy Gilliam, "Post-Furor," *Washington Post,* March 1, 1984.

56 "Jews went": Fay S. Joyce, "Jackson Outlines His Views to Jewish Group," *New York Times,* March 5, 1984.

56 "In private conversations": Rick Atkinson, "Peace with American Jews Eludes Jackson," *Washington Post,* February 13, 1984.

57 "'Hymie,'" explained: Howell Raines, "Jackson's Candor Is Praised but Remark Is Criticized," *New York Times,* February 28, 1984.

57 "I was shocked": James R. Dickenson and Kathy Sawyer, "Jackson Admits to Ethnic Slur; Candidate Makes Conciliatory Stop at Synagogue," *Washington Post,* February 27, 1984.

57 "However innocent": Ibid.

57 "You ain't got": Juan Williams, "Jackson Urges His Supporters to Unite Behind Ticket," *Washington Post,* July 19, 1984.

57 "Jesse Jackson is destroying": Haynes Johnson, "After Cheers for Jackson, Fear of White Backlash Lingers," *Washington Post,* July 19, 1984.

58 The working-class whites: Thomas B. Edsall, "Home of Reagan Democrats Holds Trouble for Bush; Voters Voice Ire in Macomb County, Mich.," *Washington Post,* March 16, 1992.

58 "The overwhelming difference": Susan Taylor Martin, "'Joy cometh' to Jackson's campaign," *St. Petersburg Times,* March 9, 1988.

59 The *New York Times:* R. W. Apple Jr., "Super Tuesday Offers a Muddled Experiment," *New York Times,* March 8, 1988.

59 "Things are not": Ibid.

59 "My message": Paul Taylor, "Jackson's Winning Ways Transform Candidate, Voters," *Washington Post,* March 15, 1988.

61 "For Ed Koch": Joyce Purnik, "Koch Defends Attack on Jackson, Saying He Evades Issue of Israel," *New York Times,* April 9, 1988.

61 Koch fired back: Gene Marlowe, "Koch Can't See Gore as No. 2," *Tampa Tribune,* April 27, 1988.

61 "What do we": Mitchell Locin, "Jackson Hits Atlanta Talking Tough," *Chicago Tribune*, July 17, 1988.

62 "[Cuomo's] mere presence": Gregory Wood, "Non-Candidate Cuomo's Still the Charismatic Draft Dodger," *Australian Financial Review*, April 15, 1988.

62 But this time: John King, "Clinton Will Nominate Dukakis," Associated Press, July 2, 1988.

63 "You're listening": "Bill Clinton Booed at Convention 1988," YouTube video, June 2, 2001, https://www.youtube.com/watch?v=J5FpRg3Tf9Y.

63 "It seems": Ibid.

63 "It was either": Maria Henson, "Clinton Speaks While Few in Atlanta or in the TV Audience Listen," *Arkansas Democrat-Gazette*, July 22, 1988.

64 "It could have": Editorial, *Arkansas Democrat-Gazette*, July 22, 1988

FOUR

65 "Physically, he is": Lee Byrd, "'How Long You Gonna Stay Around Here? I Figure It's Time to Get Out.' After 34 Years in Congress, O'Neill to Hang Up His Hat," *Los Angeles Times*, October 19, 1986.

66 "The Speaker kept": Mary McGrory, "Despite Appearances, Tip Won," *Washington Post*, December 4, 1984.

66 The *New York Times Magazine:* John M. Barry, "The Man of The House," *New York Times Magazine*, November 23, 1986.

66 "I love": Patricia O'Brien, "A Bruised but Proud O'Neill Is Leaving a Changed House," *Miami Herald*, October 5, 1986.

66 "I think he": John M. Barry, "The Man of the House," *New York Times Magazine*, November 23, 1986.

67 "He'll discipline us": Ibid.

67 "Stripped of their": Eric Pianin, "House GOP's Frustrations Intensify," *Washington Post*, December 21, 1987.

67 "There was": Charles R. Babcock, "Speaker's Royalty: 55 Percent; Friend Published Wright's Book," *Washington Post*, September 24, 1987.

68 In November: Tom Fiedler, "Gingrich Calls Wright 'Corrupt,'" *Miami Herald*, November 17, 1987.

68 "Wright is so": Tom Kenworthy, "Democrats Give New Speaker Taste of Medicine He Prescribed for Wright," *Washington Post*, January 20, 1995.

68 By December: Susan F. Rasky, "Washington Talk; The Speaker of the House; Everyone Has Something to Say About Wright," *New York Times*, December 18, 1987.

68 "It's totally without": Ibid.

68 Fred Wertheimer: Tom Kenworthy, "Group Urges Ethics Probe of Speaker Wright; Common Cause Action Could Prove Election-Year Embarrassment for Democrats," *Washington Post*, May 19, 1988.

69 The Speaker would: Tom Kenworthy, "House Ethics Committee to Investigate Wright; Charging Politics, Speaker Predicts He'll Be Exonerated," *Washington Post*, June 11, 1988.

69 Asked his personal: Tom Kenworthy, "72 Republicans Ask Panel to Probe Wright's Finances," *Washington Post*, May 27, 1988.

69 "I was just": Tom Kenworthy, "Evolving Explanations on Wright's Book," *Washington Post*, April 10, 1989.

70 The *New York Times* wrote: Susan F. Rasky, "Washington Talk: Working Profile: Representative Newt Gingrich; From Political Guerrilla to Republican Folk Hero," *New York Times*, June 15, 1988.

70 "I'm so deeply": Ibid.

70 "As a nineteen-year-old": Ken Ringle, "Memory and Anger: A Victim's Story; Watching as the Man Who Tried to Kill Her Rose to Power," *Washington Post*, May 4, 1989.

70 Feebly, Wright's fellow: Jack Nelson and Ronald J. Ostrow, "Bradley, Coelho Targets of U.S. Criminal Probe," *Los Angeles Times*, May 25, 1989.

70 "Mr. Speaker": "Turmoil in Congress; Transcript of Wright's Address to House of Representatives," *New York Times*, June 1, 1989.

72 "There's an evil": Ibid.

72 Wright's son: "Gingrich Should Cool It, GOP Minority Leader Says Whip Urged to Be More Responsible," *Los Angeles Times*, June 1, 1989.

72 From retirement: Tom Kenworthy, "Wright to Resign Speaker's Post, House Seat; Texan Again Rebuts Charges, Decries Conflict over Ethics," *Washington Post*, June 1, 1989.

72 "I think most": Steve Daley and Elaine S. Povich, "House Speaker Wright Resigns; Majority Leader Foley Seen As Likely Successor for Job," *Chicago Tribune*, June 1, 1989.

73 "Jim Wright was": Don Phillips, "Gingrich Defends Part in Wright's Fall; Evidence of Improprieties Was Overwhelming, GOP Whip Says," *Washington Post*, June 3, 1989.

73 "A professional pest": Steven V. Roberts, "One Conservative Faults Two Parties," *New York Times*, August 11, 1983.

75 "I'm a single man": "Tower Excerpts: 'I'm a Single Man; I Do Date Women'," *Los Angeles Times*, February 27, 1989.

76 Bush called Cheney: "Bush Chooses Cheney for Defense; Wyoming Congressman Picked After Rejection of Tower," *Los Angeles Times*, March 12, 1989.

76 "Everyone was stunned": Myron S. Waldman, "Republicans Jockeying for No. 2 Spot in House," *Newsday*, March 18, 1989.

76 "I'm a national": Robin Toner, "Reporter's Notebook: Race for Whip: Hyperspeed vs. Slow Motion," *New York Times*, March 22, 1989.

77 In temperament: Don Phillips, "Choice of Cheney Sparks Struggle over House GOP's Course," *Washington Post*, March 12, 1989.

77 William F. Buckley: Don Phillips, "Reps. Madigan, Gingrich Vie for GOP Post," *Washington Post*, March 16, 1989.

77 William Safire used: William Safire, "Essay: Not Mad, but Even," *New York Times*, March 13, 1989.

77 "This is not": Ibid.

77 "We are chafing": Robin Toner, "G.O.P. Focuses on Cheney Succession," *New York Times*, March 14, 1989.

78 "He came up": Steve Daley, "GOP Elects Maverick House Whip," *Chicago Tribune*, March 23, 1989.

78 "There is a fear": Myron Waldman, "GOP's New House Whip," *Newsday*, March 23, 1989.

78 He joined Michel: Michael Kranish, "Conservative Gingrich Elected House GOP Whip," *Boston Globe*, March 23, 1989.

78 "I've been one": Robin Toner, "House Republicans Elect Gingrich of Georgia as Whip," *New York Times*, March 23, 1989.

79 "I think you": Associated Press, "House GOP Elects Militant Minority Whip by 2 Votes," *Los Angeles Times*, March 23, 1989.

79 The *Los Angeles Times*': Ibid.

79 The *Washington Post*: "Gingrich Elected": Don Phillips and Tom Kenworthy, "Gingrich Elected House GOP Whip; Increased Partisan Polarization Foreseen," *Washington Post*, March 23, 1989.

79 The *New York Times*: "Aggressive Republicans Choose House Whip," *New York Times*, March 23, 1989.

79 "Suffering through": Ross Baker, "Gingrich As Whip Shows GOP Desperation," Los Angeles Times, March 31, 1989.

79 While making sure: Michael Kranish, "Conservative Gingrich Elected House GOP Whip," *Boston Globe*, March 23, 1989.

FIVE

80 "By God": "Remarks to the American Legislative Exchange Council," American Presidency Project, http://www.presidency.ucsb.edu/ws/?pid=19351.

81 "The view": Robin Toner, "Defying the Wisdom for '92, Ex-Senator Tsongas Dives In," *New York Times*, April 10, 1991.

81 The chairman: Richard Benedetto, "GOP Chief: War Stance to Be Issue," *USA Today*, April 18, 1991.

82 The *Washington Post* declared: Lois Romano, "All Quiet on the Demo-

cratic Front; Political Consultants, Stuck in the Starting Gate," *Washington Post*, March 4, 1991.

82 Since "the odds": David S. Broder, "All Eyes on '96," *Washington Post*, April 17, 1991.

83 "I have no plans": Sam Roberts, "Tick, Tick, Tick: Cuomo Can Afford to Wait on 1992," *New York Times*, June 10, 1991.

84 Hillary, he said: "Clinton sees good governor in his wife," *Arkansas Democrat-Gazette*, August 19, 1989.

84 Days before: Maralee Schwartz, "Gov. Clinton to Seek Reelection, Despite Admitted Lack of 'Fire,'" *Washington Post*, March 4, 1990.

84 "The joy": David Merriweather, "Clinton to Run Again," *Arkansas Democrat-Gazette*, March 2, 1990.

84 "I may have": Paul Taylor, "Democrats Rethinking Bush's Vulnerabilities in 1992," *Washington Post*, October 26, 1990.

85 "While we favor": Robin Toner, "Eyes to Left, Democrats Edge Toward the Center," *New York Times*, March 25, 1990.

86 For Clinton: Ellen Debenport, "Democrats' Message Aims for the Mainstream," *St. Petersburg Times*, March 25, 1990.

86 "I don't know": Tom Baxter and A. L. May, "Democratic Factions Battling for Soul of Party; Liberals React to Rise of Moderate DLC," *Atlanta Journal and Constitution*, May 5, 1991.

87 If the election: Dan Balz, "Clinton Forms Committee to Weigh Presidential Bid; Arkansas Democrat to Decide Next Month," *Washington Post*, August 16, 1991.

87 "Just what we need": Mary McGrory, "Hey, Iowa! They're Off and Running," *Washington Post*, March 24, 1991.

88 "There is no": Robert Reinhold, "Ex-California Governor to Seek the Presidency," *New York Times*, September 4, 1991.

89 Clinton and: Adam Nagourney, "Democrats in a Political Pickle," *USA Today*, August 8, 1991.

90 Harkin played up: Steve Daley, "Democratic Hopefuls Air Campaign Themes," *Chicago Tribune*, November 24, 1991.

90 "I believe": Robert Shogan and Cathleen Decker, "Clinton, Wilder Focus on Economic, Race Issues," *Los Angeles Times*, November 24, 1991.

91 "It was rather startling": Steve Daley, "Clinton Steals Show at Rorum," *Chicago Tribune*, November 24, 1991.

91 Now he reconsidered: Dan Balz, "Party Leaders Laud Clinton Performance; Arkansas Gives His 1992 Bid Boost Among State Chairmen," *Washington Post*, November 24, 1991.

91 To the Texas chairman: Steve Daley, "Clinton Steals Show at Forum," *Chicago Tribune*, November 24, 1991.

91 With a Chicago: Dan Balz, "Party Leaders Laud Clinton Performance;

Arkansas Gives His 1992 Bid Boost Among State Chairmen," *Washington Post*, November 24, 1991.

91 Clinton, the *Chicago Tribune:* Steve Daley, "Clinton Steals Show at Forum," *Chicago Tribune*, November 24, 1991.

92 "They said": Elizabeth Kolbert, "Political Memo: Cuomo Utters a 'Maybe' but Hastens to Add 'Not,'" *New York Times*, October 12, 1991.

92 "I cannot get": Sam Roberts, "An Eloquent Man and His Musings on the Presidency," *New York Times*, October 14, 1991.

93 "I don't see": Sam Howe Verhovek, "Political Memo: California's Democrats Take Cuomo at Word," *New York Times*, September 14, 1991.

94 "What makes you": Paul Tash, "Indecisive Cuomo Angering Democrats," *St. Petersburg Times*, November 16, 1991.

94 Cuomo had been: Susan Page, "He'd Be the One to Beat," *Newsday*, October 12, 1991.

94 The *Boston Globe:* Michael K. Frisby, "Democrats Say Cuomo Is on the Verge of Declaring," *Boston Globe*, October 18, 1991.

94 "It's difficult": Dan Balz and Maralee Schwartz, "Cuomo Says He Would Have to Quit as Governor to Run for Presidency," *Washington Post*, October 18, 1991.

95 "Let's say": Dan Balz and Maralee Schwartz, "Cuomo Says He Would Have to Quit as Governor to Run for Presidency," *Washington Post*, October 18, 1991.

96 Cuomo now told: Kevin Sack, "And Cuomo Chooses: (Budget? Duty? Ego? Gore?)," *New York Times*, December 10, 1991.

96 "a powerful spokesman": Kevin Sack, "The 1992 Campaign: New York; Despite Their Differences, Cuomo Gives Clinton Praise," *New York Times*, April 5, 1992.

96 "were able": Curtis Wilkie, "Clinton Takes Aim at Cuomo," *Boston Globe*, December 6, 1991.

97 If this was: Nicholas Goldberg, "Cuomo Waits on Fierce Budget Talk," *Newsday*, December 19, 1991.

97 "The more": John Riley, "Amid Fiscal Mess, Cuomo Gets N.H. Forms," *Newsday*, December 17, 1991.

97 "I always thought": Curtis Wilkie and John Milne, "In N.H., Clinton urges faster use of road fund," *Boston Globe*, December 19, 1991.

98 "The evidence": Dan Balz, "Cuomo Keeps Reporters, Staff, Electorate on Tenterhooks on Eve of Deadline," *Washington Post*, December 20, 1991.

98 Early Friday: Paul Leavitt, "Cuomo May End Speculation Today," *USA Today*, December 20, 1991.

99 "My responsibility": Morgan Lyle, "Cuomo Not Running for President," United Press International, December 21, 1991.

SIX

102 "The prospect": David Hoffman, "President Is Pushed by Deficit," *Washington Post*, May 8, 1990.

103 "Sometimes," Bush said: Myron S. Waldman, "Bush, Leaders Agree on Plan to Cut Spending, Add Taxes," *Newsday*, October 1, 1990.

104 "The naysayers": Editorial, "A Need for Political Courage . . . ," *Chicago Tribune*, October 2, 1990.

106 "Still looking": Richard L. Berke, "Shouts of Revolt Rise Up in Congressional Ranks," *New York Times*, October 1, 1990.

106 "I feel like": William J. Eaton and David Lauter, "Budget Deal Faces Revolt in House," *Los Angeles Times*, October 2, 1990.

106 "exercise leadership": Ibid.

106 "I think": Richard L. Berke, "The Budget Agreement: The Opposition; Rebellion Flares Among Republicans over Accord," *New York Times*, October 2, 1990.

107 "It is my conclusion": Ibid.

107 Chuck Douglas: Helen Dewar and Tom Kenworthy, "Conservative Republicans Assail Budget Pact," *Washington Post*, October 1, 1990.

107 To Indiana's: Richard L. Berke, "The Budget Agreement: The Opposition; Rebellion Flares Among Republicans over Accord," *New York Times*, October 2, 1990.

107 Dick Armey: Ibid.

107 "Newt Gingrich": William J. Eaton and Paul Houston, "Gingrich? Few Remain Indifferent," *Los Angeles Times*, October 3, 1990.

108 He insisted: Jack Sirica and Myron S. Waldman, "Bush Begins Hard Sell but GOP Not Buying," *Newsday*, October 2, 1990.

108 As for the idea: Mitchell Lochin and Elaine S. Povich, "Budget Deal Faces GOP Revolt," *Chicago Tribune*, October 2, 1990.

108 The compromise: Richard L. Berke, "Shouts of Revolt Rise Up in Congressional Ranks," *New York Times*, October 1, 1990.

108 Charles Schumer: Helen Dewar and Tom Kenworthy, "Conservative Republicans Assail Budget Pact; Democrats Skeptical," *Washington Post*, October 1, 1990.

108 "Obviously it is not": "Attack Puts Budget Deal Under Cloud," *St. Louis Post-Dispatch*, October 2, 1990.

108 Virginia Senator John Warner: Ibid.

108 "They should take": "Congress Blind to the Real Monster," editorial, *Chicago Tribune*, October 4, 1990.

109 "He has his reasons": Richard L. Berke, "Shouts of Revolt Rise Up in Congressional Ranks," *New York Times*, October 1, 1990.

109 "If we fail": Dan Balz and John E. Yang, "Bush Makes Appeal for

Budget Support; Voters Urged to Spare Plan's Backers," *Washington Post*, October 3, 1990.

109 "We hope": "The Budget Agreement: Democratic Leader's Response to President's Budget Address," *New York Times*, October 3, 1990.

109 This was just more: Jack Nelson, "President Was Warned to Rally the Public to Deficit Package or Risk Humiliation," *Los Angeles Times*, October 3, 1990.

109 Jerry Lewis: William J. Eaton and Paul Houston, "Gingrich? Few Remain Indifferent," *Los Angeles Times*, October 3, 1990.

110 "We're in prison": Ibid.

110 "The president's political": Richard Wolf, "Bush, Congressional Leaders Lobby to the End," *USA Today*, October 5, 1990.

110 Bush called: Reuters, "Bush Lobbies Congress to Pass Deficit Package," *Los Angeles Times*, October 3, 1990.

110 Gephardt, the House Majority Leader: Jack Nelson, "President Was Warned to Rally the Public to Deficit Package or Risk Humiliation," *Los Angeles Times*, October 3, 1990.

110 "You know": R. A. Zaldivar, "11th Hour Battle on Budget Intensifies," *Miami Herald*, October 4, 1990.

110 "Mr. Gingrich's preference": Richard L. Berke, "Gingrich, in Duel with White House, Stays True to His Role As an Outsider," *New York Times*, October 5, 1990.

111 "Every committee chairman": Richard Wolf, "Bush, Congressional Leaders Lobby to the End," *USA Today*, October 5, 1990.

112 "In past days": David Broder, "Loyalty, Ideology, Ambition—and Newt Gingrich," *Washington Post*, October 7, 1990.

SEVEN

113 Clinton, the *New York Times:* Robin Toner, "If Clinton Leads Pack, Teeth Are at His Heels," *New York Times*, January 14, 1992.

114 Most of the big boys: Howard Kurtz, "Clinton Denies Affairs," *Washington Post*, January 18, 1992.

114 "The Star": Ibid.

115 "We have been": Dan Balz and E. J. Dionne, Jr., "Clinton: Groundwork and a 'Lot of Breaks,'" *Washington Post*, January 12, 1992.

115 The *New York Post*'s: Howard Kurtz, "Clinton Denies Affairs," *Washington Post*, January 18, 1992.

115 It was embarrassing: Adam Nagourney, Richard Benedetto, David Lewis, "Clinton Forced to Deny Report of Affair—Again," *USA Today*, January 24, 1992.

116 In it, a voice: "Clinton Says New Tabloid Account of Affair Not True," Associated Press, January 24, 1992.

116 "I read the story": "Tabloid Prints New Bill Clinton Infidelity Allegations," *Nightline*, ABC News transcript, January 23, 1992.

116 The *Star* acknowledged: Alex S. Jones, "Meet the Press: Dr. Jekyll and Mr. Hyde," *New York Times*, February 9, 1992.

116 "She's obviously taken": "Clinton Says New Tabloid Account of Affair Not True," Associated Press, January 24, 1992.

117 But *Nightline*: "Tabloid Prints New Bill Clinton Infidelity Allegations," *Nightline*, ABC News transcript, January 23, 1992.

117 A *Boston Globe* editorial: Editorial, "The Star Search," *Boston Globe*, January 29, 1992.

118 Then, Pat Summerall signed off: "Bill Hillary Clinton Extramarital Affairs 1992," YouTube video, September 5, 2016, https://www.youtube.com/watch?v=uyumcdeZ4uY.

121 The *Los Angeles Times:* David Lauder and Ronald Brownstein, "Clinton Says Sex Rumors Are Irrelevant," *Los Angeles Times*, January 27, 1992.

121 "A lot of you": Curtis Wilkie, "Campaign's Fate Put in Public's Hands," Boston Globe, January 27, 1992.

121 "The man": Timothy Clifford, "Flowers: Clinton's Lying," *Newsday*, January 28, 1992.

122 Another caught: Jimmy Breslin, "Gennifer, Tammy and a Poor Dope," *Newsday*, January 28, 1992.

122 Melendez had: David Colton, "Was That a News Event, or Just an Illusion?" *USA Today*, January 31, 1992.

122 A voice: Nicholas Goldberg, "Fightin' Words," *Newsday*, January 29, 1992.

122 "I don't particularly care": Ibid.

123 The conversation: Gwen Ifill, "Clinton, Cheered by New Polls, Again Assails Bush on Economy," *New York Times*, January 29, 1992.

123 "What do you mean": Marc Humbert, "Clinton Apologizes to Cuomo for Taped Comments," Associated Press, January 29, 1992.

123 He called: Nicholas Goldberg, "Cuomo Rips Ethnic Slur by Clinton," *Newsday*, January 29, 1992.

123 Through the press: Marc Humbert, "Clinton Apologizes to Cuomo for Taped Comments," Associated Press, January 29, 1992.

124 Wynette, still a bankable star: Dennis Duggan, "Tammy: Let's Go Toe to Toe," *Newsday*, January 28, 1992.

125 She'd been: Jimmy Breslin, "Gennifer, Tammy and a Poor Dope," *Newsday*, January 28, 1992.

125 In her letter: Dennis Duggan, "Tammy: Let's Go Toe to Toe," *Newsday*, January 28, 1992.

125 No other women: Adam Pertman, "Clinton, in the South, Tries to Shift Focus Back to Platform," *Boston Globe*, January 28, 1992.

127 "To many": Associated Press, "A Letter by Clinton on His Draft Deferment: 'A War I Opposed and Despised,'" *New York Times*, February 12, 1992.

127 "Bill Clinton was able": Curtis Wilkie and Walter V. Robinson, "Clinton Letter Says He Sought ROTC Deferment to Avoid War," *Boston Globe*, February 13, 1992.

127 "No one really": R. W. Apple Jr., "The 60's; Clinton and Draft Issue: Vietnam Era Revisited," *New York Times*, February 14, 1992.

128 "You've got to": Gwen Ifill, "Vietnam War Draft Status Becomes Issue for Clinton," *New York Times*, February 7, 1992.

128 "Many of you": "A Conversation with Governor Bill Clinton," *Nightline*, ABC News, February 12, 1992.

129 Vermont Governor Howard Dean: R. W. Apple Jr., "The 60's; Clinton and Draft Issue: Vietnam Era Revisited," *New York Times*, February 14, 1992.

129 "I've got some": Susan Yoachum, "Kerrey Seems to Puzzle Voters," *San Francisco Chronicle*, February 14, 1992.

130 "I've faced death": Colin MacKenzie, "Tsongas Basks in Boomlet in Polls," *Toronto Globe and Mail*, February 11, 1992.

130 He offered himself: Karen De Witt, "The 1992 Campaign: The Victor; Tsongas Makes Bid to Widen Support," *New York Times*, February 20, 1992.

130 "Why I am here": Charles M. Madigan, "Tsongas and the Message of Survival," *Chicago Tribune*, February 14, 1992.

130 He even took: John King, "Tsongas Calls Clinton 'Pander Bear'," Associated Press, March 6, 1992.

131 "Nobody trusts us": "Address by Democratic Presidential Candidate Former Senator Paul Tsongas of Massachusetts to the United Autoworkers Convention," Federal News Service, February 3, 1992.

131 "The problem": Colin MacKenzie, "Tsongas Basks in Boomlet in Polls," *Toronto Globe and Mail*, February 11, 1992.

131 If they had to: "Tsongas Receives Praise from GOP—Sort Of," United Press International, February 18, 1992.

131 Tom Selleck: David Willman, "A Tough Sales Job for GOP in Hollywood," *Los Angeles Times*, March 9, 1992.

131 "I will tell": Adam Pertman, "On Birthday, Tsongas Has Thanks for Life," *Boston Globe*, February 15, 1992.

132 "He came here": Gary Ghiotto, "Clinton Asks for Second Chance," United Press International, February 13, 1992.

133 He kissed: Robin Toner, "Bush Jarred in First Primary; Tsongas Wins Democratic Vote," *New York Times*, February 19, 1992.

133 "Well, New Hampshire": David Von Drehle, "Tsongas Rides High with Hard Truth," *Washington Post*, February 19, 1992.

133 He was now: Robin Toner, "Bush Jarred in First Primary; Tsongas Wins Democratic Vote," *New York Times*, February 19, 1992.

134 "I'm not running": Paul Tsongas, campaign ad, 1992.

134 "No one can argue": Wire reports, "Democrats Come Out Slinging in Denver," *Chicago Tribune*, March 1, 1992.

134 Now, he was: James Kim and Linda Dono Reeves, "Does the Stock Market Care Who Is President?," *USA Today*, June 4, 1992.

134 Tsongas hated: Richard L. Berke, "The 1992 Campaign: Political Memo; Tsongas Being Haunted by Memory of Dukakis," *New York Times*, March 15, 1992.

135 "We cannot put off": David S. Broder and David Von Drehle, "2 Democrats Intensify Verbal Battle; Clinton Lays Down Attack on Tsongas," *Washington Post*, March 8, 1992.

135 The Tsongas plan: Elizabeth Kolbert, "The 1992 Campaign: Media; G.O.P. Air War Reflects Competitiveness of Georgia Contest," *New York Times*, February 28, 1992.

135 Tsongas accused him: Richard L. Berke, "Saying Clinton Is Cynical, Tsongas Goes on the Attack," *New York Times*, March 7, 1992.

135 "I must say": Dan Balz and Thomas B. Edsall, "Clinton and Bush Swamp Opponents in Sweeps of the Southern Primaries; Delegate Surge for Arkansan," *Washington Post*, March 11, 1992.

136 The president: B. Drummond Ayres, Jr., "The 1992 Campaign: Candidate's Record; Unions Are Split on Backing Clinton," *New York Times*, March 15, 1992.

137 "Every day": Marc Gunther, "Carson and Company Skewer the Candidates," *Atlanta Journal and Constitution*, February 28, 1992.

137 "I guess": Robert Samuelson, "The Recklessness of Warring in Kosovo," *Chicago Tribune*, June 4, 1999.

138 Polls showed: "CBS/N.Y. Times: Deep Dissatisfaction with All Candidates," *The Hotline*, April 1, 1992.

138 "Just when": David Nyhan, "The Arkansas Traveler Worries People," *Boston Globe*, March 15, 1992.

139 In the dying days: Robin Toner, "The 1992 Campaign: Assessment Democratic Free-for-All; Expectations of an Orderly Campaign Fade as Voters Choose One Man, Then Another," *New York Times*, February 27, 1992.

139 The result: Michael Frisby, "Resistance to Clinton May Invite Efforts to Dump His Wagon," *Pittsburgh Press*, March 25, 1992.

139 "The presumption": Martin Kasindorf, "Brown Boasts a 'Boost' from Visit to Cuomo," *Newsday*, March 26, 1992.

140 "If he does": Susan Yoachum, "What Demos Could Do If Clinton Stumbles," *San Francisco Chronicle*, March 26, 1992.

140 Since entering: George Skelton, "On the Road with Jerry—Next Stop: Madness," *Los Angeles Times*, April 13, 1992.

140 He called Clinton: George Skelton, "Brown Labels Clinton as Union-Busting Environmental Disaster for Arkansas Democrats," *Los Angeles Times*, March 26, 1992; and Robert Naylor, "Brown Denounces Questions About Evading $100 Limit," Associated Press, April 3, 1992.

140 "There are": Robert Shogan, "Rivals Seek Positive Note on New York Primary Eve," *Los Angeles Times*, April 7, 1992.

141 "I have never": Howard Kurtz, "Brown's Message Plays Despite Inconsistencies; Clinton Losing N.Y. Battle of Media Images," *Washington Post*, March 31, 1992.

141 "The answer": Wire report, "Clinton Admits to Marijuana Use in Past," United Press International, March 29, 1992.

141 With more than forty million: Linda Diebel, "Democrats' Frontrunner Looking More Like a Loser," *Toronto Star*, April 2, 1992.

142 A newspaper cartoon: "Tsongas Backers Still Campaigning," *USA Today*, April 1, 1992.

142 "A silver bullet": "'Cool' Candidate Clinton Blows His Top at Heckler," *USA Today*, March 27, 1992.

142 "I'll tell you": Gwen Ifill, "The 1992 Campaign: Democratic Primary: Brown Is Rebuked by Party Chairman for Harsh Attacks," *New York Times*, March 27, 1992.

143 "You insult": Lou Cannon, "Brown Belabored over Choice of Jackson," *Washington Post*, April 3, 1992.

143 Another activist: Maureen Dowd, "The 1992 Campaign: Brown; Candidate Is Tripped Up Over Alliance with Jackson," *New York Times*, April 3, 1992.

143 Now with Clinton: "Bill Clinton on Donahue," YouTube video, February 9, 2008, https://www.youtube.com/watch?v=bypWnBR47Rk.

144 "After a quarter-century": David Von Drehle, "Democratic Showdown in New York: Will the Party Leap from the Empire State?" *Washington Post*, April 5, 1992.

EIGHT

145 Newt Gingrich: Hedrick Smith, "Those Fractious Republicans," *New York Times*, October 25, 1987.

146 "By a single stroke": Martin Tolchin, "Conservatives First Recoil, Then Line Up Behind Bush," *New York Times*, July 18, 1980.

146 "Bush has had": Bill Thompson, "Better Keep an Eye on Your Right Flank, Mr. President!" *Miami Herald*, November 25, 1991.

147 Instead, when Bush: Steve Daley, "Bush's Tax Plan Haunts GOP Races," *Chicago Tribune*, October 25, 1990.

147 "We are in": Michael Oreskes, "G.O.P. Plagued with Democratic Problems," *New York Times*, December 15, 1990.

147 A group of disgruntled: Gwen Ifill, "Restless Conservatives Debate Future," *Washington Post*, December 24, 1990.

147 The Republicans: Martin Schram, "Some GOP Leaders Can't Handle Peace," *Newsday*, March 13, 1991.

149 "The greatest frustration": David Broder and Lou Cannon, "Libertarian Challenge?" *Washington Post*, September 17, 1991.

149 His father: James S. Kunen, "Patrick Buchanan," *People*, August 29, 1988.

149 They tried to arrest: Patrick J. Buchanan, *Right from the Beginning* (Little, Brown & Co, 1988).

150 More than half: Stephanie Mansfield, "Ringside with Braden & Buchanan," *Washington Post*, July 2, 1981.

150 *Confrontation* became: PR Newswire, June 4, 1982.

152 "We've already had": John Mashek, "Pat Buchanan Is Said to Weigh GOP Challenge," *Boston Globe*, November 15, 1990.

153 "My friends": Excerpts from Buchanan speech; Associated Press, December 10, 1991.

154 "I think God": Adam Pertman, "Buchanan Announces Presidential Candidacy," *Boston Globe*, December 11, 1991.

154 Buchanan, wrote George Will: George F. Will, "Buchanan Takes Aim," *Washington Post*, December 11, 1991.

154 "Take a look": Lally Weymouth, "Buchanan: Throwing a Hard Right at Bush," *Washington Post*, December 22, 1991.

154 "There are only": Jonathan Alter, "Is Pat Buchanan Anti-Semitic," *Newsweek*, December 23, 1991.

155 As Buchanan launched: Charles Trueheart, "Buchanan, on the Firing Line," *Washington Post*, December 12, 1991.

155 Buchanan protested: Dale W. Nelson, "Buchanan's Conservative Legacy Strong," *Jackson Sun*, January 26, 1992.

155 Michael Kinsley: E. J. Dionne, Jr., "Is Buchanan Courting Bias?" *Washington Post*, February 29, 1992.

155 Asked if he: A. L. May, "Gingrich Stumps New Hampshire for Votes for Bush," *Atlanta Journal and Constitution*, January 23, 1992.

155 "A lot of people": Chris Black, "Amazed by His Success, Buchanan Calls His N.H. Crusade Worthwhile," *Boston Globe*, February 16, 1992.

156 He still wouldn't: Susan Page, "New Hampshire Heats Up," *Newsday*, February 16, 1992.

156 "The Buchanan brigades": Ellen Warren and S. A. Paolantonio, "President

Gets Lukewarm Welcome, Attacks Buchanan's Scathing Ads," *Miami Herald*, February 16, 1992.

157 When the networks: "The '92 Vote: The New Hampshire Primary," ABC News, February 18, 1992.

157 On CBS: Update of New Hampshire Primary Results, CBS News, February 18, 1992.

157 "I think King George": Chris Black, "Bush Struggles Past Buchanan, Tsongas Outduels Clinton in N.H.," *Boston Globe*, February 19, 1992.

157 "Patrick J. Buchanan's": Ann Devroy and John E. Yang, "President's Partisans Shocked by Outcome," *Washington Post*, February 19, 1992.

157 Winning the Republican nomination: Susan Page, "Buchanan: Let's Debate," *Newsday*, February 20, 1992.

158 Moments later: Thomas Hardy and Timothy McNulty, "Chastened Bush Starts to 'Take On' Buchanan," *Chicago Tribune*, February 20, 1992.

158 "When Pat Buchanan": David Evans, "Bush Ad Becomes a Political Minefield for a Retired Marine," *Chicago Tribune*, March 13, 1992.

158 "George Bush": Steve Daley, "Stung by Bush Ad, Buchanan Gets Ferocious," *Chicago Tribune*, February 28, 1992.

159 He framed it: E. J. Dionne Jr., "Buchanan Slashes Bush for Civil Rights Action," *Washington Post*, February 21, 1992.

159 "Listen, if I": Ann Devroy, "Breaking Tax Pledge a Mistake, Bush Says," *Washington Post*, March 4, 1992.

159 "He doesn't consider": Andrew Rosenthal, "Bush Says Raising Taxes Was Biggest Blunder of His Presidency," *New York Times*, March 4, 1992.

159 "I think it": Ibid.

159 More damning: A. L. May, "Polls Show Trouble Brews for Bush, Clinton," *Atlanta Journal and Constitution*, March 4, 1992.

160 Overall, two-thirds: Ronald Brownstein, "Blacks, Blue-Collar Voters Aided Clinton, Poll Finds," *Los Angeles Times*, March 4, 1992.

160 But that was: Susan Page, "Bush Wins, Buchanan Gains," *Newsday*, March 4, 1992.

160 "We are going": Judy Wiessler, "Clinton Gets First Win; Tsongas Boosted; Bush Victories Are Bruising," *Houston Chronicle*, March 4, 1992.

160 He'd called Bush: Paul Tash, "Buchanan Now Predicts Bush Will Be Re-Elected in November," *St. Petersburg Times*, April 15, 1992.

161 If Bush didn't: Cragg Hines, "Buchanan Seeks Voice at GOP Convention," *Houston Chronicle*, April 15, 1992.

NINE

162 "President Perot?": *Newsweek*, June 15, 1992.

163 "I wouldn't bet": Steven Erlanger, "Nixon Tells Journalists He Wouldn't Bet Against Perot," *New York Times*, June 5, 1992.

163 Newt Gingrich fired: Ann Devroy, "Gingrich Warns Bush Campaign; Political Aides Are Told to Wake Up to Facts of a 'Unique' Year," *Washington Post*, June 7, 1992.

163 Desperate to siphon: Dan Balz and E. J. Dionne Jr., "Clinton Secures Party Nomination; Perot Factor Grows, Exit Polls Show; Bush Continues Unbeaten String Amid Discontent," *Washington Post*, June 3, 1992.

163 "If nothing else": David Maraniss, "For Clinton, a Chance to Start Over," *Washington Post*, June 4, 1992.

163 "Right now": Steve Daley, "As Primary Season Ends, Clinton Needs to Return Relevancy to His Campaign," *Chicago Tribune*, May 31, 1992.

164 The miniseries aired: John Carmody, "The TV Column," *Washington Post*, May 22, 1986.

164 He earned: Martin Merzer, "Tampa Crusader Wins Notice with Campaign to Oust Incumbents," *Miami Herald*, November 1, 1990.

165 "I think Perot": Jonathan Moore, "Political Activist Calls for Perot, but Texan Doesn't Want to Hold Office," *Houston Chronicle*, September 22, 1991.

165 "Unless the average": "Perot Leads Rally Against Incumbents," *Chicago Tribune*, November 3, 1991.

165 When King: John Mintz, "Perot Sampler Short on Policy Details," *Washington Post*, April 11, 1992.

167 Noted the *New York Times:* "Poll Gives Perot Clear Lead," *New York Times*, June 11, 1992.

168 "A SWIFT, STEALTHY COUP": Helen Dewar, "Swift, Stealthy Coup Raised Senate Pay," *Washington Post*, July 19, 1991.

168 To members: John Mintz, "Perot Sampler Short on Policy Details," *Washington Post*, April 11, 1992.

169 "An interview with": Ibid.

169 "Can we agree": Sonni Efron, "Angry Voters See Perot Riding in Like Cavalry Campaign: Texas Billionaire Strikes a Chord with Petitioners Weary of the Usual Political Roundup," *Los Angeles Times*, March 22, 1992.

169 Tsongas said: Renee Loth, "Perot 'Filling a Vacuum,' Tsongas Says," *Boston Globe*, June 11, 1992.

170 Perot was cleaning up: "ABC/W. Post: Bush Disapproval at 62%; Perot Keeps Lead," *The Hotline*, June 9, 1992.

170 And they were: Richard Benedetto, "Perot Attracts Support from Swing Voters," *USA Today*, June 17, 1992.

171 "Perot," the *Washington Post:* E. J. Dionne Jr., "Perot Leads Field in Poll; Bush Rating at New Low; Billionaire Appears Vulnerable on Record," *Washington Post*, June 9, 1992.

171 "I am pursuing": Richard Benedetto, "Republicans, Dems Ponder 'Perot Factor,'" *USA Today*, April 29, 1992.
172 "Sure, if I": "Clinton Says Cuomo Is Right for Court; MTV Crowd Grills Democratic Hopeful," *St. Louis Post-Dispatch*, June 17, 1992.
172 "People covering": Leslie Phillips, "TV Politicking: Novel Approach Is Now the Norm," *USA Today*, June 18, 1992.
172 Hall, who'd been critical: Jefferson Graham, "Hall's Political Party," *USA Today*, June 16, 1992.

TEN

175 He'd accused: Paul Taylor, "Jackson Says Democrats Are Pushing Blacks Out," *Washington Post*, December 7, 1985.
175 "DLC," Jackson liked: Jon Margolis, "'88 Campaign Heat on for Super Tuesday," *Chicago Tribune*, June 23, 1987.
175 The next year: Michael K. Frisby, "Jackson Counters Slight with Challenge," *Boston Globe*, April 17, 1991.
175 Jackson called it: Ibid.
175 "He's not": Dan Balz, "Jackson Struggles to Define '92 Role," *Washington Post*, June 8, 1991.
176 "It's an outrage": Howard Kurtz, "In Instant Replay, Clinton's Unvarnished Emotion; Television Captured and Tends to Amplify Democrat's Outburst at Jackson," *Washington Post*, February 29, 1992.
176 Clinton later phoned: "Jackson Brouhaha," *USA Today*, February 28, 1992.
176 "Those campaigns": Adam Nagourney, "Perot's Minority Gap: 'Big Daddy' Image, Wealth Are Concerns," *USA Today*, June 8, 1992.
177 It was early: Clinton campaign speech, C-Span video, 1:41:24, https://www.c-span.org/video/?26568-1/clinton-campaign-speech.
179 Along with some: Richard Cohen, "Don't Be a Fool; Some Rap's Not Cool," *St. Louis Post-Dispatch*, May 25, 1992.
179 "I mean": David Mills, "Sister Souljah's Call to Arms," *Washington Post*, May 13, 1992.
180 "Finally," Clinton told: Clinton campaign speech, C-Span video, 1:41:24, https://www.c-span.org/video/?26568-1/clinton-campaign-speech.
182 When Jackson: Thomas B. Edsall, "Clinton Stuns Rainbow Coalition; Candidate Criticizes Rap Singer's Message," *Washington Post*, June 14, 1992.
182 CLINTON STUNS RAINBOW COALITION: Ibid.
183 Mentioning Souljah: Sam Fulwood III, "Clinton Chides Rap Singer, Stuns Jackson," *Los Angeles Times*, June 14, 1992.
183 "The rippling effect": Timothy Clifford, "Clinton Raps Black Singer's Speech," *Newsday*, June 14, 1992.

183 "This was the best": Sam Fulwood III, "Clinton Chides Rap Singer, Stuns Jackson," *Los Angeles Times*, June 14, 1992.

183 "I did not": Sheila Rule, "The 1992 Campaign: Racial Issues: Rapper, Chided by Clinton, Calls Him a Hypocrite," *New York Times*, June 17, 1992.

183 Souljah entered: Chuck Philips, "'I Do Not Advocate . . . Murdering': Raptivist Sister Souljah Disputes Clinton Charge," *Los Angeles Times*, June 17, 1992.

184 Clinton claimed: "Jackson Has Harsh Words for Clinton," Associated Press, June 19, 1992.

184 Executing a nakedly political: R. W. Apple Jr., "The 1992 Campaign: Democrats; Jackson Sees a 'Character Flaw' In Clinton's Remarks on Racism," *New York Times*, June 19, 1992.

184 Clinton reminded reporters: Chris Black, "Clinton Rebuts Jackson," *Boston Globe*, June 20, 1992.

184 "This is a very strange": John Wagner, "Jackson Won't Rule Out Backing Perot; Rainbow Coalition Leader Indicates Clinton Lost Points at Conference," *Washington Post*, June 22, 1992.

184 "I watched two": Mary McGrory, "Backing Jackson into a Corner," *Washington Post*, June 16, 1992.

185 Clinton, he said: Mitchell Locin and Thomas Hardy, "Clinton Gears Up for Big Moment," *Chicago Tribune*, July 13, 1992.

185 "The political reality": B. Drummond Ayres, "Under the Big Top—Jesse Jackson; Jackson Faces New Reality; He Talks the Old Talk but Lacks the Old Fire," *New York Times*, July 15, 1992.

185 On ABC's prime-time newsmagazine: Mark Stencel and John Mintz, "Perot Wavers on Taxes for Schools," *Washington Post*, May 29, 1992.

186 In one instance: Bob Woodward and John Mintz, "Perot Launched Investigations of Bush," *Washington Post*, June 21, 1992.

186 Another report: Michael Wines, "Bush Aides Try to Paint Perot as a Threat to Liberties," *New York Times*, June 23, 1992.

187 "Everybody is writing": Michael Isikoff, "Perot Staff Researching Potential Running Mates," *Washington Post*, July 10, 1992.

187 Bush said: Michael Wines, "Bush Aides Try to Paint Perot as a Threat to Liberties," *New York Times*, June 23, 1992.

187 "Hitler's propaganda chief": Timothy Noah and David Rogers, "Perot Accuses Bush Campaign of 'Dirty Tricks'—Undeclared Candidate Says He Never Investigated Any of Bush's Children," *Wall Street Journal*, June 25, 1992.

187 Cracked one analyst: Kevin Phillips, "For an Angry Electorate, Perot Emerges as Answer: Populism," *Los Angeles Times*, May 31, 1992.

187 On Monday night: "Peter Jennings Reporting: Who Is Ross Perot?" ABC News, June 29, 1992.

188 A live studio audience: "Peter Jennings Reporting: A National Town Meeting," ABC News transcript, June 29, 1992.

188 Almost one in three: Ben Kubasik, "TV Spots," *Newsday*, July 1, 1992.

189 "Throughout American history": Gwen Ifill, "The 1992 Campaign: Democrats; Clinton Selects Senator Gore of Tennessee as Running Mate," *New York Times*, July 10, 1992.

189 Explained *USA Today:* Adam Nagourney, "Dem Ticket Tests New Strategy: Choice of Gore Defies Tradition," *USA Today*, July 10, 1992.

190 The Gore pick: Ibid.

190 "Financially, it's going": Reuters, "Perot Offends U.S. Blacks with 'You People' Phrase," *Toronto Star*, July 12, 1992.

190 A few minutes: John W. Mashek, "Perot Alienates NAACP Audience with 'You People' Remark," *Boston Globe*, July 12, 1992.

190 "A person of his": Ben Smith III and A. L. May, "Perot Flubs Bid for Black Votes," *Atlanta Journal and Constitution*, July 12, 1992.

190 "It never occurred": Ibid.

191 A poll showed: "ABC/Wash. Post: Clinton 42%, Bush 30%, Perot 20%, *The Hotline*, July 16, 1992.

191 "I believe": "Excerpts From Perot's News Conference on Decision Not to Enter Election," *New York Times*, July 17, 1992.

191 He said he'd concluded: "The 1992 Campaign: When Perot Said He Wouldn't Run," *New York Times*, October 2, 1992.

191 He said he wasn't endorsing: "Back to the Status Quo? . . . Well, Not Entirely," *Los Angeles Times*, July 17, 1992.

191 "In the name": "Address Accepting the Presidential Nomination at the Democratic National Convention in New York," American Presidency Project, http://www.presidency.ucsb.edu/ws/?pid=25958.

ELEVEN

193 He was raring: David Espo, "Bush Arrives; Republicans Approve Conservative Platform," Associated Press, August 17, 1992.

194 "Listen, my friends": C-Span Video Library, https://www.c-span.org/video/?31245-1/open-phones-republican-national-convention.

196 On CBS: Howard Kurtz, "TV Coverage Ignores GOP Script," *Washington Post*, August 18, 1992.

196 R. W. Apple: R. W. Apple Jr., "G.O.P. Is Flirting with the Dangers of Negativism," *New York Times*, August 19, 1992.

196 "Pat's message": Ibid.

196 On ABC's newscast: *World News Tonight with Peter Jennings*, ABC News, August 20, 1992.

196 The moment came: American Presidency Project, http://www.presidency.ucsb.edu/ws/index.php?pid=21352.

197   Clinton didn't even: "Clinton Stresses Economic Honesty," U.S. News-
      wire, August 21, 1992.

197   "Among the Republican": David Firestone, "GOP Brushes Buchanan
      Aside," *Newsday,* August 30, 1992.

198   Confronted a few weeks: E. J. Dionne Jr., "Quayle Tempers Campaign's
      'Family Values' Stance," *Washington Post,* September 9, 1992.

199   On September 28: Paul Richter and J. Michael Kennedy, "Top Clinton,
      Bush Aides Court Perot Backers on Budget," *Los Angeles Times,* Sep-
      tember 29, 1992.

199   "I thought that": Ben Smith III, "Perot Throws Hat into the Ring at
      Last," *Atlanta Journal and Constitution,* October 2, 1992.

200   "There's no great": Carl M. Cannon and Marc Gunther, "A Coy Perot:
      'Watch My Lips,'" *Miami Herald,* September 29, 1992.

200   "For George Bush": "Presidential Debate 1992 #1," YouTube video,
      April 16, 2011, https://www.youtube.com/watch?v=XD_cXN9O9ds.

204   "Well, I think": "Presidential Debate 1992 #2," YouTube video, April 16,
      2011, https://www.youtube.com/watch?v=m6sUGKAm2YQ.

205   "Mr. and Mrs. America": "Presidential Debate 1992 #3," YouTube video,
      April 16, 2011, https://www.youtube.com/watch?v=jCGtHqIwKek.

206   "This is very encouraging": Eileen Shanahan, "Economy Shows Surpris-
      ing Growth, but Pessimism Persists," *St. Petersburg Times,* October 28,
      1992.

206   After twelve years: "Presidential Debate in East Lansing, Michigan,"
      American Presidency Project, http://www.presidency.ucsb.edu/ws/index
      .php?pid=21625.

207   He provided: "Perot Alleges Threat to Ruin Wedding Day," *USA Today,*
      October 26, 1992.

209   "The next time": John Mintz, "Perot Embodied Dismay of Millions,"
      Washington Post, November 4, 1992.

209   "We have fought": Michael Kranish, "Bush Accepts Defeat Graciously,"
      Boston Globe, November 4, 1992.

TWELVE

211   Key to Clinton's: Eric Alterman and Kevin Mattson, *The Cause: The Fight
      for American Liberalism from Franklin Roosevelt to Barack Obama,* page
      369, Viking Penguin, 2012.

211   "An aggressive Democrat": Charles M. Madigan, "Clinton Elected Presi-
      dent, Bush Defeat Ends Republican Era," *Chicago Tribune,* November 4,
      1992.

212   Richard Gephardt: Dan Balz and David S. Broder, "Clinton Writing
      Agenda for Economic Change," *Washington Post,* November 5, 1992.

212   "Bill Clinton is": Steven Mufson and Eric Pianin, "The Winner's

Immediate Challenge: Jobs, Growth; Congressional Leaders Urge Speedy Action," *Washington Post*, November 5, 1992.

212 "If you listen": Greg McDonald, "Clinton Takes Different Stand on Bush Policies Toward China," *Houston Chronicle*, November 20, 1992.

212 "The simple reality": David S. Broder, "On the Campaign Trail—Again," *Washington Post*, November 18, 1992.

213 From two leaders: Dan Balz, "A Message from the Moderates; DLC Draft Urges Clinton to Quickly Rally Party Behind His Agenda," *Washington Post*, November 12, 1992.

213 "If Bill Clinton": Merrill Hartson, "Senate Democrats Cheer Victory of Four Women, Preservation of Majority," Associated Press, November 4, 1992.

214 "Bush's conservatism": Edwin J. Feulner Jr., "How Bush Has Wronged the Right," *Chicago Tribune*, November 5, 1992.

215 "I believe that": "Clinton completes cabinet, appoints first female attorney general," *Agence France-Presse* (English), December 24, 1992.

216 The chairman: David Johnston, "Clinton's Choice for Justice Dept. Hired Illegal Aliens for Household," *New York Times*, January 14, 1993.

216 "No one": Clifford Krauss, "The New Presidency: Attorney General-Designate; A Top G.O.P. Senator Backs Nominee in a Storm," *New York Times*, January 16, 1993.

216 "This is not": Michael Isikoff and Al Kamen, "Baird's Hiring Disclosure Not Seen as Major Block," *Washington Post*, January 15, 1993.

217 "She has crossed": "Baird Crossed Line, Gingrich Says," *Los Angeles Times*, January 16, 1993.

217 "She thought": Felicity Barringer, "Settling In: Around the Nation; Much Outrage, Little Sympathy on Main Street," *New York Times*, January 22, 1993.

217 "I gave too": Clifford Krauss, "The New Presidency: Attorney General; Baird Apologizes to Senate Panel for Illegal Hiring," *New York Times*, January 20, 1993.

218 "It is amazing": Joe Davidson, "Baird Tells Senate Panel She Regrets Breaking Law by Hiring Illegal Aliens," *Wall Street Journal*, January 20, 1993.

218 Alan Simpson: John Aloysius Farrell, "Baird Apologizes in Illegal-Alien Hiring," *Boston Globe*, January 20, 1993.

218 Ohio Democrat: William E. Clayton Jr., "Baird tells Senate panel she regrets hiring aliens," *Houston Chronicle*, January 20, 1993.

218 Pennsylvania's Arlen Specter: Lynne Duke and Michael Isikoff, "Baird's Illegal Hiring Raises Sharp Debate," *Washington Post*, January 21, 1993.

218 "I can't answer": David Von Drehle, "Anatomy of a Nomination's Public Way of Death," *Washington Post*, January 23, 1993.

218 "I am surprised": "Baird and Clinton Letters on Withdrawal of Nomination," *Washington Post*, January 23, 1993.

219 "It may be": Howard Kurtz, "Talk Radio's Early Word on Zoe Baird; Listeners' 'Nannygate' Reactions Signaled Trouble for Nominee," *Washington Post*, January 23, 1993.

219 The ban: Information from Rhonda Evans, "U.S Military Policies Concerning Homosexuals: Development, Implementation and Outcomes," The Center for the Study of Sexual Minorities in the Military (University of California, Santa Barbara), http://citeseerx.ist.psu.edu/viewdoc/download?doi=10.1.1.489.957&rep=rep1&type=pdf.

220 "Homosexuality": Kim I. Mills, "Opponents of Pentagon's Anti-Gay Policy Point to Harbingers of Change," Associated Press, August 24, 1992.

221 It was a basic: Melissa Healy, "Clinton Aides Urge Quick End to Military Ban on Gays," *Los Angeles Times*, January 8, 1993.

221 Clinton's victory: Andrea Stone, "Gays Hoping to End 'Years of Darkness,'" *USA Today*, November 5, 1992.

221 "My position": Thomas L. Friedman, "The Transition: The President-Elect; Clinton to Open Military's Ranks to Homosexuals," *New York Times*, November 12, 1992.

222 In Oregon: Brad Cain, "Anti-Gay Measure Rejected; Packwood Holds Lead in Senate," Associated Press, November 4, 1992.

223 "I defy you": Laurence Jolidon, "Caution Urged on Gays in Military," *USA Today*, November 16, 1992.

223 A marine general: John H. Cushman, Jr., "The Transition: Gay Rights; Top Military Officers Object to Lifting Homosexual Ban," *New York Times*, November 14, 1992.

224 "The military leaders": Ibid.

224 Liberals and gay: Donna Cassata, "Former Senator Calls for Ending Ban on Homosexuals," Associated Press, June 10, 1993.

224 Urging Clinton on: Editorial Board, "Lift the Ban on Gay Soldiers," *New York Times*, November 15, 1992.

225 "I'm well familiar": Charles Krauthammer, "Powell Needs No Lectures; The Politics Behind Clinton's Fight over Gays," *Washington Post*, January 29, 1993.

225 A poll: Eric Schmitt, "Pentagon Chief Warns Clinton On Gay Policy," *New York Times*, January 25, 1993.

225 The top Republican: Martin Fletcher, "Clinton promises to end the cold war with Congress," *Times* (London), November 17, 1992.

226 Ambushed with: Kim I. Mills, "Aspin Concerned by Opposition to Lifting Gay Military Ban," Associated Press, January 24, 1993.

226 Aspin's memo: Thomas E. Ricks, "Clinton Reiterates He'll End Military's

Ban on Homosexuals; Opposition Grows," *Wall Street Journal*, January 26, 1993.

226 "I think something": Ibid.

227 They said nothing: Eric Schmitt, "The Top Soldier Is Torn Between 2 Loyalties," *New York Times*, February 6, 1993.

227 "He's trying": Ibid.

227 Said Arizona senator: Kevin Merida, "Little Hill Support Seen for Ending Ban on Gays; Republicans Dare Clinton to Put Issue to Vote," *Washington Post*, January 26, 1993.

227 Instead, the conversation: Susan Page, "Battle Lines on Ban; Clinton May Give an Order on Gays Today," *Newsday*, January 27, 1993.

228 "I think if": Ibid.

228 "When the interests": Ruth Marcus and Helen Dewar, "Clinton Seeks Deal on Gays in Military; Nunn Urges Delay of 'Final' Action," *Washington Post*, January 28, 1993.

228 Constituent calls: "Callers Jam Capitol Switchboard," *Boston Globe*, January 28, 1993.

228 "This has touched off": Ruth Marcus and Helen Dewar, "Clinton Seeks Deal on Gays in Military; Nunn Urges Delay of 'Final' Action," *Washington Post*, January 28, 1993.

229 "I'm not a gay basher": Kathy Sawyer, "Dole Says GOP Senators Will Discuss Plan to Codify Ban," *Washington Post*, February 1, 1993.

229 It was, the *New York Times:* Gwen Ifill, "The Gay Troop Issue; Clinton Accepts Delay in Lifting Military Gay Ban," *New York Times*, January 30, 1993.

230 "Somehow he's let": Kevin Phillips, "Gay Clout, Political Dynamite: Risks—and Rewards—in the Clash of Values," *Washington Post*, January 31, 1993.

THIRTEEN

231 "Millions of Americans": Steven Greenhouse, "The 1992 Campaign: Assessment—Clinton's Retreat; Change on a Tax Cut for the Middle Class Could Forestall Accusations of Pandering," *New York Times*, June 23, 1992.

232 "I think even": "Presidential Debate in East Lansing, Michigan," American Presidency Project, http://www.presidency.ucsb.edu/ws/index.php?pid=21625.

233 "I had hoped": Richard L. Berke, "Clinton Tells Middle Class It Now Faces a Tax Increase Because Deficit Has Grown," *New York Times*, February 16, 1993.

233 "Look at this!": "Transcript of President's Address on the Economy," *New York Times*, February 16, 1993.

234 "All of our": Ruth Marcus and Ann Devroy, "Asking American to 'Face Facts,' Clinton Presents Plan to Raise Taxes, Cut Deficit," *Washington Post,* February 18, 1993.

235 There were offerings: "President Clinton's Address Wednesday Night to a Joint Session of Congress," Associated Press, February 18, 1993.

236 "The person giving": William Schneider, "The Deficit Budget Balancing: Clinton Infuriates Almost Everyone," *Los Angeles Times,* June 18, 1995.

237 "There are those": "Doing It His Way Will Cost Taxpayers," *Washington Post,* February 18, 1993.

237 Michel's remarks: Ruth Marcus and Ann Devroy, "Asking Americans to 'Face Facts,' Clinton Presents Plan to Raise Taxes, Cut Deficit," *Washington Post,* February 18, 1993.

237 "It was a good": "The Spin Cycle; 24 Hours of Very Intense," *Washington Post,* February 19, 1993.

237 "My opposition": Ibid.

238 But now Dole: Ibid.

238 From the op-ed page: Ronald Reagan, "There They Go Again," *New York Times,* February 18, 1993.

239 "It's not good": William M. Welch and Richard Wolf, "GOP Unyielding on Spending Bill," *USA Today,* April 2, 1993.

239 "There's almost unanimous": Eric Pianin, "Clinton's Economic Stimulus Suffers Early Losses in Senate," *Washington Post,* March 30, 1993.

240 "You've heard": David Von Drehle, "Clinton Targets a Few GOP Senators; President Campaigns in Specter's State for Stalled Stimulus Bill," *Washington Post,* April 18, 1993.

240 "I think this package": Ibid.

240 "I don't see": William M. Welch and Richard Wolf, "GOP Unyielding on Spending Bill," *USA Today,* April 2, 1993.

240 "It's the best": Helen Dewar, "President Unapologetic on Jobs Bill That Dole, GOP Killed; Minority Leader's Success Prompts Talk About '96," *Washington Post,* April 26, 1993.

241 "It's not our": William J. Eaton, "House Approves $1.5-Trillion Budget Plan," *Los Angeles Times,* April 1, 1993.

241 "We, the people": Dan Balz, "Perot Sharply Attacks Clinton Economic Plan," *Washington Post,* April 26, 1993.

241 "When you hear": Dan Balz, "Railing at Critics, Touring Clinton Asks Public to Support Him," *Washington Post,* May 18, 1993.

242 Now the Republican: David E. Rosenbaum, "The Clinton Budget: The Overview; House Gives Narrow Approval to Clinton Budget Proposals," *New York Times,* May 28, 1993.

242 "It's a very dangerous": Eric Pianin and Ann Devroy, "Clinton Fights

Revolt on Hill; President Visits Conservative Democrats to Defend Economic Plan," *Washington Post,* May 20, 1993.

242 "If we don't vote": Eric Pianin and David S. Hilzenrath, "House Delivers Clinton Economic Plan, 219–213; Senate Battle Threatens Fragile Pacts," *Washington Post,* May 28, 1993.

243 "America faces": "Text of President Clinton's Address," *Washington Post,* August 4, 1993.

244 "The world": "Text of Senator Dole's Response," *Washington Post,* August 4, 1993

244 "This is the Super Bowl": Kevin Merida, "For Some House Freshmen, Supporting Clinton Is Balancing Act," *Washington Post,* August 5, 1993.

245 "We've been given": Martin Kasindorf, "United the GOP Stands—for Now," *Newsday,* August 1, 1993.

246 The Clinton plan: "Budget Reconciliation Legislation," C-Span Video Library, https://www.c-span.org/video/?47894-1/budget-reconciliation -legislation.

246 "My head": "Floor Speech by Senator Bob Kerrey (D-NE) Upon Casting a Vote of Yes to the Budget Plan," Federal News Service, August 6, 1993.

246 "The message": William Schneider, "When an 800-lb Gorilla Becomes Dead Weight: Voters Now Turn Against Clinton," *Los Angeles Times,* June 13, 1993.

247 In the *Washington Post:* David Broder, "Republicans See Revival in President's Setbacks: Victories in Texas, Statehouses Hearten Party," *Washington Post,* June 21, 1993.

247 "If he fails": Michael Duffy, "That Sinking Feeling," *Time,* June 7, 1993.

FOURTEEN

248 "My style": Elaine S. Povich and Mitchell Locin, "Bye-bye, Bob: Bob Michel Returns to Illinois After 38 Years of Service in the U.S. House," *Chicago Tribune,* January 4, 1995.

248 Michel, according to: Michael Ross, "Michel's Plan to Retire Sets Up GOP Leadership Fight in House," *Los Angeles Times,* October 5, 1993.

248 David Broder: David S. Broder, "Bob Michel and the Virtues of Moderation," *Washington Post,* October 10, 1993.

248 Democrats joined: Adam Clymer, "Michel, G.O.P. House Leader, to Retire," *New York Times,* October 5, 1993.

249 "Our generation": William J. Eaton, "Gingrich Says He Has Votes to Be New House Minority Leader," *Los Angeles Times,* October 8, 1993.

249 At Gingrich's victory: Kenneth J. Cooper, "Gingrich Claims He Has Votes to Be House Minority Leader," *Washington Post,* October 8, 1993.

250 Taking stock: Thomas B. Edsall, "The Nation: Major GOP Victories Continue Year's Trend," *Washington Post*, November 4, 1993.

251 It was on March 8: Jeff Gerth, "The 1992 Campaign: Personal Finances; Clintons Joined S.&L. Operator In an Ozark Real-Estate Venture," *New York Times*, March 8, 1992.

251 Whitewater, he said: Gwen Ifill, "The 1992 Campaign: Personal Finances; Clinton Defends Real-Estate Deal," *New York Times*, March 9, 1992.

252 Dee Dee Myers: Ann Devroy and Al Kamen, "Longtime Travel Staff Given Walking Papers," *Washington Post*, May 20, 1993.

252 And she accused: Terrence Hunt, "White House Drops Arkansas Travel Agency," Associated Press, May 21, 1993.

253 The report: Thomas L. Friedman, "White House Rebukes 4 In Travel Office Shake-Up," *New York Times*, July 3, 1993.

253 "Put aside": Anthony Lewis, "Abroad at Home; The Clinton Mystery," *New York Times*, May 28, 1993.

254 There was "possibly real sleaze": David Lauter, "Clinton Relative Had Proposed She Head Travel Office Firings," *Los Angeles Times*, May 22, 1993.

254 the *Journal* complained: Editorial Board, "Who is Webster Hubbell?" Wall Street Journal, March 2, 1993.

254 The *Journal* then shifted: Editorial Board, "Who Is Vincent Foster," *Wall Street Journal*, June 17, 1993.

255 Over the next month: Ibid.

255 The *Journal* called: "A Washington Death," *Wall Street Journal*, July 22, 1993.

255 "I was not": Michael Kranish, "Note Left by Aide Reveals Distress, Makes Allegations," *Boston Globe*, August 11, 1993.

255 Conservative columnist: William Safire, "The Foster Suicide: Puzzle Still Lacks Some Crucial Pieces," *Chicago Tribune*, August 15, 1993.

256 When the president: Mary McGrory, "The Fog After Foster," *Washington Post*, August 8, 1993.

257 "All the files": Terrence Hunt, "White House Admits Papers Removed from Office of Dead Aide," Associated Press, December 21, 1993.

257 "Mr. Clinton": Editorial Board, "A Halfway Response on Whitewater," *New York Times*, December 26, 1993.

257 "I think my husband": Gwen Ifill, "First Lady, Defending President, Denounces 'Outrageous Attack'," *New York Times*, December 22, 1993.

258 Clinton and the Justice Department: "Reno Plans to Ask Court to Name Special Prosecutor," *St. Louis Post-Dispatch*, January 7, 1994.

258 Writing in *Newsday:* Susan Page, "Pressure Point; Legal Moves in Clinton Deal Raise Eyebrows," *Newsday*, January 7, 1994.

258 "Just turn it over": Gwen Ifill, "Moynihan Urges Prosecutor to Study Clinton Land Deal," *New York Times*, January 10, 1994.

258 "All the federal investigators": John Aloysius Farrell, "Clinton Says He'll Rethink Stance on Prosecutor," *Boston Globe*, January 12, 1994.

258 "I have been": Michael Isikoff, "Whitewater Special Counsel Promises 'Thorough' Probe," *Washington Post*, January 21, 1994.

259 There was much: Todd S. Purdum, "D'Amato Takes the Ethical High Ground; No Stranger to Inquiries, the Senator Seeks the Grand Inquisitor Role in Whitewater," *New York Times*, March 11, 1994.

259 "President Clinton": Editorial Board, "Slovenly White House Ethics," *New York Times*, February 27, 1994.

260 They kept the scope: "A Look at the Players in the Whitewater Scandal," *St. Petersburg Times*, March 13, 1994.

261 "There was still": Richard Whittle, "Fiske Clears Aides to President," *Dallas Morning News*, July 1, 1994.

261 "I would characterize": Paul Greenberg, "The New Ethics: Don't Get Indicted," *Arkansas Democrat-Gazette*, July 8, 1994.

261 "Was there a murder": Erik Eckholm, "From Right, a Rain of Anti-Clinton Salvos," *New York Times*, June 26, 1994.

261 On radio: Jeff Cohen and Norman Solomon, "Let's Give Rush Some Help," *Dallas Morning News*, July 5, 1994.

261 "He later told": Conor O'Clery, "Many Americans Wonder: Is Whitewater Another Watergate?" *Irish Times*, March 9, 1994.

261 The most explosive: Erik Eckholm, "From Right, a Rain of Anti-Clinton Salvos," *New York Times*, June 26, 1994.

262 Of Foster's death: Bill Nichols, "Foster's Life, Death Still in the Spotlight," *USA Today*, March 8, 1994.

262 A congressional aide: Thomas G. Watts, "Foster Case Looms over Hearings," *Dallas Morning News*, July 28, 1994.

263 A few months: David M. Shribman, "Helms' Gibe at Clinton Draws Fire," *Boston Globe*, November 23, 1994.

264 Starr himself said: "Starr Pledges 'Evenhanded' Whitewater Investigation," *Los Angeles Times*, August 7, 1994.

264 And he received: "An Even More Independent Counsel," *New York Times*, August 6, 1994.

264 But, he then said: Stephen Labaton, "Democrats Build Pressure on New Prosecutor to Quit," *New York Times*, August 9, 1994.

FIFTEEN

266 Now, Dole announced: Michael Weisskopf, "Senate Republicans Block Lobbyist Reform Measure," *Washington Post*, October 6, 1994.

266 "I think the strategy": David Dahl, "Republican Strategy: Talk Bills to Death," *St. Petersburg Times*, September 28, 1994.

266 Gingrich retorted: Adam Clymer, "Congress Winds Down: The Con-

gress; At the Capitol, Much Talk But Hardly Any Action," *New York Times*, October 8, 1994.

266 "I put my hands": David S. Broder, "Clinton's Prospects Appear Rosy in the House," *Washington Post*, December 6, 1992.

266 A rare opportunity: David S. Broder, "Clinton's Prospects Appear Rosy in the House," *Washington Post*, December 6, 1992.

267 The diagnosis: David Lauter, "Clinton Aides Hear Political Clock Ticking with Alarm," *Los Angeles Times*, June 10, 1994.

267 "The Constitution says": Robert Shaw, "Wofford Victory Would Give Democrats New Life," *Los Angeles Times*, October 30, 1991.

268 "At long last": Paul F. Horvitz, "Clinton on Health: 'Most Urgent Priority,'" *New York Times*, September 24, 1993.

269 "America is ready": Robert Dodge, "Clinton Unveils Health Care Overhaul," *Dallas Morning News*, September 23, 1993.

269 It was: David S. Broder, "GOP Health Care Strategy Emerging; Despite Divergence of Alternatives, Republicans Hope to Shape Plan," *Washington Post*, October 11, 1993.

269 Republicans, said one: Ibid.

270 "People reading": Robert Pear, "Congress Is Given Clinton Proposal for Health Care," *New York Times*, October 28, 1993.

270 "I want": "From Struggle to Straggle; Letting Clinton Speak for Himself About Health Care, Then and Now," *Boston Globe*, September 4, 1994.

270 Furious liberals: Karen Tumulty, "Odd Couple Behind Health Plan That Rivals Clinton's," *Los Angeles Times*, February 4, 1994.

271 Meanwhile, from: David S. Broder, "GOP Health Care Strategy Emerging; Despite Divergence of Alternatives, Republicans Hope to Shape Plan," *Washington Post*, October 11, 1993.

271 The health insurance lobby: Elizabeth Kolbert, "New Arena for Campaign Ads: Health Care," *New York Times*, October 21, 1993.

272 Meanwhile, on the Senate side: Dana Priest and Helen Dewar, "Mitchell's Health Bill Aims for 95% Coverage," *Washington Post*, August 3, 1994.

272 "Don't let": Dan Balz and Abigail Trafford, "Clinton Warns Against Reform 'Fearmongers,'" *Washington Post*, August 2, 1994.

273 "The combination": Ibid.

273 "For all": Dana Priest, "Democrats Pull the Plug on Health Care Reform," *Washington Post*, September 27, 1994.

273 "Thus is": Michael Ross, "103rd Congress Grinding to a Halt Amid Partisan Rage," *Los Angeles Times*, October 8, 1994.

273 From the *Washington Post:* Helen Dewar and Kenneth J. Cooper, "103rd Congress Started Fast but Collapsed at Finish Line," *Washington Post*, October 9, 1994.

273 Clinton blamed: Ann Devroy, "President Blames Republicans for Legislative Inaction," *Washington Post,* October 8, 1994.

274 When Clinton raised: David E. Rosenbaum and Steve Lohr, "With a Stable Economy, Clinton Hopes for Credit," *New York Times,* August 3, 1996.

274 "The record": Douglas Jehl, "Congress Winds Down: The President; Clinton Assails G.O.P. for Effort to 'Kill It or Just Talk It to Death,'" *New York Times,* October 8, 1994.

275 "Our government operates": David Broder, "GOP House Party," *Washington Post,* September 28, 1994.

275 "The House Republicans'": Eric Pianin, "GOP 'contract' missing its price tag, critics say," *Washington Post,* September 28, 1994.

275 "It is a vision": Editorial Board, "The G.O.P.'s Deceptive Contract," *New York Times,* September 28, 1994.

275 Referring to: Douglas Jehl, "Clinton Assails G.O.P. for Effort To 'Kill It or Just Talk It to Death,'" *New York Times,* October 8, 1994.

276 "Well, folks": "Democrats Losing Steam, Polls Say," *Chicago Tribune,* November 1, 1994.

277 At its core: "Giuliani Talks of Agonizing Over Best Man for the City," *New York Times,* October 25, 1994.

278 "They're watching you": Kevin Sack, "The 1994 Campaign: Cuomo; Clinton Campaigns Upstate with an Eye on the Nation," *New York Times,* November 4, 1994.

280 He was being challenged: "The Doctor Is In," *Seattle Times,* December 21, 2002.

280 "A bitter, bitter": "Election USA '94, Part 28," CNN transcript, November 8, 1994.

280 "If these trends": "It Was an Election Made for Television," Associated Press, November 9, 1994.

281 But, as Bruce Morton: Michael Renn, "1994 Election Night Coverage Part 2: CNN," YouTube video, September 1, 2010, https://www.youtube.com/watch?v=cejhxBQTj80.

282 "There are a lot": Michael Renn, "1994 Election Night Coverage Part 8: CBS," YouTube Video, September 17, 2010, https://www.youtube.com/watch?v=hyuIss6EPF0&t=108s.

282 "You'll recall": Michael Renn, "1994 Election Night Coverage Part 18: CNN," YouTube video, February 3, 2011, https://www.youtube.com/watch?v=k14YqLjk-g8.

283 "I'm going": Michael Renn, "1994 Election Night Coverage Part 16: CNN," YouTube video, November 2, 2010, https://www.youtube.com/watch?v=P3oBWDvrAUM.

283 "Even the most cockeyed": "ABC News Nightline #3512," ABC News transcript, November 8, 1994.

284 "It feels almost": Michael Renn, "1994 Election Night Coverage Part 11: CBS, YouTube Video, October 27, 2010, https://www.youtube.com/watch?v=BUYzYorNr0w.

284 "One of the interesting": Michael Renn, "1994 Election Night Coverage Part 14," YouTube video, October 27, 2010, https://www.youtube.com/watch?v=8mDA_L9_2sw&t=511s.

284 "This shift": Michael Renn, "1994 Election Night Coverage Part 13: ABC and NBC," YouTube video, October 22, 2010, https://www.youtube.com/watch?v=4ADLyMPHf5c&t=382s.

285 "I had high hopes": "Shelby Jumps Ship, Joins Republican Party," *USA Today*, November 10, 1994.

285 The *Boston Globe:* David M. Shribman, "Voter Rebellion Took Root at the Local Level," *Boston Globe*, November 10, 1994.

286 In his syndicated column: George F. Will, "Reagan's Third Victory," *Washington Post*, November 10, 1994.

287 "Still," he conceded: Clintonlibrary42, "President Clinton's 78th News Conference," YouTube video, July 19, 2016, https://www.youtube.com/watch?v=IeiBuUrQ0Vg.

287 "This election": R. W. Apple Jr., "Clinton's Grip On '96 Ticket Isn't So Sure," *New York Times*, November 21, 1994.

287 "It is still": Ibid.

288 "Rather than try": David E. Rosenbaum, "The Clinton Tax Plan: News Analysis; About-Face by Clinton," *New York Times*, December 16, 1994.

288 "As long as": Scot Lehigh, "Tsongas Attacks Clinton," *Boston Globe*, December 18, 1994.

SIXTEEN

290 "If only one": Al Kamen, "For Freshmen, No Harvard," *Washington Post*, November 23, 1994.

291 "Just think": Kevin Merida and Kenneth J. Cooper, "New Heritage Emerges in Orientation; Hill Newcomers Hear Conservative Gospel," *Washington Post*, December 10, 1994.

291 "I'm in awe": "Republican Freshmen Orientation," C-Span Video Library, https://www.c-span.org/video/?62105-1/republican-freshmen-orientation.

293 "This is what": Richard Wolf, "Newt's World: Leader Plans 'to Transform' Government," *USA Today*, November 16, 1994.

293 For them, he sketched: Maureen Dowd, "The 1994 Elections: G.O.P.'s Rising Star Pledges to Right Wrongs of Left," *New York Times*, November 10, 1994.

294 "Until the mid-1960's": Dale Russakoff, "Gingrich Lobs a Few More Bombs," *Washington Post*, November 10, 1994.

294 His charge: Steve Daley, "Renewing American Civilization: What Gingrich Has on Tap for 1995," *Chicago Tribune*, January 1, 1995.

294 The president: Ibid.

294 Now here: Sheryl Gay Stolberg, "For Gingrich in Power, Pragmatism, not Purity," *New York Times*, December 21, 2011.

294 "Slash and burn": Anthony Lewis, "Abroad at Home; Eye of Newt," *New York Times*, November 14, 1994.

295 The paper's editorial: Editorial Board, "Newt Gingrich, Authoritarian," *New York Times*, November 13, 1994.

295 "There was no point": "Gingrich Regrets His Nasty Jabs at Clinton," *Los Angeles Times*, November 17, 1994.

295 "I probably need": Jurek Martin, "Pugnacious Newt Takes a Fresh Tack: Bomb-Thrower of Far Right Seeks to Become a Leader," *Financial Times*, November 18, 1994.

295 It was clear: Keith Bradsher, "Many White House Employees Used Drugs, Gingrich Asserts," *New York Times*, December 5, 1994.

296 If Gingrich didn't: Ann Devroy, "Panetta Holds Gingrich's Words Against Him—and His Party," *Washington Post*, December 6, 1994.

296 A telling detail: W. Lippman and Ann Devroy, "Gingrich Takes Aim at Clinton Staff; No Specifics Offered for Claim Many Used Drugs in Recent Years," *Washington Post*, December 5, 1994.

296 HarperCollins: David Streitfield, "$4 Million Book Deal for Gingrich; Political Opponents Decry Windfall from Murdoch Firm," *Washington Post*, December 22, 1994.

297 "This is an arrogant": David Streitfield and Charles R. Babcock, "Gingrich $4.5 Million Book Deal Draws Fire; Some Democrats Say Advance Money Should be Measured Against House Ethics Standards," *Washington Post*, December 23, 1994.

297 "We're about to": Peter Applebome, "Gingrich Gives Up $4 Million Advance On His Book Deal," New York Times, December 31, 1994.

297 "He'll make": Ibid.

298 Wielding a large: Richard Wolf, "Mr. Speaker Steals Show," USA Today, January 5, 1995.

298 That morning: Michael Ross, "Gingrich Speech Debuts His Kinder, Gentler Side," *Los Angeles Times*, January 5, 1995.

298 "I know I'm a very": "Excerpts from Gingrich's Speech on Party's Agenda for the 104th Congress," *New York Times*, January 5, 1995.

298 "It almost seemed": Nolan Walter, "Softer Pitch from Master of Hardball: Gingrich Takes Control with a Kinder and Gentler Speech," *Philadelphia Inquirer*, January 5, 1995.

298 "He is the most partisan": Richard Wolf, "Mr. Speaker Steals Show,"
     USA Today, January 5, 1995.
298 What has Newt: Associated Press, "From Newt's Mom's Lips to Connie
     Chung's Ears," *Chicago Tribune,* January 4, 1995.
299 "You have no idea": "'B'-Word Controversy Laughed Off," *USA Today,*
     January 6, 1995.
299 What Gingrich: Kenneth J. Cooper, "Gingrich Fires His Pick for House
     Historian," *Washington Post,* January 10, 1995.
300 By week two: Howard Kurtz, "Gingrich Criticizes 'Nit-Picking' Media;
     Speaker to Take Break from Sunday Talk Shows," *Washington Post,* Jan-
     uary 10, 1995.

SEVENTEEN

302 Just days: Dan Meyers, "Welfare Reform More Than Just Debate in
     Many States," *Philadelphia Inquirer,* November 20, 1994.
303 "The system": Mary McGrory, "Orphanage Idea Has Many Parents,"
     Washington Post, December 13, 1994.
303 During the debate: William Claiborne, "Moynihan Presses Welfare
     Reform," *Washington Post,* January 10, 1994.
304 The criticism rained: Douglas Jehl, "Clinton Says Orphanages Can't
     Replace Strong Parents," *New York Times,* December 11, 1994.
304 Gingrich shot back: John E. Yang, "The Speaker Comes to Boys Town,"
     Washington Post, October 24, 1995.
304 "'Boys Town'": Joel Achenbach, "A Return to the Orphan Age," *Wash-
     ington Post,* December 7, 1994.
304 "I don't think": Leslie Phillips, "Critics Queasy About Orphanages,"
     USA Today, December 5, 1994.
305 "The states": Judith Havemann, "House Committee Votes to End Lunch
     Program," *Washington Post,* February 24, 1995.
305 Blowing it up: Ann Devroy and Kevin Merida, "President Hits GOP As
     Callous," *Washington Post,* February 23, 1995.
305 Words like: Editorial, "The Anti-Family Plan," *Atlanta Journal-
     Constitution,* November 18, 1994; Carl Rowan, "Both Bombast, Reason
     from GOP," *Chicago Sun-Times,* December 7, 1994; Leslie Phillips and
     Patricia Edmonds, "Critics Queasy About Orphanages," USA Today,
     December 5, 1994.
305 "It doesn't say": William Schneider, "A Clear and Present Danger: The
     Politicization of the Constitution," *Los Angeles Times,* February 26, 1995.
305 "Join us": Michael Wines, "House Votes to Cut Taxes by $189 Billion
     over 5 Years as Part of G.O.P. 'Contract,'" *New York Times,* April 6,
     1995.
306 Gingrich and his fellow: Kenneth J. Cooper and Helen Dewar, "100

Days Down but Senate to Go for Most 'Contract' Items," *Washington Post,* April 9, 1995.

306 "While we've done": "'All of Us Together . . . Must Totally Remake the Federal Government,'" *Washington Post,* April 8, 1995.

307 The president had: John F. Harris, "Clinton Offers GOP Compromise, Vetoes," *Washington Post,* April 8, 1995.

307 "In the first": Ibid.

308 "Triangulation": Peter McKay, "Ten Years of Clegg? Let's All Get Ready to Emigrate," *Daily Mail,* April 28, 2014.

308 "I do not want a pile": John F. Harris, "Clinton Offers GOP Compromise, Vetoes," *Washington Post,* April 8, 1995.

310 "Those who are lost": Tom Raum, "Clinton Prays with Anguished Nation, Outlines New Anti-Terrorism Steps," Associated Press, April 23, 1995.

310 "Our president": Lawrence Knutson, "Clintons to Children: Good Triumphs over Evil," Associated Press, April 22, 1995.

311 It was a connection: Ron Fournier, "Clinton Denounces 'Purveyors of Hatred and Division,'" Associated Press, April 24, 1995.

311 "Make no mistake": Howard Kurtz, "No Time for Dignity," *Washington Post,* April 14, 1996.

312 Before the Oklahoma City: Bennett Roth, "Appeal of Arms Ban May Be Postponed," *Houston Chronicle,* May 1, 1995.

312 "I am a gun owner": "Mr. Bush and the NRA," *Washington Post,* May 13, 1995.

313 "I want them": Ron Fournier, "Clinton Slams NRA, Praises Bush for Abandoning Group," *Chicago Sun-Times,* May 16, 1995.

313 In it, Gingrich: Steve Daley, "Gingrich Note Reveals His Vow to NRA," *Chicago Tribune,* August 2, 1995.

313 "You can see": "Clinton Blasts GOP; Threatens Veto of VA-HUD Measure," *National Journal's Congress Daily,* August 1, 1995.

EIGHTEEN

315 "It's time": Tom Raum, "Clinton Takes a Crack at Deficit," *Chicago Sun-Times,* June 14, 1995.

316 "I think some": Paul Richter, "News Analysis: Clinton Address Signals Shift in Strategy," *Los Angeles Times,* June 14, 1995.

316 On Capitol Hill: Bill Nichols, "President Decides His Best Defense Is a Good Offense," *USA Today,* June 14, 1995.

317 "We will not": Robin Toner, "Congressional Memo: G.O.P. Feels the Heat. The Furnace: Medicare," *New York Times,* May 5, 1995.

318 "Keep your tax-cutting": Todd S. Purdum, "Clinton Joins a Rally on Medicare's Birthday," *New York Times,* July 26, 1995.

318 "Those who": Ibid.

318 "I think to try": "Healthy Solutions; Waco Revisited; Whitewater Hearings; Cyberfuture; Healthy Solutions," *The MacNeil/Lehrer NewsHour* transcript, July 25, 1995.

318 "Long before": Adam Clymer, "As Demonstrators Gather, Gingrich Delays a Speech," *New York Times*, August 8, 1995.

319 "My friends": Ibid.

319 "I don't detect": Ibid.

319 It urged Republicans: Eric Pianin and John F. Harris, "President Gains Status in Budget Battle," *Washington Post*, September 3, 1995.

319 The Republican plan: Robert Pear, "G.O.P.'s Plan to Cut Medicare Faces a Veto, Clinton Promises," *New York Times*, September 16, 1995.

319 "I am not going": "An Interview with President Bill Clinton, Part 2," *NPR Morning Edition* transcript, August 8, 1995.

319 On *Meet the Press:* Robert Dodge, "Gingrich Softens GOP Threat to Force Budget on Clinton," *Dallas Morning News*, September 11, 1995.

320 "Among those": Doyle McManus, "GOP-Clinton Fiscal Standoff Is Brinkmanship Exercise," *Los Angeles Times*, September 16, 1995.

320 Clinton, Gingrich advised: John F. Harris, "Clinton Promises to Defy GOP Budget 'Blackmail,'" *Washington Post*, October 29, 1995.

320 They wanted: Edwin Chen, "Gingrich: Today's Medicare Will 'Wither,'" *Los Angeles Times*, October 26, 1995.

321 "Let's talk": Paul Taylor and John E. Yang, "Gingrich Says Democratic Ad Twists His Stance on Medicare," *Washington Post*, November 5, 1995.

321 "Now, we don't": Barbara Vobejda and John F. Harris, "Democrats Pounce on GOP Medicare Comments," *Washington Post*, October 27, 1995.

321 "I was there": "Dole Voices Pride in '65 Anti-Medicare Vote," *Washington Post*, October 26, 1995.

322 The White House's spokesman: Barbara Vobejda and John F. Harris, "Democrats Pounce on GOP Medicare Comments," *Washington Post*, October 27, 1995.

323 "As long as": Robert A. Rankin and David Hess, "Government Poised for Shutdown," *Philadelphia Inquirer*, November 14, 1995.

324 "The government": Helen Thomas, "Talks Break Down, Shutdown Lingers," United Press International, November 14, 1995.

324 At a closed-door: David Maraniss and John E. Yang, "House GOP Standing Firm on 7-Year Plan; As Drop in Poll Ratings Tests Party Unity, Freshmen Continue Strong Role," *Washington Post*, November 15, 1995.

324 "If this budget": Ibid.

325 "The president": Adam Clymer, "President Vetoes Stopgap Budget; Shutdown Looms," *New York Times*, November 14, 1995.

326 "Virtually every finding": Richard L. Berke, "Clinton's Ratings over 50% in Poll as G.O.P. Declines," *New York Times*, December 14, 1995.

NINETEEN

329 "I'm an advocate": Judy Keen, "Born to Run: Buchanan Raring to Go Again," *USA Today*, March 20, 1995.

329 "I have the most": Richard L. Berke, "April 16–22; When Good Republicans Make Bad Moves," *New York Times*, April 23, 1995.

330 "If you force": Susan Page, "'There Is a Place in Hell for Those Who Refuse to Take a Stand,'" *Newsday*, June 4, 1995.

330 "If you go back": Ronald Brownstein, "Immigration Debate Splits GOP Hopefuls," *Los Angeles Times*, May 14, 1995.

330 When Buchanan reiterated: Ibid.

332 After thanking: "United We Stand America Day 2 Part 1," C-Span Video Library, https://www.c-span.org/video/?66706-1/united-stand -america-day-2-part-1.

335 "This wasn't a victory": Richard L. Berke, "Buchanan Wins in Louisiana in Blow to Gramm Campaign," *New York Times*, February 7, 1996.

335 A stunned Gramm: William Booth, "Buchanan Stuns Gramm in Louisiana; Texan Dealt Setback in GOP Caucus Test of Conservative Field," *Washington Post*, February 7, 1996.

335 "We shocked them": Dan Balz, "Buchanan Hits Back; Gramm Endorses Dole; Establishment Panicking, Commentator Says," *Washington Post*, February 19, 1996.

336 "The poor homosexuals": Margaret Garrard Warner, "Savior on the Right?" *Newsweek*, January 19, 1987.

336 He reminded reporters: Associated Press, "Gramm Support on Dole," *Racine (Wis.) Journal Times*, February 19, 1996.

337 "I didn't think": "Remarks by Pat Buchanan, Republican Presidential Candidate, After New Hampshire Primary," *Federal News Service*, February 20, 1996.

337 "I'm telling": "New Hampshire Primary News Coverage," C-Span Video Library, https://www.c-span.org/video/?70054-1/hampshire-primary -news-coverage.

337 "We know": Richard L. Berke, "Politics: The Overview; Buchanan a Narrow Victor Over Dole in New Hampshire," *New York Times*, February 21, 1996.

339 "If Dole wins": H. Josef Herbert, "Gingrich Complains Democrats Trying to 'Demonize' Him," Associated Press, August 18, 1996.

339 Another seemed: "Clinton Unleashes a Fierce Attack on Dole," *New York Times*, September 25, 1996.

339 "I will seek": "Remarks to Reporters Prior to a News Conference,"

American Presidency Project, May 15, 1996, http://www.presidency.ucsb
.edu/ws/index.php?pid=85181.

340 Instead, he declared: William M. Welch, "Buchanan Backs 'Truce' but Doesn't Mention Dole by Name," *USA Today*, August 12, 1996.

340 "If there's anyone": Bob Edwards, "Selected Excerpts from Bob Dole's Acceptance Speech," *NPR Morning Edition* transcript, August 16, 1996.

TWENTY

341 "We said": Alison Mitchell, "Clinton, Setting Out for Chicago, Denounces G.O.P.," *New York Times*, August 26, 1996.

343 "Jim, we must": Brian McGrory and Curtis Wilkie, "Gun Control Is the First Theme," *Boston Globe*, August 27, 1996.

344 "This is the best": Peter T. Kilborn and Sam Howe Verhovek, "Clinton's Welfare Shift Ends Tortuous Journey," *New York Times*, August 2, 1996.

344 John Lewis: Richard Benedetto, "Liberal Democrats on Run with the 'End' of Welfare," *USA Today*, August 3, 1996.

344 Senator Paul Wellstone: "Excerpts from Debate in the Senate on the Welfare Measure," *New York Times*, August 2, 1996.

344 Even Moynihan: Daniel P. Moynihan, "When Principle Is at Issue," *Washington Post*, August 4, 1996.

345 "In 1968": John M. Broder and Paul Richter, "First Lady Focuses on Family," *Los Angeles Times*, August 28, 1996.

345 It was the same: William M. Welch, "Liberal Dems Rock Convention," *USA Today*, August 28, 1996.

346 "Tonight," said Dodd: Rich Hein, "Nomination Caps Triumphant Trip," *Chicago Sun-Times*, August 29, 1996.

346 Standing motionless: "Al Gore Democratic National Convention 1996 Speech," C-Span Video Library, https://www.c-span.org/video/?c4604011/al-gore-democratic-national-convention-1996-speech.

347 "Let me be": "Robert 'Bob' Dole 1996 Acceptance Speech," C-Span Video Library, https://www.c-span.org/video/?c4603908/robert-bob-dole-1996-acceptance-speech.

347 "But make no mistake": "Gore text: Clinton's American vision," United Press International, August 28, 1996.

348 The story landed: Joe Battenfeld and Gayle Fee, "Clinton's Speech Takes Back Seat to Sex Scandal," *Boston Herald*, August 30, 1996.

348 Reacting to the news: Steven Thomma and Jodi Enda, "President's Top Political Adviser Quits," *Philadelphia Inquirer*, August 30, 1996.

349 "Mr. Chairman": "Clinton's Speech Accepting the Democratic Nomination for President," *New York Times*, August 30, 1996.

350 "If Clinton": Adam Clymer, "G.O.P. Pushes Congress Strategy That Shuns Dole," *New York Times*, October 23, 1996.

351 Charlie Cook: Graham Fraser, "Battle for Congress Too Close to Call," *Toronto Globe and Mail,* October 24, 1996.

351 Gingrich himself: Dale Russakoff, "On the Stump, Gingrich Adjusts to Reduced Stature," *Washington Post,* October 24, 1996.

TWENTY-ONE

354 "Gingrich's path": Timothy Clifford, "Fresh Charge Against Newt," *New York Daily News,* December 15, 1995.

354 "If Newt": Phil Kuntz, "Democrats Sharpen Attack on Gingrich, Seeking Broader Ethics Investigation," *Wall Street Journal,* December 15, 1995.

354 "There's no doubt": Jim Dwyer, "GOP Views Speaker as a Newtron Bomb," *New York Daily News,* November 12, 1996.

355 "The party's future": David Rogers, "Republicans Break Ranks on Gingrich," *Wall Street Journal,* January 7, 1997.

355 "Let me say": "'. . . I will seek to work with every member' of the House,'" *Atlanta Journal and Constitution,* January 8, 1997.

356 "He is technically": Adam Clymer, "House, in a 395–28 Vote, Reprimands Gingrich," *New York Times,* January 22, 1997.

358 "After decades": Eric Pianin and Clay Chandler, "Clinton, GOP Both Claim Budget Victories," *Washington Post,* July 30, 1997.

358 As the votes: Jessica Lee, "Congress Revels in Budget Vote: 389 approve in House; 98 in Senate," *USA Today,* August 1, 1997.

358 He framed the act: "Budget Agreement Signing Ceremony," C-Span Video Library, https://www.c-span.org/video/?88973-1/budget-agreement -signing-ceremony.

359 From retirement: "Time for George to Get His Due," *Chicago Tribune,* August 12, 1997.

360 "Consistently, low-income": Jerry Gray, "Negative Responses Come from the Political Left and Right," *New York Times,* July 30, 1997.

360 "It is impossible": Carla Anne Robbins, "For an Ohio Congressman and His Constituents, the Issue of Bosnia Is Literally Close to Home," *Wall Street Journal,* November 20, 1995.

361 A Desert Storm: David H. Hackworth, "We Didn't Finish the Job," *Newsweek,* January 20, 1992.

361 "Mr. Clinton": Richard L. Berke and John M. Broder, "A Mellow Clinton at Ease on His Role," *New York Times,* December 7, 1997.

362 On this day: Alison Mitchell, "Gingrich Going on Tour, for Cash and the Future," *New York Times,* January 14, 1998.

TWENTY-TWO

364 It was something: Susan Schmidt, Peter Baker, and Toni Locy, "Clinton Accused of Urging Aide to Lie; Starr Probes Whether President Told

Woman to Deny Alleged Affair to Jones's Lawyers," *Washington Post*, January 21, 1998.

365 But where Willey: Richard A. Serrano, "Clinton Under Fire: Tape Excerpts Said to Reveal Explicit Talk," *Los Angeles Times*, January 24, 1998.

367 "The news of this day": "Interview with Jim Lehrer of the PBS 'News Hour,'" *Public Papers of the Presidents*, January 21, 1998.

367 "The nation": Francis X. Clines, "Public Tolerance in the Clinton Era, and Its Limits," *New York Times*, January 25, 1998.

367 "I want to say": "Clinton comment on alleged affair," Associated Press, January 26, 1998.

368 The first she'd heard: "Hillary Rodham Clinton Responds Publicly for First Time to Sex Scandal Allegations Swirling Around Her Husband," NBC News transcripts, *Today*, January 27, 1998.

368 On *The Tonight Show:* "Laugh Lines; Punch Lines," *Los Angeles Times*, January 30, 1998.

369 "I believe him": Kathy Kiely, "Defenders Work the Talk Shows," *New York Daily News*, January 26, 1998.

369 "We don't know": Alison Mitchell, "Congresswoman Angry Over Silence on Scandal," *New York Times*, February 17, 1998.

369 "Every citizen": David S. Broder, "Democrats Await Convincing Answers; Activists Express Anxiety over President's Lawyerly Rebuttal," *Washington Post*, January 23, 1998.

370 One of: Roger Simon, "Despite Polls, Clinton Aides Fear Mood Turnaround," *Chicago Tribune*, February 13, 1998.

370 She testified: Peter Baker, "Lewinsky Testifies Before Grand Jury; Ex-Intern Details Her Side of Story of Presidential Sex," *Washington Post*, August 7, 1998.

370 "The degree": Andrew Kohut, "Clinton and the Court of Public Opinion," *New York Times*, August 17, 1998.

370 "Indeed, I did": "Transcript: President Bill Clinton," CNN.com, August 17, 1998, http://www.cnn.com/ALLPOLITICS/1998/08/17/speech/transcript.html.

372 In the *New York Times:* Gloria Steinem, "Steinem: Clinton Took No for an Answer," *Toronto Globe and Mail*, March 24, 1998.

372 Another feminist writer: Francine Prose, "New York Supergals Love That Naughty Prez," *New York Observer*, February 9, 1998.

372 "What have we": Andy Seiler, "Comics Grab Bait of Another 'Gate,'" *USA Today*, January 26, 1998.

373 The *New York Times:* Editorial Board, "Bill Clinton Speaks, a Little," *New York Times*, August 18, 1998.

373 Clinton's remarks: Richard L. Berke, "Scathing and Sad, Democrats React to Clinton Speech," *New York Times*, August 19, 1998.

373 In the *Washington Post:* Dan Balz, "Hill Democrats View Speech As a Failure," *Washington Post,* August 20, 1998.

374 "In this case": "Sen. Joseph Lieberman Speaks on Clinton," CNN.com, transcript, https://www.cnn.com/ALLPOLITICS/1998/09/03/lieberman/.

374 At one point: Richard L. Berke, "Playing Safe, Republicans Offer Silence," *New York Times,* September 13, 1998.

375 "My deepest belief": Richard L. Berke, "Scathing and Sad, Democrats React to Clinton Speech," *New York Times,* August 19, 1998.

375 "I think the president": Howard Kurtz, "The Price of Admission: Clinton's Version of Truth Hasn't Set Him Free from Media Attacks," *Washington Post,* August 19, 1998.

375 "It is bad": Dan Balz and Guy Gugliotta, "Even Critics Are Cautious in Wake of President's Speech," *Washington Post,* August 19, 1998.

375 "If Republicans": Jessica Lee, "Some Republicans Feeling More Free to Speak Their Minds," *USA Today,* August 21, 1998.

376 "The president's defense": James Bennet and Don Van Natta Jr., "Clinton's Legal Team Awaits the Public's Response to His Risky Political Gamble," *New York Times,* September 13, 1998.

376 "Unconscionable overreaching": Francis X. Clines, "White House, in Rebuttal to Starr, Assails Report as 'Smear Campaign,'" *New York Times,* September 13, 1998.

376 Speaking to a group: "I Have Sinned . . . The Sorrow I Feel Is Genuine," *Washington Post,* September 12, 1998.

377 The *New York Times:* Editorial Board, "Justice or Mercy for Bill Clinton?" *New York Times,* September 14, 1998.

377 "I don't understand": Alison Mitchell, "G.O.P. Resists Speedy Action on President," *New York Times,* September 24, 1998.

377 DeLay distilled: Ibid.

377 Under Clinton: Richard L. Berke, "To Christians, G.O.P. Urges Punishment and Prayer," *New York Times,* September 19, 1998.

377 Democratic congressman: Lizette Alvarez and Eric Schmitt, "Committee Members Barely Wait for the Gavel Before They Take Off the Gloves," *New York Times,* November 20, 1998.

378 "Women will not": Mary Leonard, "Feminists Side with President," *Boston Globe,* September 25, 1998.

378 The debate: William Neikirk, "Open-Ended Impeachment Investigation Approved by House," *Chicago Tribune,* October 9, 1998.

379 The pornographer: Mark Jurkowitz, "Galled by 'Hypocrisy,' the Rogue Porn King Larry Flynt Dives into Washington's Muck," *Boston Globe,* December 29, 1998.

379 "The question asks": "Duty, Justice, Truth at Heart of Debate," *Boston Globe,* October 9, 1998.

379   As the debate: "'Follow the Truth Wherever It Leads,'" *Washington Post*, October 9, 1998.

TWENTY-THREE

381   They got a brief: The Associated Press, "Statement on Hillary Clinton," *New York Times*, August 19, 1998.

382   When Clinton went: Bill Nichols, "Trip to Ohio Provides No Reprieve for President," *USA Today*, September 18, 1998.

382   Representative Adam Smith: "Rep. Smith to Clinton: Thanks but No Thanks," *National Journal's Congress Daily*, September 3, 1998.

382   Pat Schroeder: David S. Broder and Terry M. Neal, "Democrats in Tight Races Walk a Tightrope," *Washington Post*, September 15, 1998.

382   "The attraction": Elaine Sciolino, "Hillary Clinton, Smile in Place, Hits Fundraisers," *New York Times*, September 28, 1998.

382   She hit the road: Mary Leonard, "First Lady a Campaign Warrior," *Boston Globe*, November 2, 1998.

382   She never: John F. Harris, "Clinton Team Regains Optimism As Battle Moves to Political Realm," *Washington Post*, September 25, 1998.

383   "Attacking Mrs. Clinton": Ken Fireman and William Douglas, "Clintons Relish Senate Fight," *Newsday*, October 29, 1998.

383   Standing next: Julian Borge, "Hillary's stardust to the rescue," *Guardian* (London), November 3, 1998.

384   "Does he talk": Alessandra Stanley, "Abrams-D'Amato Race Becomes a Family Affair," *New York Times*, October 27, 1992.

385   "To me": Francis X. Clines, "D'Amato Wields the Gavel, Without the Scrapper's Style," *New York Times*, July 30, 1995.

385   "History will judge": Stephen Labaton, "Whitewater Hearing Cleared the Clintons, Democrats Say," *New York Times*, June 19, 1996.

385   It was "a remarkable document": Stephen Labaton, "Report Takes Aim at Mrs. Clinton," *New York Times*, June 16, 1996.

386   The Clintons' lawyer: Stephen Labaton, "Mrs. Clinton Again Denies Knowledge of Law Files," *New York Times*, June 18, 1996.

386   That's when Hillary: Joel Siegel, "Hillary Rips D'Amato as Second Jesse Helms," *New York Daily News*, September 24, 1998.

387   A few weeks: Adam Nagourney, "Back Schumer, Hillary Clinton Tells Women Voters," *New York Times*, October 20, 1998.

387   At the end: Rick Brand, "Help to Finish Line; First Lady, Schumer Address Seniors, Gays," *New York Daily News*, October 28, 1998.

388   "He will win!": Adam Nagourney, "D'Amato and Schumer End Campaign on High, Hoarse Notes," *New York Times*, November 3, 1998.

388   "I know you": Michael Finnegan and Joel Siegel, "They're Off & Still Running," *New York Daily News*, November 3, 1998.

389 "That is the question": John F. Harris and Ceci Connolly, "Clinton: Scandal Ads Are Bid to 'Distract'; President Says Policies Are the Real Issue," *Washington Post*, October 29, 1998.

389 Gingrich was back: Lois Romano, "For Gingrich, an Easy Victory and Uneasy Future," *Washington Post*, November 4, 1998.

390 "She really feels": John F. Harris and Peter Baker, "For Clinton, the Outlook Brightens," *Washington Post*, November 4, 1998.

391 "Democrats who once": David S. Broder, "A Party Relieved, Reinvigorated," *Washington Post*, November 4, 1998.

391 "This will be": Ceci Connolly and Juliet Eilperin, "Democrats Narrow GOP's Edge in House; Midterm Gains Rare for President's Party," *Washington Post*, November 4, 1998.

391 "This was a stunning": Ceci Connolly and Juliet Eilperin, "The House: Gingrich Moves to Protect His Leadership Post," *Washington Post*, November 5, 1998.

391 From Oklahoma: Juliet Eilperin and Guy Gugliotta, "GOP Leadership Rivalries Emerge After Losses," *Washington Post*, November 6, 1998.

391 "Astonishing results": James Gerstenzang, "Decision '98 The Final Count: White House Basks in Results Afterglow," *Los Angeles Times*, November 5, 1998.

392 As *Time* wrote: John Cloud, "Give 'em Hillary," *Time*, November 16, 1998.

392 A few days later: James Barron, "Public Lives," *New York Times*, November 10, 1998.

393 "Revolutionizing takes": Edward Walsh, "For Livingston, the Time Is Now," *Washington Post*, November 7, 1998.

393 Within hours: "Gingrich says he still has 'significant role to play,'" Associated Press, November 6, 1998.

393 On a conference call: Associated Press, "Excerpts from Phone Call About Gingrich's Future," *New York Times*, November 6, 1998.

393 "For a few": Dan Balz, "Gingrich Says He's Leaving to Spare Turmoil in Party," *Washington Post*, November 10, 1998.

394 "The ideas are": Ibid.

395 "I believe": Robert G. Kaiser, "Livingston Resignation Crystallized Debate," *Washington Post*, December 20, 1998.

395 "It was twenty years": Lois Romano and Peter Baker, "'Jane Doe No. 5' Goes Public with Allegation; Clinton Controversy Lingers over Nursing Home Owner's Disputed 1978 Story," *Washington Post*, February 20, 1999.

396 "When they ask": Ibid.

396 Clinton refused: Julia Malone, "Feminists Say Rape Claim Against Clinton Is Credible," *Atlanta Journal and Constitution*, February 26, 1999.

396 "Mr. Clinton's word": Editorial, "Mrs. Broaddrick's Story," *Washington Post*, March 2, 1999.

396 Tom Daschle: Joyce Howard Price, "Lawmakers Urge Clinton to Come Clean on 1978 Incident," *Washington Times*, March 1, 1999.

TWENTY-FOUR

398 "Let me be clear": "Remarks on Signing the Continuing Resolution and an Exchange with Reporters," American Presidency Project, http://www.presidency.ucsb.edu/ws/index.php?pid=56616&st=&st1=.

398 "I'm concerned": Ibid.

398 Tom DeLay said: Bush Critical of Republican Plan to Meet Budget Targets," Associated Press, October 1, 1999.

398 "There's too much": Kathy Kiely, "House GOP Dropping Tax Credit Plan," *USA Today*, October 5, 1999.

398 He had: Dan Balz, "Bush Shows a Shadow of Clintonism; Criticism of Hill GOP Mirrors President's 'Triangulation,'" *Washington Post*, October 7, 1999.

399 "Two decades ago": Kevin Phillips, "The Rise and Folly of the GOP; As Voter Disgust Rises, So Do Clinton's Chances," *Washington Post*, August 6, 1995.

400 Surely, this was: Mark Z. Barabak, "Elizabeth Dole in N.H. For 'Major' Speech," *Los Angeles Times*, February 9, 1999.

400 They tended: Michael Duffy and Nancy Gibbs, "Who Chose George?" *Time*, June 14, 1999.

401 "It became pretty clear": Sam Howe Verhovek, "Republican Governors, Wanting One of Their Own, Decided Early on Bush," *New York Times*, November 23, 1999.

401 "It's conservative": Richard L. Berke, "Bush Tests Presidential Run with a Flourish," *New York Times*, March 8, 1999.

402 In his column: George F. Will, "Government as Therapist," *Washington Post*, February 7, 1999.

403 "We wouldn't have": Richard L. Berke, "California, Here Bush Comes, a Moderate on Immigration and Racial Quotas," *New York Times*, June 30, 1999.

403 "When I was young": Jim Yardley, "Bush, Irked at Being Asked, Brushes Off Drug Question," *New York Times*, August 19, 1999.

403 In *Newsweek:* David Brooks, "Clintonizing the GOP," *Newsweek*, February 8, 1999.

404 "I am humbled": Dick Polman, "Bush Reports a Record Haul, $36 Million, in Just 6 Months," *Philadelphia Inquirer*, July 1, 1999.

404 "George Bush's term": Tribune News Services, "Kasich Gives Up Race, Endorses Bush," *Chicago Tribune*, July 15, 1999.

404 "Too often": "Bush Hits GOP for Negative Rhetoric," *New Orleans Times-Picayune,* October 6, 1999.

405 "With this campaign": "Buchanan Campaign Announcement," C-Span Video Library, https://www.c-span.org/video/?121196-1/buchanan -announcement.

406 "The day of": Ibid.

407 "It is time": Ben MacIntyre, "Ventura Wrestles Reform from Perot: Party Delegates Elect Chairman Backed by Minnesota Governor," *Ottawa Citizen,* July 27, 1999.

408 When Trump claimed: Associated Press, "Weicker Apologizes in Feud with Trump," *New York Times,* December 5, 1993.

408 "My opposition": Hilary Waldman, "The Guv, The Donald: The War of the Words," *Hartford Courant,* December 3, 1993.

408 He issued: "Trump Ready for Call From Reform Party," *Washington Post,* October 13, 1999.

408 "I don't know anything": Mike Allen, "Reform Party Takes Decisive Turn from Perot to Ventura," *New York Times,* July 26, 1999.

408 There was common ground: Francis X. Clines, "Buchanan Nears Decision on Reform Party Bid," *New York Times,* September 13, 1999.

408 "When he asks": Bill Hutchinson, "The Body Stumps for the Donald," *New York Daily News,* September 14, 1999.

409 Buchanan wrote: Francis X. Clines, "Buchanan's Views on Hitler Create a Reform Party Stir," *New York Times,* September 21, 1999.

409 John McCain called: Alison Mitchell, "McCain Urges a Party Switch by Buchanan," *New York Times,* September 23, 1999.

410 An hour before: Kenneth Bazinet, "Trump Rips Buchanan Hitler Talk," *New York Daily News,* September 20, 1999.

410 "Well, that's a silly": "Pat Buchanan, Presidential Candidate, Talks About His Campaign," CBS News transcripts, September 19, 1999.

410 "Pat says Hitler": Francis X. Clines, "Buchanan's Views on Hitler Create a Reform Party Stir," *New York Times,* September 21, 1999.

410 "very serious": Joel Siegel, "See Donald Run," *New York Daily News,* October 8, 1999.

411 "The polls have been": "Is Donald Trump Pumped to Be Out on the Stump?" *CNN & Company,* CNN transcript, October 7, 1999.

411 On NBC's *Today*: "Donald Trump Discusses his Possible Presidential Bid," *Today,* NBC News transcript, October 7, 1999.

411 He told: Donald Lambro, "Trump, Oprah Running Mates?" *Washington Times,* October 8, 1999.

411 On the newsmagazine: "A Run for His Money," *Dateline,* NBC News transcript, October 6, 1999.

411 "I believe non-politicians": Michael Tackett and Lisa Anderson, "Trump

Hopes to Make Presidential Deal with Voters," *Chicago Tribune*, October 8, 1999.

412 "I really believe": "Donald Trump Discusses His Bid for the Reform Party Presidential Nomination," *Meet the Press*, NBC News transcript, October 24, 1999.

412 On abortion: "A Run for His Money," *Dateline*, NBC News transcript, October 6, 1999.

413 "We have to": Adam Nagourney, "A Question Trails Trump: Is He Really a Candidate?" *New York Times*, December 10, 1993.

413 But he drew: Scott Lindlaw, "Trump Renews Racism Charges Against Buchanan," Associated Press, December 7, 1999.

413 "On slow days": Donald J. Trump, "Buchanan Is Too Wrong to Correct," *Los Angeles Times*, October 31, 1999.

413 After his tour: Tom Squitieri, "Trump Snipes at Buchanan," *USA Today*, December 8, 1999.

413 Catching up: Geraldine Baum, "The Donald Erases Line Between Politics, Comedy," *Los Angeles Times*, December 6, 1999.

414 Lenora Fulani: Jonathan Alter, "Crackpot Pat Plants His Flag," *Newsweek*, October 4, 1999.

414 Fulani had been: Mike McIntire, "Fulani Loses Independence Party Role over Comments on Jews," *New York Times*, September 19, 2005.

414 John Hagelin: Mike Ferullo, "Natural Law, Reform Party Members Unite Around Hagelin in 'Coalition Convention,'" CNN, August 31, 2000, http://www.cnn.com/2000/ALLPOLITICS/stories/08/31/natural .law/.

414 "The Reform Party": Adam Nagourney, "Reform Bid Said to Be a No-Go for Trump," *New York Times*, February 14, 2000.

414 "When I held": Donald J. Trump, "What I Saw at the Revolution," *New York Times*, February 14, 2000.

415 No one put: Adam Nagourney, "Reform Bid Said to Be a No-Go for Trump," *New York Times*, February 14, 2000.

416 Pat Robertson: Peter Marks, "Keeping Up with the Joneses," *New York Times*, March 5, 2000.

416 Bush accepted: "Whitman to Endorse Bush in Connecticut, not New York," Associated Press, March 4, 2000.

418 Bush's intellect: Robert Russo, "Bush's Verbal Gymnastics Puzzle Many," *Calgary Herald*, August 3, 2000.

419 "I feel good": David Goldstein and Ron Hutcheson, "Both Bush, Gore Battle over Tennessee Electorate," *Philadelphia Inquirer*, October 25, 2000.

CREDITS

INDEX

abortion, 32, 195, 412, 416
Abraham, Spencer, 281
Abrams, Robert, 384
African Americans. *See also specific individuals*
  after Civil War, 22
  Civil Rights and Voting Rights Acts, 24–25
  Bill Clinton and, 135, 184, 190
  election of 1992, 175
  importance of, in New York, 142–143
  importance to Democratic Party by 1988, 26, 54, 55
  Jackson candidacy and, 54–55, 56, 61
  Mondale and, 55
  Perot and, 190
  Reagan and, 22, 25
  Revels, 23
After Hours, 150
Aid to Families with Dependent Children (AFDC), 302–305, 308–309, 343–345
Alexander, Lamar, 336, 337, 338, 404
Alfred P. Murrah Federal Building (Oklahoma City) bombing, 309–310, 311

Altman, Roger, 259–260, 385
"America first" theme, 153
American Legion, 224
Andersen, Kurt, 5
Andrews, Tom, 279
anti-Semitism
  Buchanan, 154–155, 412, 413
  Fulani, 414
  Jackson, 56–57
Apple, R. W., 59, 196, 287
Arafat, Yasser, 55–56
Arkansas and election of 1992, 136–137
Arkansas Democrat-Gazette, 51–52
armed forces
  in Bosnia, 360, 361
  Clinton and draft, 125–129, 222–223, 227, 360
  homosexuals in, 219–221, 222–223, 224–230
Armey, Dick, 107, 237–238, 356
Arpaio, Joe, 337
Aspin, Les, 226, 227

Babbitt, Bruce, 49–50
Baird, Zoë, 215–219
Baker, Howard, 282
Baker, Ross, 79

Balanced Budget Act of 1997,
　　357–360
Balz, Dan, 91, 373, 398–399
Barbour, Haley, 350
Bauer, Gary, 304
Bayh, Evan, 345
Beasley, David, 338
Begala, Paul, 267, 369
Bennett, Robert, 264
Bennett, William, 290
Bentsen, Lloyd, 62, 82
Biden, Joseph R., 49, 54, 216, 218
Boehner, John, 249
Boland, Edward, 37, 39
"Boll Weevils," 22
Bonior, David, 297, 354
Boren, David, 239, 279
Borger, Gloria, 410
Bork, Robert, 405
Bosnia, 360, 361
Boston Globe, 79, 94, 117, 137, 139,
　　228, 285–286, 382
Boston Herald, 114
Boxer, Barbara, 390
Braden, Tom, 150
Bradley, Bill, 82, 418
Bradley, Tom, 25
Brady, James, 242–243
Brady, Sarah, 242–243
Brady Bill (1993), 312
Branch Davidians, 310–311
Breaux, John, 239
Breslin, Jimmy, 125
Brinkley, David, 157, 284
Broaddrick, Juanita, 395–396
Broder, David
　　on Clinton, 212, 247
　　on Democratic Party identity
　　　　crisis, 14–15
　　election of 1996, 82
　　on election of 1998, 391

on Gingrich, 112
on Michel, 248
on Republicans' health care
　　reform strategy, 269
on Robb, 47
Brokaw, Tom, 2, 3, 4, 63, 200, 284
Brooks, David, 403–404
Brooks, Jack, 72
Brown, Jerry, 88–89, 90, 136, 139,
　　140–144, 169
Brown, Ron, 86
Brown, Willie, 171
Brownback, Sam, 324
Bryan, Richard, 239
Buchanan, Patrick J.
　　background, 6, 149–151
　　election of 1988, 151–152
　　election of 1992, 152, 153, 161,
　　　　193–196
　　election of 1996, 329, 330, 332–
　　　　340
　　election of 2000, 405–407, 410,
　　　　412, 415, 421
　　on U.S. fight against Hitler, 409
Buchanan, Shelley, 194
Buckley, William F., 77, 155
Bumpers, Dale, 48
Burton, Dan, 107, 262, 264
Bush, Barbara, 194
Bush, George H. W.
　　Cheney and, 75–76
　　economy and, 91–92, 96, 101
　　election of 1988, 146, 156–160,
　　　　209, 261
　　Gulf War and, 83
　　movement to right by, 146
　　NAFTA, 136
　　NRA and, 312–313
　　Operation Desert Storm, 80, 81
　　as politician, 104–105
　　popularity, 97

primaries of 1980, 145–146
Saddam Hussein and, 360–361
taxes and, 101–111, 148, 196–197, 236, 359
Tower nomination, 75–76
Bush, George H. W. and election of 1992
debates, 200–202, 203–204, 205–206
loss as vindication of Gingrich, 213–214
Perot and, 163, 167, 170–171, 186–187
Republican Convention, 193–197
television and, 172–173
Bush, George W.
background, 1, 334, 416
budget of 1999 and, 398
characteristics, 401
election of 2000, 1–4, 398, 400–402, 403–405, 416–417, 419
Bush, Jeb, 402
Bush, Prescott, 201–202

Cable Satellite Public Affairs Network (C-SPAN), 35–37
California, Proposition 13, 31
Callaway, Bo, 29
Campbell, Carroll, 269, 337–338
Campbell, Tom, 355
Carr, Bob, 281
Carson, Johnny, 137
Carter, Jimmy, 13, 22, 31, 81, 171, 213
Carvey, Dana, 151
Carville, James, 267
Cellucci, Paul, 402
Chafee, John, 240, 269
Chicago Tribune, 91, 163, 211, 302

Children's Health Insurance Program (CHIP), 358
Choate, Pat, 412
Christie, Jeff, 291
Chung, Connie, 298–299
Clinton, Bill
Arkansas gubernatorial election of 1990, 83–84
background, 10–12, 13–14, 26, 210
Democratic Convention of 1984, 9, 14–16
Democratic Convention of 1988, 62–64
Democratic Leadership Council and, 46, 85, 87
Dukakis and, 62
election of 1988, 46, 48, 49, 50, 51–52
on Hillary's use of maiden name, 13
marital fidelity and, 50–51, 367–368, 381–383, 392, 395–396
personality, 138
as savior of Democratic Party, 5
as Slick Willie, 138, 141, 160
Vietnam War and draft status of, 125–129, 360
Whitewater Development, 250–252
Clinton, Bill, as president. See also health-care reform during Clinton presidency
AFDC and, 308–309, 343–345
Baird nomination, 215–219
Bosnia and, 360, 361
budget of 1997 and, 357–360
budget of 1999 and, 397–399
budget showdown with Gingrich, 315–323
on Contract with America, 275–276

Clinton, Bill, as president (*cont.*)
election of 1994, 275–276, 278,
284–285, 286–287
election of 1996, 339, 341–342,
348–350, 351
on election of 1998, 391
first budget, 241–247
government shutdown, 324–327,
341, 347–348
gun control and, 312
homosexual rights, 219, 220–221,
222, 224–230
impeachment of, 377–379, 388–
390, 391, 394, 395
Iraq and, 360–361
Jackson and, 213
Jones settlement, 391–392
Lewinsky affair, 364–376
lobbying reform bill, 265–266
on Michel, 248
middle-class bill of rights, 287–
288
moves to counter Gingrich,
307–309
Oklahoma City bombing and,
310
plurality win and, 212, 213
popularity, 211, 274
as reformer of Democratic Party,
230
Republican strategy toward, 249
role of government, 211
stimulus package, 232–233, 238–
240
taxes, 231, 241–245, 287–288,
359
weapons ban and, 313
White House Travel Office
firings, 252–254
Whitewater and, 250–252, 256–
260, 263–264, 363

Clinton, Bill and election of 1992
Chicago Palmer House Hilton
and, 89–92
Cuomo and, 95, 96, 97, 134
debates, 200–202, 203, 204–205,
206
Democratic Convention, 190–192
exploratory committee, 87
extramarital affairs reported,
114–122
fundraising, 113
Gulf War position, 137–138
homosexuals and homosexuality,
195, 219, 220
Jackson and, 174, 175–185
past marijuana use, 141–142
Perot and, 162, 163, 167, 170,
171, 189
primaries and caucuses, 113–114,
117–118, 125, 131–136, 139,
142–144
public opinion of, 137, 138
Sister Souljah and, 180, 182–184
taxes, 130, 232
use of television, 171–172
Whitewater Development, 251–
252

Clinton, Hillary Rodham
background, 12–13, 252, 380
Bill's extramarital affairs and,
367–368, 381–383, 392
D'Amato and, 385–387
election of 1988, 51
election of 1992, 118, 119, 120,
123–124
election of 1998, 382–383, 386–
387, 388
election of 2000, 422
as elitist, 124–125
as first lady, 6
Gingrich's mother on, 298–299

on Gingrich's proposed changes to AFDC, 304
health-care system reform task force, 268
as polarizing, 380–381
as radical feminist, 195
run for Senator from New York, 392
White House Travel Office firings, 253–254
Whitewater and, 250–252, 257–260, 363
Coalition for Democratic Values, 86
Coelho, Tony, 19, 70, 73
Common Cause, 68
Confrontation, 150
Connecticut, election of 1992, 139
Conservative Opportunity Society, 35
Contract with America, 274–276, 306
Conyers, John, 377–378
Cook, Charlie, 351
Cooper, Jim, 270, 280
Cornelius, Catherine, 252–253
corruption/scandals
    Jerry Brown and election of 1992, 88
    Clinton staffers' past drug use, 295–296
    Clinton White House Travel Office firings, 252–254
    Clintons and Whitewater, 250–254, 256–260, 263–264, 363, 383–384, 385
    Clinton's extramarital affairs, 50–51, 114–122, 364–376, 381–383, 392, 395–396
    Clinton's past marijuana use, 141–142
    Coelho and junk bonds, 70

D'Amato and federal housing program, 384
Democratic perks for big donors, 351
Diggs, 34
during election of 1992, 168
Gingrich's book deal, 296–297
Gingrich's college course, 353–356
Gingrich's race against Flynt, 30
Hastert's sexual molesting of students, 395
Hyde's extramarital affair, 379
Morris and prostitute, 348
Wright's book deal, 67–71, 78
Cranston, Alan, 19
Crenna, Richard, 164
Crossfire (formerly *Confrontation*), 150, 151
Crowley, Candy, 3–4
Crystal, Billy, 141–142
culture wars
    abortion, 32, 195, 412, 416
    election of 2000 Republican primaries, 416
    feminism, 32
    homosexuality, 185–186, 188, 195, 219, 220–221, 222, 224–230
    religion, 195–196
    two-parent families, 195
Cuomo, Andrew, 278
Cuomo, Mario
    Bill Clinton and, 191
    election of 1988, 21–22, 52, 60, 62
    election of 1992, 83, 92–100, 113, 122–123, 134
    election of 1994, 276–279
    as governor of New York, 52–53
    keynote speaker at 1984 Democratic Convention, 16–20
    labor support, 136

Cuomo, Mario (*cont.*)
  as "old-style" liberal, 17–19, 21
  welfare reform and convention of
    1996, 345
Cutler, Lloyd, 261

Dale, Billy, 252, 253
Daley, Richard J., 242
D'Amato, Alfonse Marcello
  background, 384–385
  Hillary Clinton and, 385–387
  election of 1992, 97, 259
  on Fiske, 264
  New York gubernatorial election
    of 1994, 277
  Senate election of 1998, 383–385,
    386–388, 390, 391
  Whitewater and, 383–384, 385
D'Amato, Antoinette "Mama," 384,
  387–388
Darman, Richard, 106, 290
Daschle, Tom, 325, 396
Dateline, 295
Davis, Gray, 391
Dean, Howard, 129
DeConcini, Dennis, 281
DeLay, Tom, 78, 305, 375
Demjanjuk, John, 154–155
Democratic Leadership Council
  (DLC)
  Bill Clinton and, 85, 87, 213
  Jackson and, 174–175
  liberals and, 86
  origins, 45–46
  Super Tuesday, 54
Democratic Party. *See also specific*
  members
  after Civil Rights and Voting
    Rights Acts, 24–25
  George H. W. Bush and taxes,
    102–103

  Clinton and Gingrich budget
    proposals and, 315–316, 317–
    318, 321
  Democratic control of executive
    and legislative branches after
    1992 election, 211–212, 266
  election of 1968, 242
  election of 1984, 9, 14–20
  election of 1988, 47–54, 62–64
  election of 1992, 82–83, 86–89,
    139–140, 190–192
  election of 1994, 274–286, 279–
    286
  election of 1996, 341–343, 345–
    350, 351–352
  election of 2000, 418–422
  health care reform and, 270–271,
    272
  importance of African
    Americans to, 26, 54, 55
  importance of homosexuals to,
    220, 221
  Lewinsky affair and, 373–374,
    376, 377–378
  nomination process, 46–48, 54,
    81–82
  Operation Desert Storm and,
    80, 81
  southern defections and Reagan
    Democrats, 26, 45, 58
  southern domination through
    1968, 22–23, 28
  tax and spend reputation of, 231,
    238, 245–246
DeWine, Mike, 279
Diggs, Charlie, 34, 214
Dodd, Christopher J., 346
Dole, Bob, 247, 335–340
  budget showdown with Clinton,
    317, 320, 321–322
  Bush taxes and, 104, 108

on Clinton plurality win, 213
Clinton stimulus package, 238,
    239, 240
Clinton taxes, 244
election of 1976, 327
election of 1988, 327
election of 1996, 247, 317, 335–
    340, 350–351
election of 2000, 400, 404, 405
on election of 1994, 282
on Fiske, 264
Gingrich and, 328, 356
health care reform and, 269, 273
homosexuals in armed forces,
    225, 230
lobbying reform bill and, 265–266
Domenici, Pete, 317
Donahue, Phil, 143, 172
"Don't Ask, Don't Tell" policy,
    219–230
Douglas, Chuck, 107
Dowd, Maureen, 293
The Drudge Report, 364, 366
Dukakis, Michael
    election of 1988, 49, 54, 59–60,
        61–62, 64, 81
    Jackson and, 178
Duke, David, 336, 414

economy
    1992 election, 91–92, 96, 104,
        168, 206
    Balanced Budget Act of 1997
        and, 357–360
    boom in 1997, 356–357
    budget showdown between
        Clinton and Republicans,
        315–323
    Bush taxes and, 101–111
    Clinton stimulus package, 232–
        233, 238–240

deficits and Bill Clinton, 357
election of 1992 and, 91–92, 96,
    104, 168, 170, 214
election of 1994 and, 274
Jackson campaign of 1988, 58,
    59
Edelman, Peter, 344–345
Edsall, Thomas, 182
Edwards, John, 390
election of 1964, 29
election of 1968, 242
election of 1972, 29
election of 1974, 29, 30
election of 1976, 31, 81, 327
election of 1978, 31–33
election of 1980, 21, 145–146,
    384–385
election of 1984
    Bloody Eighth district in
        Indiana, 42–43
    Democratic Convention, 9,
        14–20, 57
    Jackson, 25–26, 54–58, 59, 61
    Reagan, 9, 13, 22
    results, 20, 26
election of 1986, Senate, 47
election of 1988
    Buchanan, 151–152
    George H. W. Bush, 146, 156–160,
        209, 261
    Clinton, 46, 48, 49, 50, 51–52
    Cuomo, 21–22, 52, 60, 62
    Democratic Convention, 62–64
    Dole, 327
    Dukakis, 49, 54, 59–60, 61–62,
        64
    Gore, 59, 61
    Jackson, 49, 58–62
    potential Democratic candidates,
        47–54
    Rand Paul, 148

election of 1988 (*cont.*)
    results, 81
    Super Tuesday, 59
    Trump, 408–409
election of 1990
    Arkansas gubernatorial, 83–84
    Republican House losses, 147
    THRO wins, 164–165
election of 1991, 267
election of 1992, 384–385. *See also*
        Bush, George H. W. and
        election of 1992; Clinton, Bill
        and election of 1992; Tsongas,
        Paul and election of 1992
    Buchanan, 152, 153, 161, 193–
        196
    Cuomo, 83, 92–100, 113, 122–
        123, 134
    D'Amato, 97, 259
    Democratic candidates, 87–89
    Democratic Convention, 190–192
    Democratic potential candidates,
        82–83, 86–87
    economy and, 91–92, 96, 104,
        168, 206
    Gulf War (Operation Desert
        Storm) and, 81, 82, 91
    Harkin and, 175
    homosexuality as issue, 185–186
    Jackson, 174, 175–185
    Kerrey, 87, 90, 129, 139, 201
    Rand Paul, 148–149
    Perot, 162–163, 167, 169–171,
        186–187, 189, 198–201, 208–
        209
    Republican Convention, 161,
        193–197, 222
    Republican primaries, 155–160
    results, 207–209, 211–212
    television and, 171–173, 200–206
election of 1993, 249–250
election of 1994

    Cuomo, 276–279
    economy and, 274
    lack of Democratic progress on
        issues, 273–274
    Republican strategy, 274–276
    results, 279–286
election of 1996
    Buchanan, 329, 330, 332–340
    Democratic Convention, 341–343,
        345–350
    Dole, 247, 317, 335–340, 350–351
    Perot, 331–334, 351, 406
    polls after government shutdown,
        326
    Republican Convention, 247,
        339–340, 347
    Republican hopefuls, 317, 321,
        325–330
    results and beginning of Blue
        America, 351–352
    unofficial start, 82
election of 1998
    Hillary Clinton and, 382–383,
        386–387, 388
    D'Amato and Whitewater, 383–
        384, 385
    impeachment of Clinton and,
        388–390, 391
    results, 390–391
election of 2000
    Buchanan, 405–407, 410, 412,
        415, 421
    Bush, 1–4, 398, 400–402, 403–
        405, 416–417, 418–419
    Hillary Clinton, 422
    Bill Clinton factor, 1, 418
    colors chosen for political parties,
        5, 419
    election night, 1–4
    Gore, 1–4, 246, 404, 418–422
    Nader, 421
    Perot, 405, 408, 412

potential candidates and primaries, 398, 399, 400–405, 415–416, 418
as product of partisan war, 422
results, 419–422
Trump, 407–408, 410–411, 412–415
Weicker, 407, 408
Empower America, 290
Engler, John, 402
Ensign, John, 391
evangelical Christians in Republican Party. *See also* culture wars
Buchanan, 195–196
Foster death and, 261–262
homosexuals in armed forces, 224
Lewinsky affair and, 377
power of, 220, 285

Faircloth, Lauch, 263–264, 390
Falwell, Jerry, 261–262, 416
Farrakhan, Louis, 56
Feinstein, Dianne, 373
feminism, 124, 195
Ferraro, Geraldine, 20
Fish, Hamilton, 41
Fiske, Robert, 258–259, 260–261, 262, 263
Fitzwater, Marlin, 103, 152, 172
Flanagan, Michael Patrick, 283
Florida
election of 1992, 134–135
election of 2000, 2–4
Florio, Jim, 250
Flowers, Gennifer, 115–116, 117, 118–119, 121–122, 372
Flynt, John J., 29, 30, 31
Flynt, Larry, 379
Foley, Tom, 103, 108, 272, 283
Ford, Gerald, 30, 81, 110
Foster, Vincent, 156, 254–256, 260, 261–262

Fox, Jon, 282, 390
Frank, Barney, 227, 298
Frist, Bill, 280
Fulani, Lenora, 412–415
Fulbright, J. William, 126

Gargan, Jack, 164–165, 407, 414
Georgia and Republican primary of 1992, 158–160
Gephardt, Richard A.
Bush taxes and, 103–104, 110
on Democratic control of executive and legislative branches, 266
election of 1988, 49, 54, 59
election of 1992, 82, 212
election of Gingrich as speaker, 297–298
Gewirtz, Paul, 215
Gibbons, Gene, 205
Gibbons, Sam, 271
Gigot, Paul, 290
Gilmore, Jim, 402
Gingrich, Kathleen, 298–299
Gingrich, Newt
background, 27–29
Baird nomination, 216–219
Buchanan and, 155
Buckley on, 77
Cheney on, 73
Clinton and Whitewater, 258
Bill Clinton stimulus package, 237
on Coelho, 73
conservation movement, 29–30
Contract with America as Republican strategy for election of 1994, 274–275
culture of politics and, 293
determination to end Permanent Democratic Congress, 33, 67, 73

Gingrich, Newt (*cont.*)
  Diggs and, 34
  on Dole's intransigence, 240
  election of 1974, 29, 30
  election of 1978, 32
  election of 1992, 105
  on Foster death, 262
  homosexuals in armed forces,
    227–228
  as leader of Republican
    Revolution, 146–147
  Michel and, 34–35
  on militancy of Republican Party
    in House, 43–44
  move up to Republican whip,
    74–75, 76–79
  on Perot, 163
  Republican primary of 1992,
    159
  showdown with and rebuke of
    O'Neill, 38–41
  Special Order speeches, 35–37
  Wright book deal and, 67–71, 78
  Young Turks and, 35
Gingrich, Newt, as Speaker
  attempts to remove, 354–355,
    356
  Bill Clinton's moves to counter
    Republicans and, 307–309
  book deal, 296–297
  budget of 1997 and, 357–360
  budget showdown with Clinton,
    306, 315–323
  counterculture of boomers and,
    293–294
  election and talk about
    nonpartisanship, 297–298
  election of 1994, 283–284
  election of 1996, 243
  election of 1998, 389, 391
  ethics probe of, 353–356

  government shutdown, 324–327,
    351
  gun control and, 313
  as having mandate, 301–302
  impeachment of Clinton, 378
  Jeffrey appointment and, 299–300
  as leader of Republican Party,
    5–6, 314, 399
  Lewinsky affair and, 369, 374–375
  media and, 295
  orientation for freshmen, 289–
    292
  power of, 306–307
  public opinion of Republican
    Congress, 319
  resignation from House, 392–394
  taxes and, 306
  welfare and, 302–305
Gingrich, Newt and taxes
  Bush's loss as vindication, 213–
    214
  Bush's taxes and, 105–106, 107–
    110, 111–112, 146–147
  position, 32, 236–237
Giuliani, Rudolph, 277–278
Goldberg, Lucianne, 365
Goldwater, Barry, 29, 224
Goodling, William, 305
GOPAC, 354, 355
Gore, Albert
  background, 49, 189
  debate with Perot, 331
  on Democratic control of
    executive and legislative
    branches, 212
  election of 1988, 48, 59, 61
  election of 1992, 82, 189–190
  election of 1996, 346–348
  election of 2000, 1–4, 346, 404,
    418–422
  as vice president, 243, 246

Gramm, Phil, 131, 321, 329–330, 334, 335, 336
Gramm-Rudman-Hollings Act, 102
Green Party, 421
Greenberg, Stanley, 58
Greenfield, Jeff, 2, 284
Greenspan, Alan, 233
Grow, Galusha A., 283
Grunwald, Mandy, 117
Gulf War (Operation Desert Storm), 80, 81, 82, 84–85, 91, 137–138, 147–148
gun control, 242–243, 312–313

Hagelin, John, 414
Hall, Arsenio, 172–173
Hammerschmidt, John Paul, 10–11
Harkin, Tom, 86–87, 90, 136, 175
HarperCollins, 296–297
Harris, Katherine, 421
Hart, Gary, 17, 50, 52, 53
Hartman, Phil, 151
Hastert, J. Dennis, 395
Hatch, Orrin, 216
health-care reform during Clinton presidency
  CHIP, 358
  Clinton call for, 234–235, 270
  Clinton plan, 268–269, 269–270
  Democratic Party proposals, 270–271, 272
  groups fighting, 271–272, 273
  public fear of, 271–273
  public support for, 267, 268–269
  Republican response, 268–269, 272, 273
  task force established, 268
Heritage Foundation, 290
Holmes, Eugene, 125–127, 128
Holt, Rush, 391

homosexuals and homosexuality
  AIDS crisis, 220
  armed forces ban, 219–221, 222–223, 224–230
  Buchanan, 336
  Bill Clinton as candidate and, 195, 219, 220–221
  as issue in election of 1992, 185–186, 188, 195
  as power in Democratic Party, 220, 221
  public opinion about rights of, 221–222, 225, 228–230
Hubbell, Webster, 254
Hume, Brit, 196
Humphrey, Hubert, 24
Hunt, Al, 219
Hussein, Saddam, 360–361
Hyatt, Joel, 279
Hyde, Henry, 40, 72–73, 77, 78, 378–379
"Hymie" and "Hymietown," 56–57, 61

immigration
  Baird's employment of illegal immigrants, 215–219
  Buchanan, 154
  election of 1996, 330, 334
Inhofe, Jim, 279
Inslee, Jay, 382
Iraq, 360–361
Ireland, Patricia, 378
Isikoff, Michael, 364, 365
Israel, 56, 60–61, 154

Jackson, Jesse, 6
  Jerry Brown and, 142–143
  Bill Clinton and, 190, 213
  election of 1984, 25–26, 54–58, 59, 61

Jackson, Jesse (*cont.*)
  election of 1988, 49, 58–62
  election of 1992, 174, 175–185
  tensions between African
    Americans and American Jews,
    56–57
  welfare reform and convention of
    1996, 345
Jagt, Guy Vander, 110
Javits, Jacob, 384
Jeffords, Jim, 240
Jeffrey, Christina, 299–300
Jennings, Peter, 4, 187–188, 284
John F. Kennedy School of
  Government, 289–290
Johnson, Haynes, 19
Johnson, Lyndon, 24
Johnson, Nancy, 78
Jones, Paula, 364, 365–366, 391–392
Jordan, Hamilton, 167, 189
Jordan, Vernon E., Jr., 364, 366

Kaplan, Richard, 116
Kasich, John, 241, 404
Keating, Charles, 168
Keating, Frank, 310
Keating Five scandal, 168
Kelley, P. X., 158
Kemp, Jack, 32, 290
Kennedy, John F., 210
Kennedy, Ted, 52, 281, 318
Kerrey, Bob
  Clinton stimulus package, 239
  election of 1992, 87, 90, 129,
    139, 201
  on election of 1994, 287
  first Clinton budget, 246
Kilborn, Craig, 372
King, Coretta Scott, 55
King, Larry, 163–164, 165–166,
  172, 199, 331, 411

King, Peter, 354
Kinsley, Michael, 155
Kirkpatrick, Jeane, 290
Koch, Ed, 17, 60–61, 99
Koppel, Ted, 117, 128–129, 383
Koresh, David, 311
Kramer, Marcia, 141
Kroft, Steve, 118–121
Kyl, John, 281

labor, 20, 136, 140
LaPierre, Wayne, 312
Largent, Steve, 354–355
Larry King Live, 163–164, 165–166,
  331
Lauer, Matt, 368
Lazio, Rick, 417
Leach, Jim, 260, 355
Leavitt, Mike, 401
Lehrer, Jim, 205, 366–367
Leno, Jay, 368
Letterman, David, 422
Lewinsky, Monica, 364–376, 377–
  378
Lewis, Anthony, 253, 294–295
Lewis, Jerry, 77, 109–110
Lewis, John, 318, 344
liberalism
  backlash of "silent majority," 33
  Cuomo's as old-style, 17–19, 21
  domination of Washington
    since Roosevelt, Franklin D.,
    33
Lieberman, Joe, 373–374, 376, 418
Limbaugh, Rush, 261, 280, 291–
  292, 311, 375, 416
Livingston, Bob, 393, 394–395
Los Angeles Times, 65, 70, 121, 199,
  248, 273, 320, 413
Lott, Trent, 39–40, 74
Lugar, Richard, 196

Lungren, Dan, 390–391
Luntz, Frank, 383

Mack, John, 70, 71
Madigan, Edward R., 77–78
Madison Guaranty, inquiry into,
    256, 259, 260
The Man from Hope (television
    program), 210
Margolies-Mezvinsky, Marjorie,
    244–245, 282
Marino, John, 93–94, 95–96, 98
Martin, Lynn, 77
McCain, John, 227, 404, 409,
    415–416
McCarthy, Carolyn, 243
McCloskey, Frank, 42–43
McCurdy, Dave, 279
McCurry, Mike, 322
McDermott, Jim, 270
McDougal, James, 251, 257, 258
McDougal, Susan, 251
McGovern, George, 25, 29
McGrory, Mary, 21, 36, 66, 87, 256,
    302–303
McInturff, Bill, 269
McIntyre, Richard, 42–43
McLaughlin, John, 151
The McLaughlin Group, 151
McVeigh, Timothy, 309–310, 311
Medicare, 315–319, 321–323, 325,
    344, 350
Meet the Press, 295, 319, 412
Melendez, "Stuttering John," 122
Metzenbaum, Howard, 86, 218,
    279
Mfume, Kweisi, 298
Miami Herald, 50
Michel, Robert
    Bush taxes and, 104, 109, 214
    characteristics, 34–35

Clinton taxes and, 242
elected Republican House leader,
    74
election of 1984 in Bloody eight,
    43–44
Gingrich and, 68, 74–76, 78–79,
    249
O'Neill and, 41, 65
retirement, 248–249
taxes and, 237
Mills, David, 179
Minge, David, 245
Mintz, John, 169
Mitchell, George
    Bush taxes and, 103, 109
    election of 1992 and, 82
    health care reform and, 272, 273
    lobbying reform bill and, 266
    retirement, 279
Mitchell, Henry, 11
Mixner, David, 220–221
Moakley, Joe, 39–40
Mondale, Walter, 9, 20, 26, 55,
    178
Moore, Carlos, 67
Moorer, Thomas, 223
Morris, Dick, 308, 348
Morton, Bruce, 281
Moynihan, Daniel Patrick, 99, 258,
    303, 344, 392
Murray, Charles, 290
Murray, Patty, 391
Murtha, John P., 111
Myers, Dee Dee, 252

Nachman, Jerry, 115
Nader, Ralph, 421
NAFTA, 136
Nagourney, Adam, 89
National Gay and Lesbian Task
    Force, 221

National Rifle Association (NRA),
    312–313, 399, 412
New Hampshire
    election of 1988, 54, 58
    election of 1992, 95, 113–114,
        117, 121, 125, 129, 131–133,
        152, 155, 156–157
New Orleans Declaration (DLC),
    85–86
The New Republic, 186–187
New York, election of 1992, 140
New York Daily News, 141, 325, 387,
    410
New York Post, 98, 115
New York Times, 41, 47, 57, 59, 69,
    70, 81, 93, 110, 113, 133, 146,
    167, 185, 196, 215, 216, 217,
    223, 224–225, 227, 238, 257,
    259–260, 264, 270, 275, 288,
    294–295, 361, 367, 372, 373,
    377, 385–386, 392, 414–415
New York Times Magazine, 66
Newsday, 258
Newsweek, 364, 403–404
Nichols, Larry, 114–115
Nichols, Terry, 309–310
Nightline, 117, 128–129
Nixon, Richard
    Buchanan and, 149–150
    Bush taxes and, 110
    election of 1972, 29
    impeachment hearings, 379
    pardoning of, 30
    on Perot, 162–163
    "silent majority" and, 33
    Vietnam War draft and, 125
North, Oliver, 340
Nunn, Sam
    background, 47
    election of 1988, 47–48
    election of 1992, 82
    first Clinton budget, 246

homosexuals in armed forces,
    226, 228, 229
reason for Democratic
    Leadership Council, 46
Nussbaum, Bernard, 256–257, 261
Nyhan, David, 137, 139

Oakar, Mary Rose, 67
Obey, David, 316
Oklahoma City bombing, 309–310,
    311
O'Neill, Tip
    C-SPAN and, 37
    described, 65
    Diggs and, 34
    retirement, 65–66
    showdown with Gingrich and
        rebuke of, 38–41
    on Wright, 72
Operation Desert Storm (Gulf
    War), 80, 81, 82, 84–85, 91,
    137–138, 147–148

Page, Susan, 258
Panetta, Leon, 296, 305
Pappas, Mike, 391
Pataki, George E., 277–278, 402
PBS *NewsHour*, 366–367
Pelosi, Nancy, 93, 356, 369, 378
Permanent Democratic Congress,
    33, 34, 70, 102, 105, 207, 211,
    234, 280
Perot, Henry Ross
    African Americans and, 190
    background, 163–164, 165, 170
    campaign suspension, 191
    characteristics, 168–169
    debates, 200–203, 331
    election of 1992, 162–163, 167,
        169–171, 186–187, 189, 198–201,
        208–209
    election of 1996, 331–334, 351, 406

election of 2000, 405, 408, 412
first Clinton budget, 241
homosexuality as issue, 185–186, 188
on television, 163–164, 165–166, 185–186, 187–188, 207
Trump and, 6
use of private investigators by, 186–187, 188
Washington scandals, 168
Phelan, Richard J., 70
Phillips, Kevin, 229–230, 399–400
Powell, Colin L., 223–224, 225, 227
Price, Christina, 299–300
Pryor, David, 11

Quayle, Dan, 106, 195, 198, 404

race and racial politics
election of 1984 and, 55, 57
Nixon's southern strategy, 25
Reagan and states' rights, 25
Reagan Democrats and, 58
of Reagan voters, 22
radio
Limbaugh, 261, 280, 291–292, 311
WRC in Washington, D.C., 150
Rainbow Coalition, 175, 176–182
Rangel, Charlie, 61, 379
Rather, Dan, 1–2, 157, 196
Rayburn, Sam, 65
Reagan, Ronald
AIDS crisis, 220
Clinton stimulus plan, 238
debt and, 101–102
election of 1980, 21, 145–146
election of 1984, 9, 13, 20, 22
on election of 1994, 286
government as problem, 211
on O'Neill, 65
states' rights, 25
Reagan Democrats, 45, 58, 211

Reed, Ralph, 290
Reflections of a Public Man (Wright), 67–69
Reform Party, 6, 406–410, 411–415
Reid, Harry, 391
Renewing American Civilization, 353–356
Reno, Janet, 256, 258, 263, 311
A Republic, Not an Empire (Buchanan), 409
Republican Party. *See also* evangelical Christians in Republican Party; *specific members*
after Civil Rights and Voting Rights Acts, 24–25
after Civil War, 22
"big tent" image, 314
budget of 1999 and, 397–399
budget showdown with Clinton, 317–323
George H. W. Bush's loss as vindication of Gingrich, 213–214
Democrats move to, 27
dominance in presidential elections, 81
election of 1964, 29
election of 1980, 145–146
election of 1990, 147
election of 1992, 155–160, 161, 193–197, 222
election of 1994, 274–286
election of 1996, 247, 317, 321, 328–330, 335–340, 347, 351–352
election of 2000, 398, 399, 400–405, 415–416, 418–422
election of 1993 resurgence of, 249–250
family values and, 195
Gingrich as leader of, 314, 399
health care reform and, 268–269, 272, 273

Republican Party (*cont.*)
　Hispanic vote and, 403
　immigration in platform, 154
　Lewinsky affair and, 369, 374–375,
　　377
　NRA and, 312
　popular opinion of budget and
　　government shutdown and,
　　325
　public opinion about after
　　impeachment, 399
　recriminations after election of
　　1998, 392–393
　response to Clinton health-care
　　reform, 268–269, 272, 273
　"southern strategy," 21, 25
　taxes and, 237, 245–246
Revels, Hiram, 23
Rhodes, John, 34
Ribicoff, Abraham, 242
Rice, Donna, 50
Richards, Ann, 402
Ridge, Tom, 402
Riegle, Don, 281
Robb, Charles, 46–47
Roberts, Steven, 41
Robertson, Pat, 261, 377, 416
Rockefeller, Jay, 82
Rockefeller, Nelson, 28
Rodham, Hillary. *See* Clinton,
　Hillary Rodham
Rollins, Ed, 167, 189
Romney, Mitt, 281
Rose Law Firm, 254–255
Rostenkowski, Dan, 212, 266, 271,
　283
Rove, Karl, 404
Rowlands, Sherry, 348
Royko, Mike, 88
Ruby Ridge, Idaho standoff, 310
Russakoff, Dale, 293

Russert, Tim, 2, 99, 280, 412
Ryan, George, 402

Safire, William, 77, 255
Sahl, Mort, 150
Salon, 379
Santorum, Rick, 282
Sasser, Jim, 280
Sasso, John, 184–185
Saturday Night Live, 82–83, 151, 224,
　411
savings and loan (S&Ls) crisis and
　bailout, 168, 251
scandals. *See* corruption/scandals
Schneider, Bill, 200, 246, 280
Schneider, Claudine, 78
school lunch program, 304–305
Schroeder, Pat, 382
Schumer, Charles, 108, 242, 383,
　386–388
Schwarzenegger, Arnold, 156
Scowcroft, Brent, 76
Selleck, Tom, 131
Sentelle, David, 263, 390
Seven Dwarfs, 49–50
Shales, Tom, 19–20
Shapard, Virginia, 31
Shaw, Bernard, 2, 282, 337
Shays, Chris, 391
Shelby, Richard, 285
Simon, Paul, 49
Simpson, Alan K., 104, 216, 218
Simpson, Carole, 204
Sister Souljah, 179–180, 182–184
60 Minutes, 118–121, 143
Slouching Towards Gomorrah (Bork),
　405
Small, Pamela, 70, 71
Smeal, Eleanor, 378
Smith, Adam, 382
Snowe, Olympia, 78, 279

Sosnik, Doug, 370
Southern Baptist Convention,
    224
Special Order speeches, 35–37
Specter, Arlen, 206–207, 218, 240
Squires, Jim, 209
Star, 114–116, 121, 348
Starr, Kenneth W.
    attacks on, 378
    background, 263
    expansion of inquiry by, 364
    Lewinsky and, 366–376
    Whitewater and, 263–264, 363
Steinem, Gloria, 372
Stennis, John C., 74
Stephanopoulos, George, 218
Sununu, John, 106–107, 106–108
Super Tuesday, 54, 59, 135, 136–
    137
Supreme Court, 4, 421

Taft, Bob, 402
talk-show campaign (1992), 171–
    173
taxes. *See also* Gingrich, Newt and
    taxes
    Jerry Brown, 142
    George H. W. Bush, 101–111,
        146, 148, 196–197, 236, 359
    Bill Clinton as candidate, 130,
        232
    Clinton as president, 231, 241–245,
        287–288, 359
    Democratic Party as tax and
        spend, 231
    as foundational conservative issue,
        147
    Gingrich, 317
    New Jersey election of 1993, 250
    Perot, 241
    Proposition 13 in California, 31

    Republican Party position, 237,
        245–246
    Trump, 412
Tsongas, 130–131
television
    Buchanan on, 150, 151, 410
    Bill Clinton's extramarital
        affairs, 116–117, 118–121,
        143
    Bill Clinton's past marijuana use,
        141–142
    Bill Clinton on, 128–129, 243–
        244, 367
    Hillary Clinton on, 367–368
    election of 1992, 171–173, 200–
        206
    Gingrich on, 295–296, 306–307,
        319
    Gingrich's mother on, 298–299
    Gore on, 331
    homosexuals in armed forces as
        topic, 224, 226
    The Man from Hope, 210
    Perot on, 163–164, 185–186,
        187–188, 199, 207, 241, 331
    Trump on, 411, 412
This Week with David Brinkley,
    153–154
Thomas, Bill, 42, 110
Thomason, Harry, 253
Thompson, Fred, 280
Thompson, Tommy, 402
Thornton, Ray, 245
Throw the Hypocritical Rascals Out
    (THRO), 164–165
Thurmond, Strom, 24, 337
Time, 164, 247, 401
To Renew America (Gingrich), 296–
    297
Today, 367–368, 411
The Tonight Show, 137, 368

Tower, John G., 75–76
Travelgate, 253–254
"Triangulation" strategy, 307–309, 315–316
Tripp, Linda, 364–366
Trump, Donald J.
  background, 6, 408, 409
  on Buchanan, 410
  election of 1988, 408–409
  election of 2000, 407–408, 410–411, 412–415
  Reform Party and, 6
Tsongas, Paul, 236, 288
Tsongas, Paul and election of 1992
  background, 87
  Connecticut primary, 139
  Florida primary, 134–135
  New Hampshire primary, 95, 113, 114, 117, 125, 129–131, 133–134
  Perot and, 169
  Super Tuesday, 136–137
  withdrawal, 138
20/20, 185–186

Union Leader, 155
United We Stand America Convention, 331–334
USA Today, 89, 189–190, 316, 358, 375

Van Drehle, David, 144
Ventura, Jesse, 407, 408, 414
Veterans of Foreign Wars, 224
Vinson, Carl, 47

Waco, Texas standoff, 310–311
Walker, Robert, 36, 37, 40
Wall Street Journal, 125, 127, 128, 254, 255, 411
Wallace, Chris, 63

Wallace, George, 47
Walters, Barbara, 185–186
Warner, John, 108
Washington, Harold, 25
Washington Post, 11, 14, 19, 31, 34, 56, 57, 67, 70, 77, 79, 82, 91, 102, 144, 150, 157, 163, 168, 169, 179, 182, 186, 212, 235, 237, 247, 250, 273, 275, 304, 307, 324, 364, 373, 393, 395, 396, 398–399
Washington Times, 256, 411
Watts, J. C., 391
Waxman, Henry, 270
Weaver, Randy, 310
Weber, Vin, 36, 77, 108–109, 291
Weicker, Lowell, 407, 408
welfare system, 302–305, 308–309, 343–345
Wellstone, Paul, 270, 344, 360
Wertheimer, Fred, 68
Wexler, Robert, 378
White, Frank, 13–14
White, John C., 94
White, Mark, 17
Whitewater, 250–252, 254, 256–260, 263–264, 363, 383–384, 385. *See also* Starr, Kenneth W.
Whitman, Christine Todd, 250
Who Is Ross Perot? (television program), 187–188
Wilder, Doug, 87, 90
Will, George, 154, 286, 402
Willey, Kathleen, 365
Williamson, Lisa (Sister Souljah), 179–180, 182–184
Williamson, Pat, 245
Wilson, Charlie, 242
Wilson, Pete, 403
Wofford, Harris, 267, 281–282
Wright, Betsey, 51

Wright, James C., Jr.
   background, 66
   described, 66–67
   Ethics Committee investigation,
      69–71
   Gingrich and book deal of,
      67–71, 78
   as majority leader, 67
   resignation of, 70–73
Wright, Jim, 37, 44
Wynette, Tammy, 124–125

Young, Andrew, 55
Young Turks, 35